Advanced Oxidation Processes for Micropollutant Remediation

Advanced Oxidation Processes for Micropollutant Remediation covers current state-of-the-art in advanced oxidation processes (AOPs) for the removal of micropollutants from industrial and pharmaceutical wastewater. It covers various AOP-based techniques like Fenton process, ozonation, hydrogen peroxide, ultraviolet radiation, electrochemical oxidation, ultrasonic irradiation, and other hybrid technologies. It focuses on aspects like impact of micropollutants on the ecosystems, different types of AOPs, their principles, applications, and challenges in implementing these techniques and their future perspectives.

Features:

- Presents state-of-the-art in advanced oxidation processes (AOPs) along with hybrid technologies.
- Covers latest advances in technological know-how for the remediation of wastewater soiled with micropollutants.
- Discusses pressing environmental pollution issues associated with AOPs needed for wastewater remediation.
- Presents future perspective as well as techno-economic analysis in implementing various AOPs.
- Reviews strategies to limit the micropollutants in water bodies.

This book is aimed at graduate students and researchers in chemical and industrial engineering, environmental science, and wastewater treatment.

Advanced Oxidation Processes for Micropollutant Remediation

Edited by
Mohammad Khalid, Yuri Park,
Rama Rao Karri and Rashmi Walvekar

CRC Press
Taylor & Francis Group
Boca Raton London New York

CRC Press is an imprint of the
Taylor & Francis Group, an **informa** business

Designed cover image: © Shutterstock

First edition published 2024
by CRC Press
6000 Broken Sound Parkway NW, Suite 300, Boca Raton, FL 33487–2742

and by CRC Press
4 Park Square, Milton Park, Abingdon, Oxon, OX14 4RN

CRC Press is an imprint of Taylor & Francis Group, LLC

ISBN: 978-1-032-16291-1 (hbk)
ISBN: 978-1-032-16292-8 (pbk)
ISBN: 978-1-003-24791-3 (ebk)

DOI: 10.1201/9781003247913

Typeset in Times LT Std
by Apex CoVantage, LLC

*To my beloved parents, M. S. Siddiqui and Badarun Nisa
Siddiqui, for being the source of inspiration and motivation
My wife, Rashmi, for her support and encouragement
And my wonderful kids, Tawheed, Tawfeeq and
Zainab, for their love and understanding.*

Prof. Dr. Mohammad Khalid

*I dedicate this book, first and foremost, to God, who has provided
all that was needed to complete this work for which it was
undertaken for. There was never lack or want.
I also dedicate this work to my lovely family and friends, who have
encouraged me all the way and have continually prayed for my
success. Thank you. My love for you all can never be quantified.*

Dr. Yuri Park

*I dedicate this book to the memory of my
beloved Father Karri Sri Ramulu
I also dedicate this to my mother, Karri Kannathalli, who protected,
guided and supported me all these years. She is a superwoman and
my inspiration for driving me to achieve the best.
I also thank my lovely wife, Soni, my lovely children, Yajna and Jay,
and my in-laws L.V. Rao and Prameela without whose support and
understanding, this book as well as my research
achievements were not possible.*

Dr. Rama Rao Karri

*In memory of my beloved Father, Gangasa Walvekar,
and my loving Mother, Kasturi Walvekar, for
her unconditional love and support.*

Dr. Rashmi Walvekar

Contents

About the Editors ix
List of Contributors... xiii
Preface...xvii
Acknowledgments...xix

Chapter 1   Introduction to Micropollutants 1

                 Khalid Ansari and Moonis Ali Khan

Chapter 2   Homogeneous Advanced Oxidation Processes 13

                 Muhammad Asam Raza, Umme Farwa,
                 Muhammad Waseem Mumtaz, Hamid Mukhtar
                 and Umer Rashid

Chapter 3   Heterogeneous Advanced Oxidation Process 47

                 Abdul Sattar Jatoi, Zubair Hashmi, Nabisab Mujawar
                 Mubarak and Rama Rao Karri

Chapter 4   Photochemical and Sonochemical Strategies in Advanced
                 Oxidation Processes for Micropollutants' Treatments 65

                 Franklin Vargas, Edmanuel Lucena-Mendoza,
                 Tamara Zoltan, Yesica Torres, Miguel León,
                 Beatriz Angulo and Gabriel Ibrahin Tovar

Chapter 5   Kinetic Modelling of the Photodegradation Process
                 of Water-Soluble Polymers 99

                 Nadeem A. Khan, Afzal Husain Khan, Viola Vambol,
                 Sergij Vambol, Nabisab Mujawar Mubarak and
                 Rama Rao Karri

Chapter 6   Application of UV/TiO$_2$ and UV/H$_2$O$_2$ Systems
                 for Micropollutants' Treatment Process 107

                 Yuri Park and Allison L. Mackie

Chapter 7   Emerging Materials in Advanced Oxidation Processes
                 for Micropollutant Treatment Process.. 133

                 Anup Singh, Jashandeep Singh, Ajay Vasishth, Ashok Kumar
                 and Shyam Sundar Pattnaik

Chapter 8   Fenton, Photo-Fenton, and Electro-Fenton Systems
for Micropollutant Treatment Processes 157

Anjan Deb, Jannatul Rumky and Mika Sillanpää

Chapter 9   Techno-Economic Assessment of the Application of
Advanced Oxidation Processes for the Removal of Seasonal
and Year-Round Contaminants from Drinking Water.....................187

Reece Lima-Thompson and Stephanie Leah Gora

Chapter 10 Current Challenges and Future Prospects in AOPs 215

Pratibha Biswal, Umit Gunes and Mohamed M. Awad

Index.. 229

About the Editors

Dr. Mohammad Khalid is Professor and Head of Graphene and Advanced 2D Materials Research Group at the School of Engineering and Technology, Sunway University, Malaysia. He completed his bachelor's degree in Chemical Engineering from Visvesvaraya Technological University, India; master's in Chemical and Environmental Engineering from Universiti Putra Malaysia; and PhD in Engineering from International Islamic University, Malaysia. Prior to joining Sunway University, he was Associate Professor in the Department of Chemical & Environmental Engineering at the University of Nottingham Malaysia Campus. His primary research interests are in advanced nanomaterial synthesis for energy harvesting, storage, and conversion. He has over 200 peer-reviewed scientific articles to his credit, as well as six edited books and five patents. For the year 2021, he was named as one of the top 2% of the world's most influential scientists in the field of material science. He is also Fellow of the Higher Education Academy (FHEA), UK.

Dr Yuri Park holds a PhD degree in Chemistry from the Queensland University of Technology, Queensland, Australia (2013). After earning the degree, she joined the Department of Green Chemistry (DGC), Lappeenranta University of Technology (LUT), Finland, where she worked as Assistant Professor. Later, she was inducted into the Brain Pool Program (BP Program) in Korea and is currently Research Professor at Seoul National University of Science and Technology, Institute of Environmental Technology. She has core competence in the synthesis of sustainable nanomaterials and their application for environmental remediation application, using advanced water treatment methods, particularly with advanced oxidation processes (AOPs), including UV, ozone, and photo(electro)catalysis, and chemical analysis of environmental contaminants including analytical method development/validation. The main outlets for her articles in the past seven years are quartile one (Q1) international journals in the fields of chemistry, material science, earth and planetary science, and environmental science and engineering. Examples of high-impact journals in which she published her work include *Environmental Science & Technology*, *Journal of Colloid and Interface Science*, *Science of the Total Environment*, *Journal of Chromatography A*, *Chemical Engineering Journal*, *Chemosphere*, *Journal of Chromatography A*, *Environmental Science: Water Research & Technology*, and *Journal of Environmental Engineering*.

Dr. Rama Rao Karri is Professor (Sr. Asst) in Faculty of Engineering, Universiti Teknologi Brunei, Brunei Darussalam. He has obtained his PhD in Chemical Engineering from Indian Institute of Technology (IIT) Delhi, master's from IIT Kanpur, and bachelor's from Andhra University College of Engineering, Visakhapatnam. He has worked as Post-Doctoral Research Fellow at NUS, Singapore, for about six years and has over 18 years of working experience in academics, industry, and research. He has led the water resources cluster team at UTB (2015–2017), focusing on Water-Energy Nexus, Smart Water and Environmental Management. His research activities mainly focus on developing modelling methods for systems with inherent uncertainties in multidisciplinary fields. He has experience of working in multidisciplinary fields and has expertise in various evolutionary optimization techniques and process modeling. He has published 150+ research articles in reputed journals, book chapters, and conference proceedings with a combined impact factor of 479.2 and has an H-index of 24 (Scopus citations: 2,100+) and 26 (Google Scholar citations: 2,500+). Among 75 journal publications, 60 articles published are Q1 and high IF journals. He is an editorial board member in 10 renowned journals and a peer-review member for more than 93 reputed journals, and he peer reviewed more than 410 articles. Also, he handled 112 articles as an editor.

He also has the distinction of being listed in the top 2% of the world's most influential scientists in the area of environmental sciences and chemical for year 2021. The List of the Top 2% Scientists in the World compiled and published by Stanford University is based on their international scientific publications, number of scientific citations for research, and participation in the review and editing of scientific research. He is also a recipient of Publons Peer Reviewer Award to be listed as top 1% of global peer reviewers for Environment & Ecology and Crossfield categories for the year 2019. He is also delegated as advisory board member for many international conferences. He held a position as Editor-in-Chief (2019–2021) in *International Journal of Chemoinformatics and Chemical Engineering*, IGI Global, USA. He is also Associate Editor in Scientific Reports, Nature Group (IF:4.996) and *International Journal of Energy and Water Resources* (IJEWR). He is also Managing Guest editor for Special Issues: 1) *"Magnetic nano composites and emerging applications"* in *Journal of Environmental Chemical Engineering* (IF: 7.968); 2) *"Novel CoronaVirus (COVID-19) in environmental engineering perspective"*, in *Journal of Environmental Science and Pollution Research* (IF: 5.19); and 3) *"Nanocomposites for the sustainable environment"* in *Applied Sciences Journal* (IF: 2.679). He, along with his mentor, Prof. Venkateswarlu, has authored the book, *Optimal State Estimation for Process Monitoring, Diagnosis and Control*. He is also co-editor and managing editor for nine edited books.

Dr. Rashmi Walvekar is Associate Professor in the Department of Chemical Engineering at Xiamen University Malaysia Campus. She did her PhD in Engineering from International Islamic University, Malaysia; master's in Chemical and Environmental Engineering from the University Putra Malaysia; and her bachelor's in Chemical Engineering from Visvesvaraya Technological University, India. Her research includes nanofluid synthesis for heat transfer application, deep eutectic solvents as green alternative solvents, recycling of waste rubber tires, nanomaterial production from biomass, nano-lubricants and solar energy harvesting, and storage. She has published more than 170 research articles in peer-reviewed journals and has an H-index of 34. Her other research output includes two edited books and four patents. She is also Chartered Chemical Engineer and Chartered Member of the Institution of Chemical Engineers (IChemE), UK.

Contributors

Beatriz Angulo
Laboratorio de Hidrogeoquímica,
PDVSA—Intevep.
Los Teques, Venezuela

Khalid Ansari
Department of Civil Engineering,
Yeshwantrao Chavan College of
Engineering
Nagpur, India

Mohamed M. Awad
Mechanical Power Engineering
Department, Faculty of Engineering,
Mansoura University
Mansoura, Egypt

Pratibha Biswal
Department of Chemical Engineering,
Indian Institute of Petroleum and
Energy
Visakhapatnam, Andhra Pradesh, India

Anjan Deb
Department of Chemistry, University of
Helsinki, A. I. Virtasen aukio 1
Helsinki, Finland

Umme Farwa
Department of Chemistry, University of
Gujrat
Gujrat, Punjab, Pakistan

Stephanie Leah Gora
Department of Civil and Resource
Engineering, Dalhousie University,
1360 Barrington St, Bldg D, Rm
D215
Halifax, Nova Scotia, Canada

Umit Gunes
Department of Naval Architecture
and Marine Engineering, Yildiz
Technical University
Besiktas, Istanbul, Turkey

Zubair Hashmi
Department of Chemical Engineering,
Dawood University of Engineering
and Technology
Karachi, Sindh, Pakistan

Abdul Sattar Jatoi
Department of Chemical engineering,
Dawood University of Engineering
and Technology
Karachi, Sindh, Pakistan

Rama Rao Karri
Petroleum and Chemical Engineering,
Faculty of Engineering, Universiti
Teknologi Brunei
Bandar Seri Begawan, Brunei
Darussalam

Afzal Husain Khan
Civil Engineering Department, Jazan
University
Jazan, Saudi Arabia

Moonis Ali Khan
Chemistry Department, College of
Science, King Saud University
Riyadh, Saudi Arabia

Nadeem A. Khan
Civil Engineering Department, Jamia
Millia Islamia
New Delhi, India

Ashok Kumar
Department of Applied Sciences,
    National Institute of Technical
    Teachers Training and Research
Chandigarh, Punjab, India

Miguel León
Laboratorio de Fotoquímica,
    Laboratorio de Fisicoquímica
    Orgánica, Centro de Química
    "Dr. Gabriel Chuchani", Instituto
    Venezolano de Investigaciones
    Científicas (IVIC)
Caracas, Venezuela

Reece Lima-Thompson
Department of Civil and Resource
    Engineering, Dalhousie University,
    1360 Barrington St, Bldg D, Rm
    D215, Halifax
Nova Scotia, Canada

Edmanuel Lucena-Mendoza
Laboratorio de Fotoquímica,
    Laboratorio de Fisicoquímica
    Orgánica, Centro de Química
    "Dr. Gabriel Chuchani", Instituto
    Venezolano de Investigaciones
    Científicas (IVIC)
Caracas, Venezuela

Allison L. Mackie
Department of Engineering, School
    of Science and Technology, Cape
    Breton University
Sydney, Nova Scotia, Canada

Nabisab Mujawar Mubarak
Petroleum and Chemical Engineering,
    Faculty of Engineering, Universiti
    Teknologi Brunei
Bandar Seri Begawan, Brunei
    Darussalam

Hamid Mukhtar
Institute of Industrial Biotechnology,
    Government College University
Lahore, Pakistan

Muhammad Waseem Mumtaz
Department of Chemistry, University of
    Gujrat
Gujrat, Punjab, Pakistan

Yuri Park
Institute of Environmental Technology,
    Department of Environmental
    Engineering, Seoul National
    University of Science and Technology
Seoul, South Korea

Shyam Sundar Pattnaik
Media Engineering, National Institute
    of Technical Teachers Training and
    Research
Chandigarh, India

Umer Rashid
Institute of Nanoscience and
    Nanotechnology (ION2), Universiti
    Putra Malaysia
Serdang, Selangor, Malaysia

Muhammad Asam Raza
Department of Chemistry, University of
    Gujrat
Gujrat, Punjab, Pakistan

Jannatul Rumky
Department of Separation Science, LUT
    University, Yliopistonkatu 34, FI-53850
Lappeenranta, Finland

Mika Sillanpää
Zhejiang Rongsheng Environmental
    Protection Paper Co. LTD,
    NO. 588 East Zhennan Road, Pinghu
    Economic Development Zone
Zhejiang, P.R. China

Anup Singh
Department of Physics, Chandigarh
 University
Gharuan, Mohali, Punjab, India

Jashandeep Singh
Department of Physics, Gulzar Group
 of Institutes Khanna
Khanna, Punjab, India

Yesica Torres
Laboratorio de Fotoquímica,
 Laboratorio de Fisicoquímica
 Orgánica, Centro de Química
 "Dr. Gabriel Chuchani", Instituto
 Venezolano de Investigaciones
 Científicas (IVIC)
Caracas, Venezuela

Gabriel Ibrahin Tovar
Universidad de Buenos Aires, Facultad de
 Farmacia y Bioquímica, Departamento
 de Ciencias Químicas, Buenos Aires,
 Argentina; CONICET—Universidad
 de Buenos Aires, Instituto de
 Química y Metabolismo del Fármaco
 (IQUIMEFA)
Buenos Aires, Argentina

Sergij Vambol
Department of Life Safety, State
 Biotechnological University
Kharkiv, Ukraine, Kharkiv, Ukraine

Viola Vambol
Department of Environmental
 Engineering and Geodesy,
 University of Life Sciences
 in Lublin, Lublin, Poland;
 Department of Applied Ecology and
 Environmental Sciences, National
 University "Yuri Kondratyuk
 Poltava Polytechnic"
Poltava, Ukraine

Franklin Vargas
Laboratorio de Fotoquímica,
 Laboratorio de Fisicoquímica
 Orgánica, Centro de Química
 "Dr. Gabriel Chuchani", Instituto
 Venezolano de Investigaciones
 Científicas (IVIC)
Caracas, Venezuela

Ajay Vasishth
Department of Physics, Chandigarh
 University
Gharuan, Mohali, Punjab, India

Tamara Zoltan
Laboratorio de Fotoquímica,
 Laboratorio de Fisicoquímica
 Orgánica, Centro de Química
 "Dr. Gabriel Chuchani", Instituto
 Venezolano de Investigaciones
 Científicas (IVIC)
Caracas, Venezuela

Preface

Water is a fundamental element throughout the lifecycle of human beings and the ecosystem. With the increasing population and rapid industrialization–urbanization, mitigating water scarcity is now one of the leading challenges for the sustainable development of society. Water plays a crucial role in accomplishing sustainable livelihood, and achieving clean water is one of the main sustainable development goals (SDGs). Water scarcity is expected to be intensified due to climate change, which has already affected every continent. In the last few decades, the occurrence and fate of micropollutants, which are highly found in untreated sewage and terrestrial and aquatic environments, have been a growing concern due to their pseudo-persistent nature. Micropollutants are broadly defined as a group of biochemical substances included in trace amounts (less than parts per billion, $\leq$ μg/L) in surface and subsurface water. Perhaps this is why wastewater treatment plants (WWTPs) face challenges from variable substances with various chemical properties that remain unresolved. Although developed countries have realized the need for highly effective solutions, which are generally adopted in wastewater treatment processes, they are usually energy intensive and not cost-effective for the application in the real context. In the same timeframe, advanced oxidation processes (AOPs) have emerged as alternative technologies for the treatment of trace micropollutants present in waterways.

The book *Recent Trends in Advanced Oxidation Process for Micro-pollutants Removal* is proposed and edited to address these issues. This book offers information on the sources and prevalence of micropollutants in the environment, as well as the most recent methods for removing them from aquatic environments. Accordingly, the chapters of this book are organized by covering the basic fundamentals of the latest technologies to mitigate the micropollutants from water bodies. The first chapter provides a detailed introduction to the presence of micropollutants in the environment, their ill effects, and sources, followed by the chapters "Homogeneous Advanced Oxidation Processes" (Chapter 2), "Heterogeneous Advanced Oxidation Process" (Chapter 3), and "Photochemical and Sonochemical strategies in Advanced Oxidation Processes for Micropollutants' Treatments" (Chapter 4), covering the introduction of current AOPs such as photo-assisted and other catalyzed processes under either homogeneous or heterogeneous regimes towards micropollutants treatment. Special attention has been paid to identifying the most critical concerns that have prevented more application of AOPs in this field: the necessity in the design of photocatalytic reactors since the degradation of pollutants is highly associated with the integration of nanomaterials and nanocomposites to establish sustainable water treatment systems. The next chapter, entitled "Kinetic modelling of the Photodegradation Process of Water-Soluble Polymers" (Chapter 5), discusses the integration of mechanical kinetic modelling to determine the photodegradation rate of water-soluble polymers under UV/H_2O_2, which offers an important parameter for constructing industrial photo-reactors. Further, up-to-date heterogeneous AOP application studies, including main process conditions and operational parameters for the removal of micropollutants, are presented in the chapters "Application of

UV/TiO$_2$ and UV/H$_2$O$_2$ systems for Micropollutants' Treatment Process" (Chapter 6), "Emerging materials in Advanced Oxidation Processes for Micropollutants' Treatment Process" (Chapter 7), and "Fenton, Photo-Fenton, and Electro-Fenton System for Micropollutant Treatment Processes" (Chapter 8).

The book finishes with the chapters "Techno-economic Assessment of the Application of Advanced Oxidation Processes for the Removal of Seasonal and Year-round Contaminants from Drinking Water" (Chapter 9) and "Current Challenges and Future Prospects in AOPs" (Chapter 10). These last two chapters are dedicated to the current challenges associated with AOPs, which are mostly developed on the laboratory scale to scale up the application of AOPs to industrial-scale treatment. Although there are a few installations of AOP on an industrial scale, the development of AOP is still in the infancy stage. Therefore, to implement the AOP treatment process in water treatment fields, both economy and energy efficiency should be improved. In addition, the hybridization of different developed and emerging AOPs can be considered to manage a wide range of micropollutants, which further reduces the energy cost by using solar and other renewable energy in the treatment process. Moreover, cost curves and subsequent analyses are inevitable to determine important capital and operational cost drivers for each process and predict when AOPs are installed on a pilot and industrial scale.

This book gives researchers working/affiliated with environmental, wastewater management, micropollutants mitigation, and advanced oxidation process applications with a one-stop solution. We hope that this book will encourage scientists, managers, and decision-makers at industrial and public health levels to embrace AOPs with the ultimate goal of practically applying green and efficient technological alternatives to supplement drinking water treatment plants (WTPs) in the near future.

Editors
Mohammad Khalid
Yuri Park
Rama Rao Karri
Rashmi Walvekar

Acknowledgments

I gratefully acknowledge the International Research Network Grant Scheme (STR-IRNGS-S ET-GAMRG-01–2022) from Sunway University to support this work. Also, I owe a special thanks to my co-editors and authors, without whom this book would not have been possible.

Prof. Dr. Mohammad Khalid

I sincerely appreciate the entire academic staff and AWTL research team members at Seoul National University of Science and Technology. I offer my deepest gratitude to Dr. Yuhoon Hwang and Dr. Tae-Hyun Kim, who have been of tremendous help to me. I say a big thank you to all. I also appreciate the efforts of all the authors and contributors who have given everything to make this work possible.

Dr. Yuri Park

I thank Prof. Zohrah, Vice Chancellor, Universiti Teknologi Brunei; Prof. Ramesh; and higher management for the support. I offer special thanks to my co-editors; without their support and cooperation, this book would not have been possible. Finally, I thank all the authors who contributed high research value chapters.

Dr. Rama Rao Karri

I thank the co-editors and all the authors for their efforts and collaboration in making this book possible.

Dr. Rashmi Walvekar

1 Introduction to Micropollutants

Khalid Ansari and Moonis Ali Khan

CONTENTS

1.1   Introduction 1
1.2   Classification.. 2
    1.2.1   Major Classes of Organic Compounds in Water 2
1.3   Exposure Pathways 4
1.4   Potential Risk of Micropollutants to the Living Beings and Environment 5
    1.4.1   Surfactant and Personal Care Product.. 6
    1.4.2   Pharmaceuticals.. 6
    1.4.3   Steroid Hormones 6
    1.4.4   Industrial and Consumer Application.. 7
    1.4.5   Pesticides 7
    1.4.6   Heavy Metals.. 7
1.5   Conclusion 7
References... 8

1.1   INTRODUCTION

Over the last century, the population boom and economic developments have ensued a 500% increase in water usage [1]. The estimated global water demand at present is about 4600 km^3/yr, expected to rise due to community increase and industrialization in the course of the next decades [2]. Continuous socio-economic developmental activities and the use and discarding of a myriad of offered chemicals through human beings' day-to-day routine activities have resulted in the widespread occurrence of many such substances (in traces) in surface and sub-surface water, commonly termed micropollutants (MPs). Thus, MPs are broadly defined as a group of biochemical substances that are included in trace amounts ($\leq$ µg/L) in surface and sub-surface water. They are derived from natural and anthropogenic sources. Some of the major anthropogenic sources include municipal, domestic, hospital, and industrial wastewater; agricultural runoff; livestock; and aquaculture [3]. The MPs are made up of a wide range of compounds, including pharmaceuticals (such as antibiotics, hormones, and endocrine disrupters), artificial sweeteners, pesticides, plasticizers, surfactants, industrial chemicals, and personal care products (including UV filters, fragrances) [4]. The major categories of MPs, along with their sub-classes and sources in an aquatic environment, are displayed in Table 1.1.

In recent times, the occurrence of MPs has raised global concern while challenging the fate of clean water availability. Most of the MPs are tenacious, can deposit in

DOI: 10.1201/9781003247913-1

TABLE 1.1
Some Common Categories and Sources of Micropollutants Present in Water [5].

Category	Sub-classes	Major Sources
Detergent, surfactants	Cationic, anionic, and nonionic surfactants, perfluorinated compounds.	Bathing, laundry, dishwashing, household dilutants, and dispersants.
Industrial chemicals	Plasticizers, fire retardants, solvents, hydrocarbons.	Leaching, garden and road runoff, etc.
Pharmaceuticals	Antibiotics, anticonvulsants, lipid regulators, β-blockers, stimulants, and veterinary drugs.	Excretion, hospital effluents, runoff from concentrated animal feeding operations, and aquaculture [3].
Personal care products	UV filters, insect repellents, disinfectants, fragrances, synthetic musk.	Showering, swimming, bathing, shaving.
Steroids and hormones	Estrogens, endocrine, disruptive chemicals.	Excretion, hospital effluents, runoff from concentrated animal feeding operations, and aquaculture [3].
Pesticides	Insecticides, herbicides, and fungicides.	Agricultural runoff, runoff from gardens, lawns, roadways, etc. [3].

the aquatic environment, and could interfere with hormonal functions, thus, emerging as a potential threat to both living beings and the environment.

1.2   CLASSIFICATION

The huge database on organic chemicals in water must be logically structured for effective usage, focusing on classifying individual organics into an array relevant to health researchers and analytical chemists. Chemically related chemicals frequently have similar health effects and are usually accessible to the same analytical procedures; therefore, classification by chemical function makes the most sense. Toxicologists should review the database more efficiently and prioritize chemical classes for health effects investigations if chemicals are classified into optimally sized classes. As a result of this procedure, the classes should be prioritized based on the necessity for analytical methods development. In light of these considerations, the following organics in the water classification scheme are proposed. This is an analytical chemist's system based on chemical function similarities and is broken down into narrow enough sub-classes so that only one or two analytical methods are needed to analyze all members of each class. Because of so little knowledge in this field, health impacts are forced to take a back seat [6].

1.2.1   MAJOR CLASSES OF ORGANIC COMPOUNDS IN WATER

All chemicals are classified into one of the 24 major classes listed alphabetically. Except for miscellaneous nonvolatile compounds, all classes are functional groups.

Alphabet	Micropollutants
A	Alcohols, aldehydes, alkane, hydrocarbons, alkene, alkyne, alkyne, terpenoid hydrocarbons, amides, amines, and amino acids.
B	Benzenoid hydrocarbons.
C	Carbohydrates, carboxylic acids, and anhydrides.
E	Esters, ethers, and heterocyclic oxygen compounds
H	Halogenated aliphatic compounds, halogenated aromatic compounds
K	Ketones
N	Nitro-compounds, nitrogen compounds, nonvolatile compounds, miscellaneous.
O	Organometallic compounds
P	Phenols and naphthols, phosphorus compounds, polynuclear organic compounds
S	Steroids, sulfur compounds

These MPs arise from various sources, and their harmful effects on humans and aquatic species, such as toxicity, bioaccumulation, and endocrine disruption, vary widely. Furthermore, their physiological and chemical features differ in their functional groups, molar mass, octanol–water partition coefficient, and water content. Their collective elimination in wastewater treatment is frequently influenced by their solubility, among other factors. Furthermore, a cocktail of chemicals with a synergistic impact on aquatic species may be more lethal by the same or a different way of action. [7]. As a result, it is critical to look into eliminating various things. Table 1.2 classified some of the common MPs, their types, applications, and major concerns.

Nanomaterials are materials with morphological properties in at least one dimension smaller than one-tenth of a micrometer. Cosmetics, sunscreen, and joint replacement materials all contain engineered nanoparticles [14]. When nanomaterials are consumed by cells or organisms, they cause a wide range of toxic cell damage, which has far-reaching implications for animal health and environmental risk [15]. Pesticides, on the other hand, are the most dangerous pollutants to the ecology. The pesticide level of drinking water in Europe is 0.1 μg/L [16, 17]. Chlorinated solvents such as tetrachloroethene (PCE), trichloroethene (TCE), and 1,1,1-trichloroethane (1,1,1-TCA), petroleum hydrocarbons, especially polyaromatic hydrocarbons (PAH), and other industrial additives and by-products contaminate groundwater because they're water-soluble located nearer to improper subsurface storage locations [18, 19]. Furthermore, food additives are the most significant pollutant in the environment. According to the World Health Organization (WHO), food additives are substances added to food to maintain its safety, freshness, flavor, texture, or appearance [20]. Camphor, eucalyptol 1, 8-cineole, citral, citronellal, cis-3-hexanol, heliotropin, and hexanoic acid are food additives that include triacetin, menthol, and terpineol. Some of them, as well as endocrine disruptors and oxidants, could be involved [21].

TABLE 1.2
Classification, Applications, and Major Concerns of Micropollutants [8].

Micropollutant	Type	Uses	Concerns
Estrone (E1)	Steroid hormone (natural)	As medication in hormone replacement therapy	Feminizing male fish can lead to population collapse [9].
17 α-ethinylestradiol (EE2)	Steroid hormone (natural)	In contraceptive pills	Feminizing male fish can lead to population collapse [9].
PBDEs: Industrial, Chemical	Brominated flame retardants (BFRs)	Plastics, polyurethane foam, and textiles contain flame retardants.	Persistent, Bioaccumulative, and toxic reproduction in humans and animals [10, 11].
Perfluorinated compounds (PFCs)	Industrial chemical	Antimicrobial ingredients in toothpaste, soaps, cosmetics, and other items	Endocrine disruptor, toxic [12].
PAHs	Industrial chemical	Product of incomplete combustion of organic materials, a by-product in the processing of raw materials	Carcinogenic and mutagenic compounds to humans and animals [13]

1.3 EXPOSURE PATHWAYS

The MPs enter the environment through several pathways such as (bio) chemical conversion, suspension in water, and/or irrevocable to solids [22]. Exposure to the environment degrades only fractions of MPs. In contrast, major fractions are diffused into the surroundings where their hazards depend on the physical and chemical properties of the compound, such as its solidity, aqueous phase solubility, vapor pressure, water partitioning, and environmental sector in which the microbial metabolic activities of MPs have been released [23]. Chemicals with low water solubility are more persistent, hazardous, and bio-accumulative than those with higher solubility. As a result, they can travel to far-flung destinations. On the other hand, compounds with higher solubility and transformation rates spread quickly. Then, through a variety of routes, metabolic activities can transform them from less available forms to more available forms (Figure 1.1) (phase I and phase II) [24]. Most of these MPs are not completely digested by the animals, resulting in the excretion of the parent molecules, as well as their metabolites and/or conjugates, into the environment via urine and feces [25]. Similarly, the bulk of antibiotics is eliminated without altering their molecular structure [26]. As a result, most MPs end up in either wastewater treatment plants (WWTPs) or food chains, where they often go unnoticed. Direct emission from households, animal feedlots, agricultural regions, hospitals, and industries adds to the MP burden in surface waterways [27, 28].

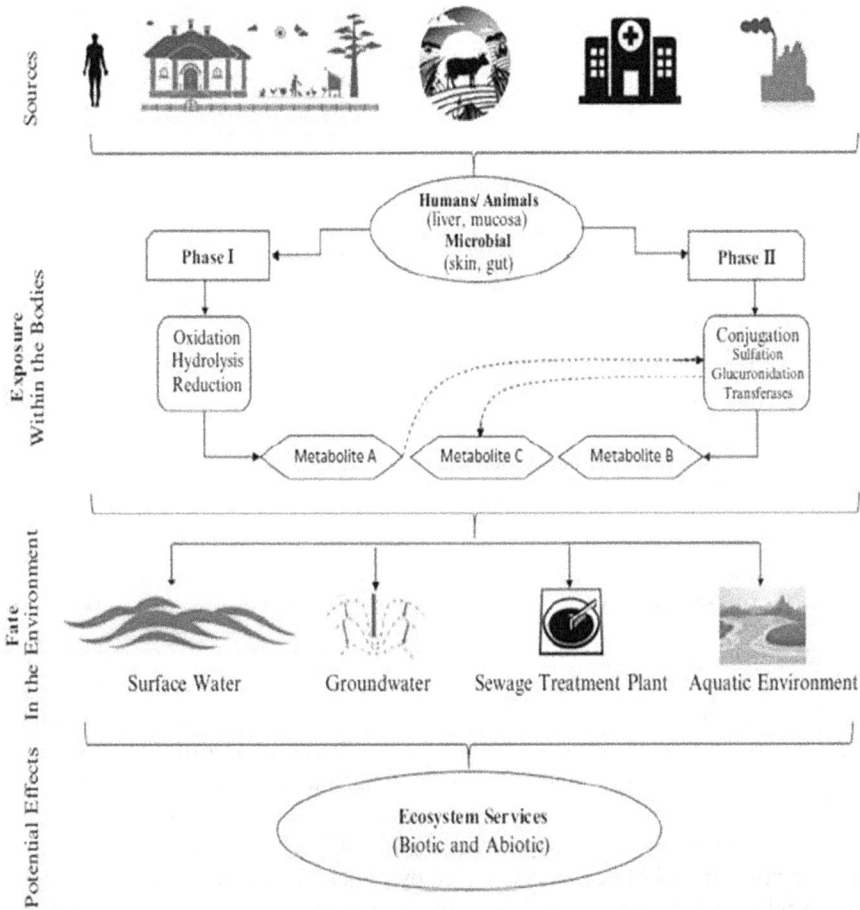

FIGURE 1.1 Scheme illustrating micropollutant biotransformation, destiny, and environmental impacts [29].

1.4 POTENTIAL RISK OF MICROPOLLUTANTS TO THE LIVING BEINGS AND ENVIRONMENT

Even at trace levels, most MPs are bioactive and may be hazardous to living beings [30]. Numerous researches have evaluated the acute and chronic impacts of MPs. Mostly, acute MP toxicity is less likely to develop at its current levels in the environment; however, chronic toxic effects are well-known and are the subject of substantial research in developed countries, strikingly active in complex mixtures [31]. For example, triclosan, a widely employed biocide, has been continuously detected in trace amounts of wastewater, streams, and seawater and is found to be highly toxic to green [32]. Higher toxicity levels of pharmaceuticals prolonged exposure to carbamazepine, clofibric acid, diclofenac, ofloxacin, propranolol, and sulfamethoxazole [33]. MPs are associated with ecotoxicity and endocrine-disrupting effects, including a reduction in fertility and physiological disorders in aquatic organisms [34]. Gee et al. reported that feminization and decline in male fish fertility were associated with steroidal hormones [35]. A study by Diamanti-Kandarakis and co-workers on human beings found the adverse effect of endocrine disruptors on fertility, breast development, neuroendocrinology, thyroid, metabolism and obesity, and prostate [36]. Some of the related case studies are given in the next sections.

1.4.1 Surfactant and Personal Care Product

Detergents, polymers, textiles, and paper sectors utilize surfactants extensively. According to a previous study by [37], nonylphenol (NP) is more harmful to aquatic organisms than the breakdown product, nonylphenol ethoxylate (NPE), which indicates that environmentally relevant quantities of NP can inhibit the growth of the mysid crustacean *Americamysis bahia*, which is used for personal hygiene and beauty products [38]. It comprises skin care products, shampoos, soaps, dental care products, etc. This product contains many synthetic chemical compounds and a large variety of harmful parabens and triclosan (TCS, $C_{12}H_7Cl_2O_2$).

1.4.2 Pharmaceuticals

Many pharmaceutical compounds are released into the environment after use by humans and animals. Experiments with a single drug indicated that the acute toxicity of most medications on aquatic creatures is improbable at environmentally relevant doses. It can have acute or persistent effects on flora and wildlife. NSAIDS (non-steroidal anti-inflammatory medications), DCF (diclofenac), GEM (gemfibrozil), IBF (ibuprofen), and other pharmacological chemicals. For 28 days, a DCF concentration of 5 g/L can harm the kidneys and liver of fish. DCF, IBF, and GEM bioaccumulation in rainbow trout blood was discovered by [39]. Antibiotics and their metabolites are also released into the environment, increasing the possibility of bacterial resistance to antibiotics in aquatic ecosystems. According to [40], the most commonly discharged chemicals into aquatic ecosystems include sulfonamides, tetracyclines, fluoroquinolones, macrolides, lactams, and aminoglycosides. Toxic substances can harm organs and cause reproductive problems, hormonal imbalances, stunted or abnormal growth, and behavioral disorders in children.

1.4.3 Steroid Hormones

The endocrine and immunological systems are regulated by steroid hormones, which are a class of chemicals. Natural hormones are divided into four categories: estrogens, androgens, progestagens, and corticoids. In addition to natural hormones, synthetic hormones such as ethinylestradiol, mestranol, and dexamethasone have been developed. The steroid hormone molecules have a higher level of endocrine disruption. Even at quantities close to detection limits, they are hazardous. They can affect fish, amphibians, reptiles, and invertebrates. Plant growth was altered by estrone (E1) and estradiol (E2) [41]. In humans, ethinylestradiol (EE2) causes breast cancer [42].

1.4.4 INDUSTRIAL AND CONSUMER APPLICATION

Perfluorinated compounds (PFCs) have been made since 1950, and they have been utilized in a vast array of industrial and consumer applications due to their physico-chemical features. The enhanced toxicity of perfluorooctane sulfonate (PFOS) and perfluorooctanoic acid (PFOA) results in carcinogenic and hormonal consequences. According to research conducted on animals, this exacerbates the effects on serum, liver, and kidney. Furthermore, due to their toxicity, PFCs have the potential to disrupt the food chain. PFCs have also been identified in blood and tissue samples from people exposed to PFCs in occupational and nonoccupational contexts around the globe [43–46].

1.4.5 PESTICIDES

Pesticide is a substance used to eradicate undesirable insects, pests, or organisms. Insecticides, herbicides, urea, and many others fall under the category of pesticides. Mammals are moderately poisoned by organochlorine insecticides, whereas aquatic organisms are severely poisoned. According to a prior study, pesticides thin the eggshell of black ducks [47]. Pesticides containing organophosphorus are hazardous to various organisms. An organophosphorus chemical called chlorpyrifos is a major environmental threat. Organophosphates and oxon metabolites harm sperm DNA, according to [46, 47]. Harmful to sperm DNA, the examined chemicals and their metabolites were more toxic than the parental compounds, according to the researchers. Male farmers' sperm quality and fertility have declined, while female farmers' breast cancer risk has increased because of atrazine contamination in their water, according to [47].

1.4.6 HEAVY METALS

Heavy metals, such as zinc (Zn), copper (Cu), nickel (Ni), mercury (Hg), cadmium (Cd), lead (Pb), and chromium (Cr), pose a significant threat to aquatic ecosystems due to their toxicity and non-biodegradability. These metals are of particular concern in wastewater, where their accumulation can have detrimental effects on aquatic organisms. Moreover, arsenic (As) consumption leads to skin, lung, and bladder cancer, according to [47].

1.5 CONCLUSION

MPs continue to pose new and serious threats to aquatic ecosystems as well as human health, even though the generation of new MPs is changing due to global economic growth. Implementing integrated MP safety management requires modern technology, risk assessment methodologies, conservation measures, and remediation technologies. In addition, a complete analysis of the toxicity of microorganisms, their metabolites, and their fate in the environment is necessary to successfully forecast the impacts of MPs on the receiving environment.

REFERENCES

1. Wada, Yoshihide, M. Flörke, N. Hanasaki, S. Eisner, G. Fischer, S. Tramberend, Y. Satoh et al. "Modeling global water use for the 21st century: The water futures and solutions (WFaS) initiative and its approaches." *Geoscientific Model Development* 9, no. 1 (2016): 175–222.
2. Burek, P., Y. Satoh, G. Fischer, M.T. Kahil, A. Scherzer, S. Tramberend, L.F. Nava, Y. Wada, S. Eisner, M. Flörke, et al. *Water Futures and Solution-Fast Track Initiative (Final Report)*. Laxenburg: IIASA, 2016.
3. Ribeiro, Ana R. Lado, Nuno F. F. Moreira, Gianluca Li Puma, and Adrián M. T. Silva. "Impact of water matrix on the removal of micropollutants by advanced oxidation technologies." *Chemical Engineering Journal* 363 (2019): 155–173.
4. Schäfer, Andrea I., Ime Akanyeti, and Andrea J. C. Semião. "Micropollutant sorption to membrane polymers: A review of mechanisms for estrogens." *Advances in Colloid and Interface Science* 164, no. 1–2 (2011): 100–117.
5. Muriuki, Cecilia, Pius Kairigo, Patrick Home, Elijah Ngumba, James Raude, Anthony Gachanja, and Tuula Tuhkanen. "Mass loading, distribution, and removal of antibiotics and antiretroviral drugs in selected wastewater treatment plants in Kenya." *Science of The Total Environment* 743 (2020): 140655.
6. Garrison, Arthur W., Lawrence H. Keith, and Walter M. Shackelford. "Occurrence, Registry, and Classification of Organic Pollutants in Water, with Development of a Master Scheme for their Analysis." In *Aquatic Pollutants*, pp. 39–68. Pergamon, 1978. doi:10.1016/B978-0-08-022059-8.50009-1.
7. Schwarzenbach, René P., Beate I. Escher, Kathrin Fenner, Thomas B. Hofstetter, C. Annette Johnson, Urs Von Gunten, and Bernhard Wehrli. "The challenge of micropollutants in aquatic systems." *Science* 313, no. 5790 (2006): 1072–1077.
8. Luo, Yunlong, Wenshan Guo, Huu Hao Ngo, Long Duc Nghiem, Faisal Ibney Hai, Jian Zhang, Shuang Liang, and Xiaochang C. Wang. "A review on the occurrence of micropollutants in the aquatic environment and their fate and removal during wastewater treatment." *Science of the Total Environment* 473 (2014): 619–641.
9. Liu, Yongze, Haowan Sun, Liqiu Zhang, and Li Feng. "Photodegradation behaviors of 17β-estradiol in different water matrixes." *Process Safety and Environmental Protection* 112 (2017): 335–341.
10. Kidd, Karen A., Paul J. Blanchfield, Kenneth H. Mills, Vince P. Palace, Robert E. Evans, James M. Lazorchak, and Robert W. Flick. "Collapse of a fish population after exposure to a synthetic estrogen." *Proceedings of the National Academy of Sciences* 104, no. 21 (2007): 8897–8901.
11. Gorga, Marina, Elena Martínez, Antoni Ginebreda, Ethel Eljarrat, and Damià Barceló. "Determination of PBDEs, HBB, PBEB, DBDPE, HBCD, TBBPA and related compounds in sewage sludge from Catalonia (Spain)." *Science of the Total Environment* 444 (2013): 51–59.
12. Lozano, Nuria, Clifford P. Rice, Mark Ramirez, and Alba Torrents. "Fate of triclocarban, triclosan and methyl triclosan during wastewater and biosolids treatment processes." *Water Research* 47, no. 13 (2013): 4519–4527.
13. Chen, Xiaoyang, Zhiyong Xue, Yanlai Yao, Weiping Wang, Fengxiang Zhu, and Chunlai Hong. "Oxidation degradation of rhodamine B in aqueous by treatment system." *International Journal of Photoenergy* 2012 (2012).
14. Colvin, Vicki L. "The potential environmental impact of engineered nanomaterials." *Nature Biotechnology* 21, no. 10 (2003): 1166–1170.
15. Stuart, M. E., K. Manamsa, J. C. Talbot, and E. J. Crane. "Emerging contaminants in groundwater." *British Geological Survey Open Report* OR/11/013 (2011): 111. https://doi.org/10.1089/109287503768335887.

16. Kolpin, Dana W., Jack E. Barbash, and Robert J. Gilliom. "Occurrence of pesticides in shallow groundwater of the United States: Initial results from the national water-quality assessment program." *Environmental Science & Technology* 32, no. 5 (1998): 558–566.
17. Lindinger, Helga, and A. Scheidleder. "Indicator fact sheet: (WEU1) Nitrate in groundwater." In *European Environmental Assessment.* Copenhagen: European Environment Agency, 2004. https://www.eea.europa.eu/data-and-maps/indicators/nitrate-in-ground water-1/weu1_ nitrategroundwater_110504.pdf
18. Miller, W. W., H. M. Joung, C. N. Mahannah, and J. R. Garrett. "Identification of water quality differences in Nevada through index application." *American Society of Agronomy, Crop Science Society of America, and Soil Science Society of America* 15, no. 3 (1986).
19. Verliefde, Arne, Emile Cornelissen, Gary Amy, Bart Van der Bruggen, and Hans Van Dijk. "Priority organic micropollutants in water sources in Flanders and the Netherlands and assessment of removal possibilities with nanofiltration." *Environmental Pollution* 146, no. 1 (2007): 281–289.
20. Vogel, Timothy M., and Perry L. Mccarty. "Biotransformation of tetrachloroethylene to trichloroethylene, dichloroethylene, vinyl chloride, and carbon dioxide under methanogenic conditions." *Applied and Environmental Microbiology* 49, no. 5 (1985): 1080–1083.
21. Stuart, Marianne, Dan Lapworth, Emily Crane, and Alwyn Hart. "Review of risk from potential emerging contaminants in UK groundwater." *Science of the Total Environment* 416 (2012): 1–21.
22. Metcalfe, Chris D., Sonya Kleywegt, Robert J. Letcher, Edward Topp, Purva Wagh, Vance L. Trudeau, and Thomas W. Moon. "A multi-assay screening approach for assessment of endocrine-active contaminants in wastewater effluent samples." *Science of the Total Environment* 454 (2013): 132–140.
23. Corcoran, Jenna, Matthew J. Winter, and Charles R. Tyler. "Pharmaceuticals in the aquatic environment: A critical review of the evidence for health effects in fish." *Critical Reviews in Toxicology* 40, no. 4 (2010): 287–304.
24. Boxall, Alistair B. A., Murray A. Rudd, Bryan W. Brooks, Daniel J. Caldwell, Kyungho Choi, Silke Hickmann, Elizabeth Innes et al. "Pharmaceuticals and personal care products in the environment: What are the big questions?" *Environmental Health Perspectives* 120, no. 9 (2012): 1221–1229.
25. Zhang, Yongjun, Sven-Uwe Geißen, and Carmen Gal. "Carbamazepine and diclofenac: Removal in wastewater treatment plants and occurrence in water bodies." *Chemosphere* 73, no. 8 (2008): 1151–1161.
26. Kümmerer, Klaus. "Antibiotics in the aquatic environment–a review–part I." *Chemosphere* 75, no. 4 (2009): 417–434.
27. Eggen, Trine, Monika Moeder, and Augustine Arukwe. "Municipal landfill leachates: A significant source for new and emerging pollutants." *Science of the Total Environment* 408, no. 21 (2010): 5147–5157.
28. Verlicchi, Paola, Alessio Galletti, Mira Petrovic, and Damiá Barceló. "Hospital effluents as a source of emerging pollutants: An overview of micropollutants and sustainable treatment options." *Journal of Hydrology* 389, no. 3–4 (2010): 416–428.
29. Arslan, Muhammad, Inaam Ullah, Jochen A. Müller, Naeem Shahid, and Muhammad Afzal. "Organic micropollutants in the environment: Ecotoxicity potential and methods for remediation." *Enhancing Cleanup of Environmental Pollutants* (2017): 65–99.
30. Thongprakaisang, Siriporn, Apinya Thiantanawat, Nuchanart Rangkadilok, Tawit Suriyo, and Jutamaad Satayavivad. "Glyphosate induces human breast cancer cells growth via estrogen receptors." *Food and Chemical Toxicology* 59 (2013): 129–136.
31. La Farre, Marinel, Sandra Pérez, Lina Kantiani, and Damià Barceló. "Fate and toxicity of emerging pollutants, their metabolites and transformation products in the aquatic environment." *TRAC Trends in Analytical Chemistry* 27, no. 11 (2008): 991–1007.

32. Tatarazako, Norihisa, Hiroshi Ishibashi, Kenji Teshima, Katsuyuki Kishi, and Koji Arizono. "Effects of triclosan on various aquatic organisms." *Environmental Sciences: An International Journal of Environmental Physiology and Toxicology* 11, no. 2 (2004): 133–140.

33. Ferrari, Benoît, Raphael Mons, Bernard Vollat, Benoît Fraysse, Nicklas Paxēaus, Roberto Lo Giudice, Antonino Pollio, and Jeanne Garric. "Environmental risk assessment of six human pharmaceuticals: Are the current environmental risk assessment procedures sufficient for the protection of the aquatic environment." *Environmental Toxicology and Chemistry: An International Journal* 23, no. 5 (2004): 1344–1354.

34. Baynes, Alice, Christopher Green, Elizabeth Nicol, Nicola Beresford, Rakesh Kanda, Alan Henshaw, John Churchley, and Susan Jobling. "Additional treatment of wastewater reduces endocrine disruption in wild fish. A comparative study of tertiary and advanced treatments." *Environmental Science & Technology* 46, no. 10 (2012): 5565–5573.

35. Gee, David, Philippe Grandjean, Steffen Foss Hansen, Malcolm MacGarvin, Jock Martin, Gitte Nielsen, David Quist, and David Stanners. "Late lessons from early warnings: Science, precaution, innovation." *European Environment Agency. EEA Report* 2013, no. 1 (2013). https://doi.org/10.2800/70069.

36. Diamanti-Kandarakis, Evanthia, Jean-Pierre Bourguignon, Linda C. Giudice, Russ Hauser, Gail S. Prins, Ana M. Soto, R. Thomas Zoeller, and Andrea C. Gore. "Endocrine-disrupting chemicals: An endocrine society scientific statement." *Endocrine Reviews* 30, no. 4 (2009): 293–342.

37. Kalutharage, Nishantha K. "Effects of major Micropollutants in Aquatic Ecosystem." *Chemistry in Sri Lanka ISSN 1012–8999* 34, no. 1 (2017): 9.

38. Kanno, Sanae, Seishiro Hirano, Hideaki Kato, Mamiko Fukuta, Toshiji Mukai, and Yasuhiro Aoki. "Benzalkonium chloride and cetylpyridinium chloride induce apoptosis in human lung epithelial cells and alter surface activity of pulmonary surfactant monolayers." *Chemico-Biological Interactions* 317 (2020): 108962.

39. Brown, Jeffrey N., Nicklas Paxéus, Lars Förlin, and DG Joakim Larsson. "Variations in bioconcentration of human pharmaceuticals from sewage effluents into fish blood plasma." *Environmental Toxicology and Pharmacology* 24, no. 3 (2007): 267–274.

40. Zhu, Ting-ting, Zhong-xian Su, Wen-xia Lai, Yao-bin Zhang, and Yi-wen Liu. "Insights into the fate and removal of antibiotics and antibiotic resistance genes using biological wastewater treatment technology." *Science of the Total Environment* 776 (2021): 145906.

41. Shore, Laurence S., Yoram Kapulnik, Bruria Ben-Dor, Yechezkial Fridman, Smadar Wininger, and Mordechai Shemesh. "Effects of estrone and 17 β-estradiol on vegetative growth of Medicago sativa." *Physiologia Plantarum* 84, no. 2 (1992): 217–222.

42. Prentice, A., B. J. Randall, A. Weddell, A. McGill, L. Henry, C. H. W. Horne, and E. J. Thomas. "Ovarian steroid receptor expression in endometriosis and in two potential parent epithelia: Endometrium and peritoneal mesothelium." *Human Reproduction* 7, no. 9 (1992): 1318–1325.

43. Hess-Wilson, J. K., and K. E. Knudsen. "Endocrine disrupting compounds and prostate cancer." *Cancer Letters* 241, no. 1 (2006): 1–12.

44. Calafat, Antonia M., Zsuzsanna Kuklenyik, Samuel P. Caudill, John A. Reidy, and Larry L. Needham. "Perfluorochemicals in pooled serum samples from United States residents in 2001 and 2002." *Environmental Science & Technology* 40, no. 7 (2006): 2128–2134.

45. Olsen, Geary W., Jean M. Burris, David J. Ehresman, John W. Froehlich, Andrew M. Seacat, John L. Butenhoff, and Larry R. Zobel. "Half-life of serum elimination of perfluorooctanesulfonate, perfluorohexanesulfonate, and perfluorooctanoate in retired fluorochemical production workers." *Environmental Health Perspectives* 115, no. 9 (2007): 1298–1305.

46. Martin, Jonathan W., Scott A. Mabury, Keith R. Solomon, and Derek CG Muir. "Bioconcentration and tissue distribution of perfluorinated acids in rainbow trout (Oncorhynchus mykiss)." *Environmental Toxicology and Chemistry: An International Journal* 22, no. 1 (2003): 196–204.
47. Darko, Godfred, and Samuel Osafo Acquaah. "Levels of organochlorine pesticides residues in dairy products in Kumasi, Ghana." *Chemosphere* 71, no. 2 (2008): 294–298.

2 Homogeneous Advanced Oxidation Processes

Muhammad Asam Raza, Umme Farwa,
Muhammad Waseem Mumtaz,
Hamid Mukhtar and Umer Rashid

CONTENTS

2.1 Introduction 14
2.2 Classification of Homogeneous AOPs 15
    2.2.1 Homogeneous AOPs Using Energy 16
        2.2.1.1 UV Radiation 16
            2.2.1.1.1 UV Radiation and Ozone (UV/O_3)................... 16
            2.2.1.1.2 Ultraviolet Radiation and Hydrogen
                     Peroxide (UV/H_2O_2).. 16
            2.2.1.1.3 Ultraviolet Radiation, Hydrogen Peroxide
                     and Ozone (O_3/H_2O_2/UV) 17
            2.2.1.1.4 Photo-Fenton (Fe^{2+}/H_2O_2/UV) 17
        2.2.1.2 Electrical Energy 18
            2.2.1.2.1 Electrochemical Oxidation 18
    2.2.2 Electrochemical Advanced Oxidation Processes–Persulfate System 18
    2.2.3 EAOPs–Peroxy Mono Sulfate System.................................... 19
    2.2.4 EAOPs–Peroxone Systems 19
        2.2.4.1 Anodic Oxidation... 20
        2.2.4.2 Electro-Fenton 21
    2.2.5 Ultrasound Energy.. 22
        2.2.5.1 Ultrasound Radiation and Ozone......................... 22
        2.2.5.2 Ultrasound Radiation and Hydrogen Peroxide ... 23
        2.2.5.3 Ultrasound–Persulfate Systems 23
2.3 Homogeneous AOPs Without Harness of Energy 24
    2.3.1 Ozone/Alkaline Medium 24
    2.3.2 Ozone and Hydrogen Peroxide (O_3/H_2O_2) 24
    2.3.3 Hydrogen Peroxide and Catalyst.. 24
2.4 Parameters Impacting the Homogeneous AOPs............................... 25
    2.4.1 Operating pH 25
    2.4.2 Concentrations of Oxidizing Species 26
    2.4.3 Catalyst Concentration... 27
    2.4.4 Pollutants' Concentration 27
    2.4.5 Effects of Inorganic Species and Salts 27
    2.4.6 Intensity of Radiation 27

DOI: 10.1201/9781003247913-2

2.5   Advantages and Limitations of Prior Methods to New Homogeneous AOPs 28
2.6   Application in Removal of Micropollutants 29
      2.6.1   In Urban Wastewater 29
      2.6.2   Treatment of Leather and Tannery Wastes 29
      2.6.3   In Pharmaceutical Industry 30
      2.6.4   In Landfill Leachate 30
2.7   Future Prospective and Challenges 36
      2.7.1   Toxicity Estimation 36
      2.7.2   Design of Reactor 36
      2.7.3   Existence of Ionic Moieties 37
      2.7.4   Catalyst as an Instance of Limitation 37
      2.7.5   Cost of Hydrogen Peroxide 37
      2.7.6   By-products in Reaction Process 38
2.8   Conclusion 38
References 38

2.1   INTRODUCTION

In the last few decades, the progressive aquatic pollution due to numerous biological pollutants from agronomic, breeding activity, industrial production, and anthropogenic discharges has been a concern for more attention (Alfonso-Muniozguren et al., 2020). Although several conventional physicochemical methods of coagulation (Gan et al., 2019), adsorption (Iram, Guo, Guan, Ishfaq, & Liu, 2010), and Fe-C microelectrolysis (Lai, Zhou, & Yang, 2012) have been utilized for the removal of pollutants from water. The present method exhibits limited efficacy in producing satisfactory results, owing to significant drawbacks such as catalyst renewal difficulties, reduced treatment efficiency, and catalyst passivation, which impede its practical implementation. Similarly, alternative technologies for wastewater treatment have failed to ensure compliance with national quality standards due to the intricate composition of effluents derived from textile industries. So, a more effective system is required to overwhelm the current challenges in wastewater treatment (Tony, Purcell, & Zhao, 2012). Many skills have been planned and established to encounter the present and predicted action necessities. The new treatment procedures must promise to eradicate pollutants to meet the general effluent quality standards (Hammami, Oturan, Oturan, Bellakhal, & Dachraoui, 2012). The insight of innovative oxidation-based techniques was recognized by Glaze, Kang, and Chapin (1987). AOP involves the production of very reactive hydroxyl radicals ($^\bullet$OH), which have grown as an essential tool for oxidation and degradation of refractory substances in the effluent (Tony, Purcell, & Zhao, 2012).

Homogeneous processes take place in only one phase and exploit ozone (O_3) (Katsoyiannis, Canonica, & von Gunten), hydrogen peroxide (H_2O_2), ultraviolet (UV), and Fenton reagent to produce hydroxyl free radicals (Loures et al., 2013). The Fenton process can be improved by UV treatment. UV rays' irradiation can completely degrade several biological effluents in sewage water. The efficacy of UV irradiation is credited to the photo-reduction of Fe (III) to Fe (II), which generates H_2O_2 (Navarro, Ichikawa, & Tatsumi, 2010). The photo-Fenton reaction is founded on the production of $^\bullet$OH by the reaction of H_2O_2 with Fe^{2+} (catalyst) in the presence of UV–visible radiation (Kortangsakul & Hunsom, 2009). The main light-absorbing

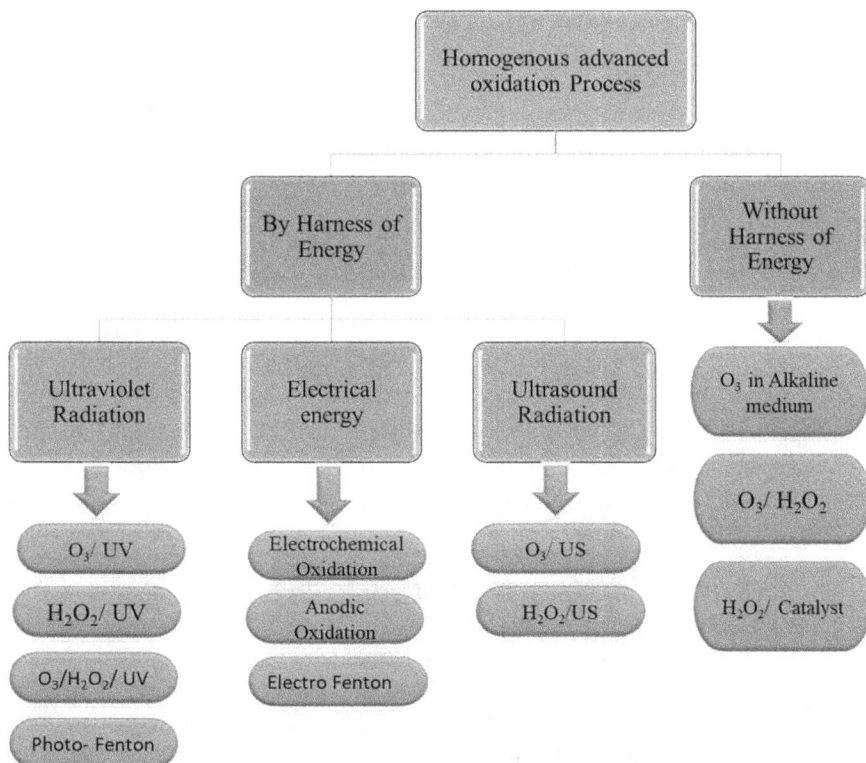

FIGURE 2.1 Classification of homogeneous advanced oxidation process.

substances in the system are iron complexes, which yield extra Fe^{2+} by reactions of photo-induced ligands with metal charge transference. Photo-Fenton oxidation is appropriate for giving aromatic chemicals because of its usefulness in destroying carbon-based compounds (Giroto, Teixeira, Nascimento, & Guardani, 2008), and the classification of homogeneous advanced oxidation processes is given in Figure 2.1.

Oxidizing power is the most vital feature in choosing an oxidizing species for the oxidation processes. H_2O_2 is a low price and efficient oxidant extensively used in water treatment. The blends of H_2O_2 with ultraviolet, O_3, and Fe (II) are emphatic AOPs used to ruin refractory organics by the production of •OH (Fatta-Kassinos, Dionysiou, & Kümmerer, 2016). H_2O_2 is also the second strongest oxidizing agent, with a comparative oxidation power of 2.8 eV. This nonselective principal oxidant is proficient in degrading nearly all organic moieties and altering them to fewer detrimental products. Mostly, OH radical is chiefly found in the breakdown processes of H_2O_2 and O_3 in the presence of Co (II), Cu (II), and Fe (II) (Vujević, Papić, Koprivanac, & Božić, 2010).

2.2 CLASSIFICATION OF HOMOGENEOUS AOPs

The homogeneous advanced oxidation process (HAOP) is classified mainly into two main methods, accomplished by the harness of energy or without the harness of energy:

2.2.1   Homogeneous AOPs Using Energy

2.2.1.1   UV Radiation

HAOPs can be characterized as photo and non-photo-AOPs. In the photo-oxidation reaction, mainly ultraviolet radiation is used, but it is not economically viable due to the pH operating conditions and high electricity demand (Rizzo et al., 2019).

2.2.1.1.1   UV Radiation and Ozone (UV/O₃)

In a struggle to raise the efficacy of the ozone method, ultraviolet radiation has been combined with O_3 for wastewater treatment (Scheme 2.1). As a result, the photolysis of dissolved ozone takes place. H_2O_2 is a principal agent for preparing hydroxyl radicals and undergoes additional photolysis to form further hydroxyl radicals by itself (Ikehata & Li, 2018).

This process has been extensively valuable for the treatment of water having a variety of contaminants. However, stand-alone ozonation has some drawbacks, such as slow ammonia elimination, ozone budget, and toxicity of intermediary products like bromate (Mishra et al., 2017).

2.2.1.1.2   Ultraviolet Radiation and Hydrogen Peroxide (UV/H₂O₂)

The breakdown mechanism of compounds by UV/H_2O_2 involves both direct photolysis and oxidative degradation mediated by OH radicals. The OH radicals initiate the breakdown process by engaging in a second-order reaction with the compounds, leading to their breakdown. A pseudo-first-order response could be presumed by keeping a persistent concentration of radicals in the entire reaction process. Numerous factors, such as basicity, pH, and inorganic and organic matter in the water medium disturb the speed of degradation. The ultraviolet radiation system supports a twin role in the water decontamination process wherever the UV radiation can support the photolysis method of the peroxide (Scheme 2.2) for producing •OH and also

$$O_3 + H_2O + hv \rightarrow H_2O_2 \tag{2.1}$$

$$H_2O_2 + hv \rightarrow 2\,^{\bullet}OH \tag{2.2}$$

SCHEME 2.1   Synthetic process for hydroxyl-free radical

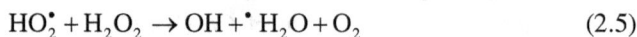

$$H_2O_2 + hv\,(UV) \rightarrow OH^{\bullet} + OH^{\bullet} \tag{2.3}$$

$$OH^{\bullet} + H_2O_2 \rightarrow HO_2^{\bullet} + H_2O \tag{2.4}$$

$$HO_2^{\bullet} + H_2O_2 \rightarrow OH +\,^{\bullet}H_2O + O_2 \tag{2.5}$$

$$HO_2^{\bullet} + HO_2^{\bullet} \rightarrow H_2O_2 + O_2 \tag{2.6}$$

SCHEME 2.2   Photolysis of hydrogen peroxide

physically deactivating the microbes by itself. The optimal concentration of H_2O_2 is a critical consideration, as excessive concentrations can lead to the scavenging of free radicals generated by peroxide molecules due to an increased concentration of H_2O_2 sources (Mierzwa, Rodrigues, & Teixeira, 2018).

2.2.1.1.3   Ultraviolet Radiation, Hydrogen Peroxide, and Ozone (O_3/H_2O_2/UV)

Several studies cited that the combination of O_3, H_2O_2, and UV light shows the best outcomes for removing impurities. The decolorization efficiencies of unlike processes like UV/O_3 and UV/O_3/H_2O_2 for eliminating RR45 dye were studied. The results revealed that all systems, with the exception of the UV treatment, exhibited complete decolorization. Besides other systems, the UV/O_3/H_2O_2 system displayed the best performance. In partial mineralization systems, the order of degree of eradication of pollutants was UV < UV/H_2O_2 < UV/O_3 < UV/O_3/H_2O_2. The UV/O_3/H_2O_2 was four times quicker than the UV/H_2O_2 process (Peternel, Koprivanac, & Kusic, 2006). Various UV methods carried out the degradation of phenol. Again, the outcomes recommended that UV/H_2O_2 process was the most effective one, with 58% total organic carbon (TOC) elimination in contrast to 44.3% by the UV/O_3 process. It also highlighted that the combination of O_3, UV, and H_2O_2 not only marks higher degradation of organic compounds but also is the furthermost economical one (Kusic, Koprivanac, & Bozic, 2006).

2.2.1.1.4   Photo-Fenton (Fe^{2+}/H_2O_2/UV)

In this process, the production of HO$^\bullet$ free radical occurred by the photolytic process of H_2O_2 along with the Fenton reaction (Figure 2.2), as the Fenton process is accelerated by ultraviolet (UV) light. This process not only fastens the reaction process but also increases the elimination of moieties in contrast to the old Fenton process. The main reason for the higher efficacy of this reaction is a conversion of Fe^{3+} to Fe^{2+} via a photocatalytic process.

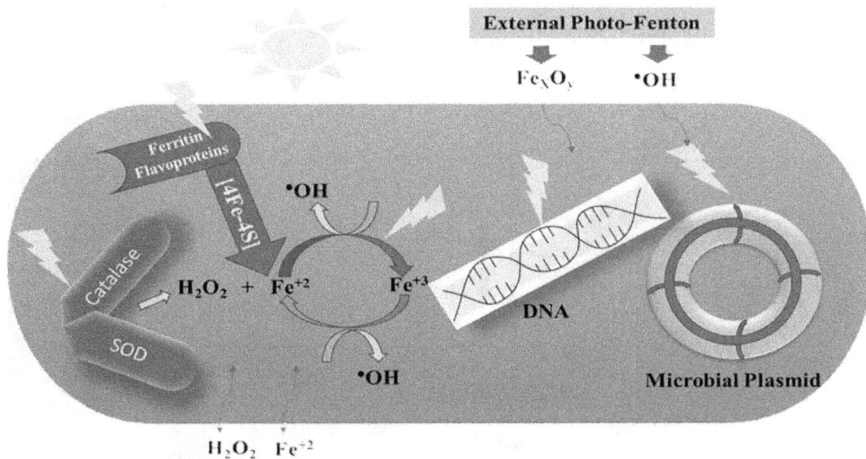

FIGURE 2.2   Reactions' pathways involved during microbial deactivation by photo-Fenton.

2.2.1.2 Electrical Energy

2.2.1.2.1 Electrochemical Oxidation

Electrochemical oxidation processes are unique and innovative (Figure 2.3) due to their eco-friendly, high catalytic ability, no secondary toxins, and mild working conditions (Qin, Sun, Liu, & Qu, 2015). Specifically, the electrochemical advanced oxidation processes (EAOPs) for the decay of organic compounds might be exploited as the pretreatment procedure before the bio-treatment process (Feng & Li, 2003). Moreover, EAOPs have the potential to efficiently decompose various pesticides, azo dyes, and some minor acids. Particularly, the electro-Fenton reaction was the most known EAOP accompanying the Fenton process. It embraces the production of H_2O_2 on the cathode via the usage of O_2. The additional Fe^{2+} reacted with H_2O_2 to give OH radical. The reduction of Fe^{3+} to Fe^{2+} quickens the Fenton process and free radical formation. Due to the supremacy of the electrochemical Fenton system, joint technologies with oxidizing agents such as persulfate (PS), peroxy mono sulfate (PMS), and O_3 are planned. Though homogeneous AOPs have trouble with catalyst reproduction and separation of catalyst, a high cost is required.

2.2.2 ELECTROCHEMICAL ADVANCED OXIDATION PROCESSES–PERSULFATE SYSTEM

The homogeneous EAOPs in the presence of persulfate along the other homogeneous catalysts are used to degrade diverse organic moieties. In this, the main electrodes are usually the cathode of titanium, stainless steel, iron sheet, and anode of assorted metal oxides, i.e., $Ti/RuO_2\text{-}IrO_2$, carbon felt, and iron sheet used (Zhang et al., 2014). In coupling processes of EAOPs–PS, catalyst Fe (II)/Fe (III) was used in the electro/persulfate process. During such a reaction, electron transfer produced a sulfate radical on the cathode. While on the other side, the Fe (II) activator stimulates PS to

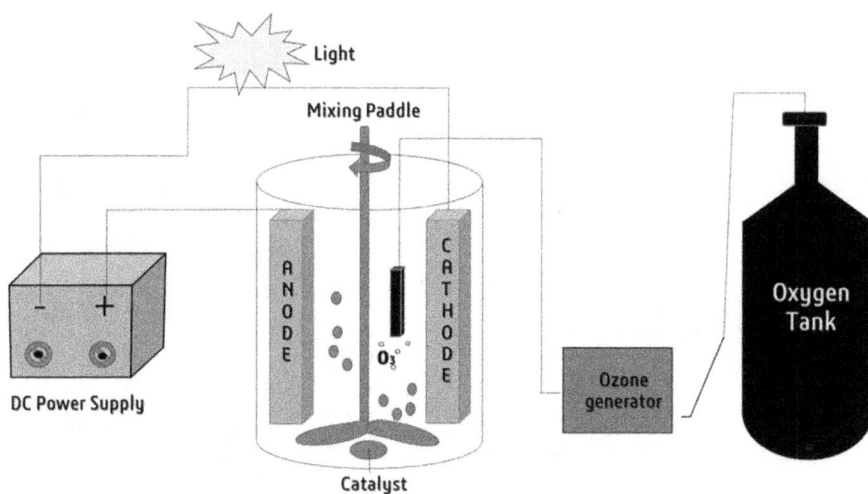

FIGURE 2.3 Setup for electrochemical AOPs.

$$Fe^{2+} + S_2O_8^{2-} \rightarrow SO^{4-} + SO_4^{2-} + Fe^{3+} \tag{2.7}$$

$$Fe^{3+} + e^- \rightarrow Fe^{2+} \tag{2.8}$$

$$Fe - 2e^- \rightarrow Fe^{2+} \tag{2.9}$$

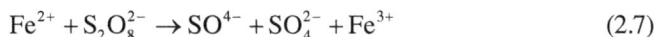

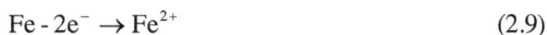

SCHEME 2.3   Reactions involved in processes of EAOPs–PS

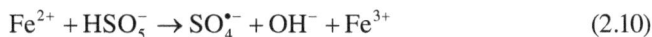

$$Fe^{2+} + HSO_5^- \rightarrow SO_4^{\bullet-} + OH^- + Fe^{3+} \tag{2.10}$$

$$HSO_5^- + e^- \rightarrow SO_4^{\bullet-} + OH^- \tag{2.11}$$

$$M(H_2O) + e^- \rightarrow M(OH^\bullet) + H^+ + e^- \tag{2.12}$$

SCHEME 2.4   Reactions involved in processes of EAOPs–PMS

give sulfate radical, and then Fe (III) formed by Fe (II) oxidation can be additionally reduced at the cathode to release Fe^{2+} (Scheme 2.3). Fe^{2+} was produced from the surface of the iron (anode). It stimulates the PS activation process when the anode is made up of iron material (Silveira, Cardoso, Barreto-Rodrigues, Zazo, & Casas, 2018).

2.2.3   EAOPs–Peroxy Mono Sulfate System

Normally, the EAOPs–PMS has also been investigated for decontamination. Recently, Fe^{2+} and Fe^{3+} as homogeneous catalysts have been added to the EAOPs–PMS system (Lin, Wu, & Zhang, 2014; Wang & Chu, 2013). The oxidation state of Fe (II) was produced from Fe (III) on the cathode, triggering the PMS to give sulfate radicals. In the meantime, the sulfate radical was also prepared from electron transfer over the cathode; additionally, hydroxyl radical formed on the anode (Scheme 2.4).

2.2.4   EAOPs–Peroxone Systems

Electro-peroxone system is an energetic EAOP coupling technique, and reported studies to explain that the EAOPs–peroxone systems utilized for numerous organic compounds, e.g., rhodamine B, oxalic acid, amoxicillin, diethyl phthalate degradation, and ibuprofen. In the electro-peroxone systems, a mixture of ozone and oxygen from the ozone generator was expunged into the reaction solution according to the concentration required. The oxygen molecule in the sparged gas transformed into H_2O_2 on the cathode (Scheme 2.5) in Eq. (2.13) (Qu et al., 2019). The conjugated base $HO_2^\bullet$ photo-Fenton from Eq. (2.14) reacts with ozone in the solution to yield HO_5 (Eqs. 2.15–2.18) (Wang et al., 2015). The recent findings indicate that

$$O_2 + 2H + 2e^- \rightarrow H_2O \qquad (2.13)$$

$$H_2O_2 \rightarrow H^+ + HO_2^- \qquad (2.14)$$

$$HO_2^- + O_3 \rightarrow HO_5^- \qquad (2.15)$$

$$HO_5^- \rightarrow HO_2 + O_3 \qquad (2.16)$$

$$O_3 \rightarrow O_2 + O^{\bullet -} \qquad (2.17)$$

$$H_2O + O^{\bullet -} \rightarrow HO^{\bullet} + OH^- \qquad (2.18)$$

$$2O_3 + H_2O_2 \rightarrow 2HO^{\bullet} + 3O_2 \qquad (2.19)$$

$$HO_5^- \rightarrow 2O_2 + OH^- \qquad (2.20)$$

$$O_3 + H_2O + e \rightarrow HO + O_2^{\bullet} + OH^- \qquad (2.21)$$

SCHEME 2.5 Reactions involved in processes of EAOPs–peroxone

the HO creation efficiency was half of what is achieved in Eq. (2.19) due to the decomposition of HO_5 into O_2 and OH as described in Eq. (2.20) (Fischbacher, von Sonntag, von Sonntag, & Schmidt, 2013). Ozone molecules in sparged gas can also be converted into OH through a certain reaction Eq. (2.21) (Zhang et al., 2019). Overall, the formation of H_2O_2 on the cathode is hazardous to the operation of the EAOPs–peroxone system, but it is necessary for pollutant removal (Wang, Yu, Deng, Huang, & Wang, 2018).

2.2.4.1 Anodic Oxidation

The anodic oxidation (AO) process finds a direct method for the electrochemical generation of •OH radicals without forming any side pollutants. The electrode material is important for the performance of AO, and numerous blends of anode and cathode have been experienced (Du, Zhang, Li, & Lai, 2020). Metal complexes can be distressed in the anodic area by any anode directly or in situ formation of •OH (Martinez-Huitle & Ferro, 2006). The anodic oxidation (AO) progression includes the direct transfer of an electron to the anode surface represented by (M), H_2O, or OH^- if the effluent is accountable for producing influential physisorbed •OH at the anode surface, denoted M(•OH), generated via Scheme 2.6 (Eq. 2.22).

The anodic oxidation process has received growing attention for high degradation flexibility, proficiency, and versatility (Du et al., 2020). This process eliminates not only ammonia but also restores nickel. The novel electrochemical reactor was developed with a rotating mesh disc as a cathode to remove the Cr,

$$M + H_2O \rightarrow M(^\bullet OH) + H^+ + e^-$$ (2.22)

$$M(^\bullet OH) + L \rightarrow M + LO + H^+ + e^-$$ (2.23)

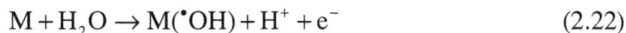

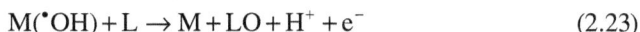

SCHEME 2.6 Reactions involved in anodic oxidation

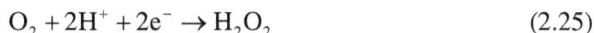

$$Fe \rightarrow Fe^{2+} + 2e^-$$ (2.24)

$$O_2 + 2H^+ + 2e^- \rightarrow H_2O_2$$ (2.25)

SCHEME 2.7 Reactions involved in electro-Fenton process

Ni, and Cu complexes (Wang, Chen, Yao, & Huang, 2015). It actively removed up to 99% within 2 hours. However, the oxygen progression reaction of the anode contests the oxidation process of the organic compounds (Panizza & Cerisola, 2009). This essential drawback limits actual applications of the AO process, whose undesirable impact can be diminished by innovative electrode material production.

2.2.4.2 Electro-Fenton

Electro-Fenton (EF) reaction comprises the formation of Fe^{2+} ions on the anode (Eq. 2.24) and the production of H_2O_2 on the cathode. The in-situ production of H_2O_2 results from the cathodic reduction of dissolved O_2 (Eq. 2.25). Afterwards, Fe^{2+} counters with H_2O_2 to cause strongly oxidative •OH for the decomposition of heavy metal, and both formations are given in Scheme 2.7.

EF owns several merits compared to the traditional Fenton reaction. It covers

(1) In situ and constant generation of H_2O_2, storage, transport, and handling.
(2) Advanced exclusion efficacy by supporting the production of Fe^{2+}.
(3) Insignificant secondary pollutants and no sludge formation.
(4) Work with the highest concentration of wastewater (He, Liu, Ye, & Wang, 2018).

The EF process is recognized as an active way to degrade heavy metal complexes (Wen, Niu, Zhang, Li, & Chen, 2018). To synthesize Co-N-doped MoO_2, improved carbon felt as cathode brings about the high H_2O_2 concentration by oxygen reduction in the EF process. To explore the electro-Fenton method, Cu/Ni-EDTA was used to eliminate the pollutants. The results disclosed that the decomplexation of Cu/Ni-EDTA occurred in the presence of •OH, and the Cu^{2+} and Ni^{2+} together promoted the joint decomposition of H_2O_2, which directed to noteworthy removal (Zhao et al., 2018) and anodic oxidation along the electrochemical Fenton reaction exhibited in Figure 2.4.

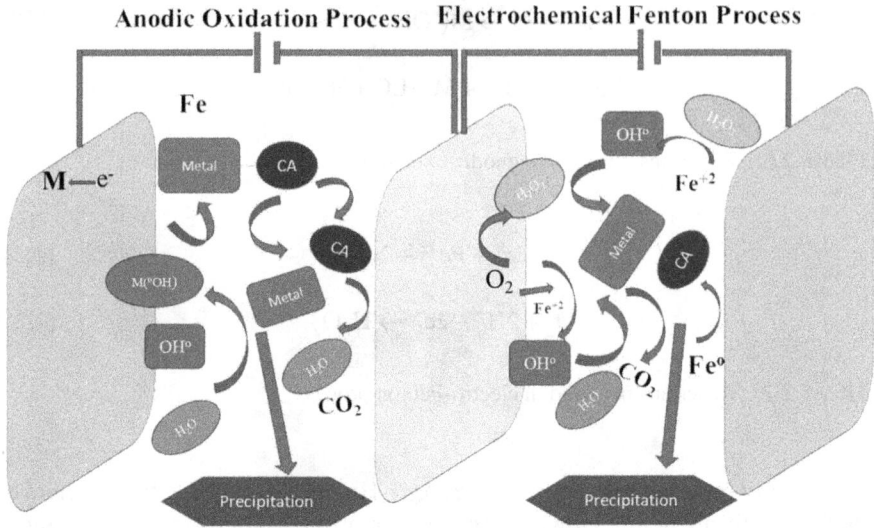

FIGURE 2.4 EAOPs by anodic oxidation and electrochemical Fenton for heavy metal complexes.

2.2.5 Ultrasound Energy

2.2.5.1 The Ultrasound (US) Energy-based pollutant decay methods encompass ultrasonic waves in the water to form free radicals and sensitive zones via cavitation, which is important for the production of microbubbles (Ince, 2018). In US systems, the inside and interface of gas solution in a cavitation bubble play a distinct role in initiating pollutants and oxidants degradation. The degradation in US arrangements is relatively unlikely from the UV-based homogeneous AOPs. Few reviews for pollutant decontamination have been reported, but a comprehensive explanation of the combinations of US and oxidants has not been provided (Wang et al., 2019). The commercial charge of the US is relatively high when considering an effective elimination process (Mahamuni & Adewuyi, 2010).

2.2.5.1 Ultrasound Radiation and Ozone

Sonolysis is an effective and operative method of catalytic approaches to improve the ozonation for the degradation of micropollutants. As ozone comprising gas bubbles passes in water, the transmission of ozone from gas medium to solution is the precondition to degrade effluents in water (Figure 2.5). The process of the US increasing ozonolysis primarily involves the increase of ozone mass transfer (Zhang, Duan, & Zhang, 2007). The ozone comprising gas bubbles was smashed in the US presence, which significantly enlarged the area of mass transmission.

The breakdown of O_3 in cavitation bubbles is the source of greater generation of •OH and H_2O_2. The two free radicals recombined to form hydrogen peroxide again. Furthermore, the produced •OH and H_2O_2 reacted with the ozone to form •OH. The

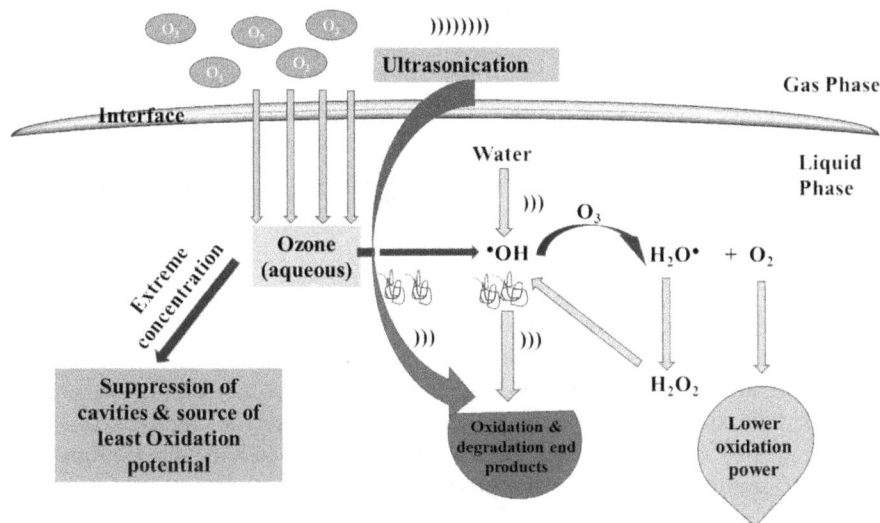

FIGURE 2.5   Combined process of ozonation and ultrasonication.

higher concentration of $^\bullet$OH also damages the ozone-refractory oxidation and is the main reason for the decay of micropollutants (Gültekin & Ince, 2006).

2.2.5.2   Ultrasound Radiation and Hydrogen Peroxide

H_2O_2 can also break down in the existence of the US to give $^\bullet$OH. H_2O_2 along the US was studied for cyanide ion dyes and the removal of herbicides. However, the elevation of H_2O_2 concentration in the US reduces the decontamination frequency of impurities owing to the inhibitory characteristics of $^\bullet$OH (Bagal & Gogate, 2012). Although the formation of $^\bullet$OH is improved and there exist no complications of waste. The pollutants' decay by US/H_2O_2 is restricted due to the US activation of hydrogen peroxide. In the meantime, the self-consumption of $^\bullet$OH was increased with the attendance of extreme $^\bullet$OH produced by the H_2O_2 decay at the gas–solution boundary. As well, hydrogen peroxide decays to water and oxygen in the process of hydrogen peroxide moving from solution to the gas–solution interface, which also decreases the consumption of hydrogen peroxide.

2.2.5.3   Ultrasound–Persulfate Systems

Sulfate-based peroxide comprises peroxymonosulfate (PMS) and peroxydisulfate (PDS). It was noted that in the last decades, PDS and PMS had become study hotspots owing to their possible applications in in-situ chemical oxidation (ISCO) as well as in wastewater refinement (Matzek & Carter, 2016). When PDS and PMS are not present in inactivated form, they produce $SO_4^{\bullet-}$, a strong oxidant that degrades numerous refractory organics in a nonselective manner. As per $SO_4^{\bullet-}$, electron transmission is predicted during the reaction mechanism, while $^\bullet$OH includes electron transference and hydrogen atom removal. The earlier discussed difference governs that both $^\bullet$OH and $SO_4^{\bullet-}$ are excellent at destroying diverse kinds of pollutants. Furthermore,

$SO_4^{\bullet-}$ has more life compared to $^{\bullet}OH$ in water solution (Ghanbari & Moradi, 2017). Of all the activation approaches, the US is one of the most updated activators for persulfate. Therefore, studies on US (PDS and PMS) for pollutant degradation have rapidly increased.

2.3 HOMOGENEOUS AOPS WITHOUT HARNESS OF ENERGY

2.3.1 Ozone/Alkaline Medium

Ozonation at high pH is a established AOP because it favors the production of $^{\bullet}OH$ (Buffle, Schumacher, Meylan, Jekel, & Von Gunten, 2006). Still, the treated water's pH directly affects the efficacy of the ozonation process through dissociated target organic compounds (Calderara, Jekel, & Zaror, 2002). Additionally, the increase of hydroxide ions rightly influences the generation of $^{\bullet}OH$ and thus indirect ozonation. If the water used for decontamination has a pH greater than 8, ozonation functional as an AOP might be an auspicious process if the formation of the precipitate of $CaCO_3$ is not considered.

2.3.2 Ozone and Hydrogen Peroxide (O_3/H_2O_2)

During the peroxone reaction, O_3 is treated with peroxide anion (HO_2^{-}) to produce $^{\bullet}OH$ as a precursor that is next reacted with $^{\bullet}OH$. A detailed mechanistic depiction of the peroxone process was reported by Merenyi, Lind, Naumov, and Sonntag (2010) (Merenyi, Lind, Naumov, & Sonntag, 2010). Residual H_2O_2 possibly had to be destroyed earlier to discharge the water into the aqueous environment. The optimal molar ratio value for the peroxone process is $H_2O_2/O_3 = 0.5$ mol/mol (Katsoyiannis, Canonica, & von Gunten, 2011). Distinctive ozone dosages are typically used in 1–20 mg/L for the peroxone process. Peroxide can also be generated from the ozone process with the aquatic matrix, but its influence on overall $^{\bullet}OH$ creation during sewage water ozonation is not noteworthy (Nöthe, Fahlenkamp, & Sonntag, 2009). O_3/H_2O_2 is a well-developed method in freshwater treatment and water reuse. However, current studies have exposed that its assistances for its application in contaminated water are restricted due to more competitive reactions and already competent radical formation with ozone alone (Hübner, Zucker, & Jekel, 2015).

2.3.3 Hydrogen Peroxide and Catalyst

Fenton's reagent, discovered by H.J.H. Fenton a century ago, comprises ferrous salt and H_2O_2. It was rapidly comprehended that Fenton's reagent is a powerful oxidant for numeral organic substances (Walling, 1975). Later, the OH radicals were generated by Scheme 2.8 by the following series of reactions (2.26–2.30).

In a homogeneous Fenton reaction, starting material and catalyst occur in a single phase, which neglects the mass transmission restrictions. The episodic addition of H_2O_2 is one of the chief disadvantages of the Fenton process that elevate

$$Fe^{2+} + H_2O_2 \rightarrow Fe^{3+} + {}^{\bullet}OH + OH^- \tag{2.26}$$

$$OH + H_2O_2 \rightarrow HO_2^{\bullet} + H_2O \tag{2.27}$$

$$Fe^{2+} + {}^{\bullet}OH \rightarrow Fe^{3+} + OH^- \tag{2.28}$$

$$Fe^{3+} + HO_2^{\bullet} \rightarrow Fe^{2+} + O_2 + H^+ \tag{2.29}$$

$$OH^{\bullet} + OH^{\bullet} \rightarrow H_2O_2 \tag{2.30}$$

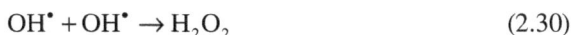

SCHEME 2.8 Possible reaction mechanism

the operational price. Also, the existence of dissolved and suspended iron catalysts designed after the reaction significantly hinders its drinking water uses. The main disadvantage of the Fenton process (Figure 2.6) is that the reformation of Fe^{2+} is less than consumed, and the process is very good only at acidic pH (2.0–3.0). Additionally, the precipitate formation of iron as iron hydroxide at basic pH decreases the accessibility of the catalyst for the reaction, and mud is produced at the end of the reaction (Divyapriya, Nambi, & Senthilnathan, 2016).

2.4 PARAMETERS IMPACTING THE HOMOGENEOUS AOPs

The reaction's efficacy is significantly reliant on the formation of ${}^{\bullet}OH$. Therefore, it is crucial to know the connection between operating constraints along with ${}^{\bullet}OH$ formation and consumption. Based on literature studies, the factors that immensely affect AOPs comprise pH of the system, type of catalysts along concentration and oxidizing agent, temperature, retention time (RT), partial pressure of O_3, catalyst, agitation speed, wavelength/intensity of light, matrix of pollutant, pH of the buffer, radiant flux, the of radiation, aeration, and reactor design (Beltrán, Masa, & Pocostales, 2009).

2.4.1 OPERATING pH

These processes have occurred efficiently in a wide range of pH values. For example, ozonation needs higher pH for nonselective reaction with all the refractory pollutants (Ikehata & El-Din, 2004). In distinction, the catalytic disintegration of H_2O_2, a Fenton, requires low pH for the formation of ${}^{\bullet}OH$ (Babuponnusami & Muthukumar, 2013). The best pH for the Fenton reaction is about 3 (Kusic, Koprivanac, & Srsan, 2006). At advanced pH, Fe^{3+} formulates $Fe(OH)_3$, which responses slowly with H_2O_2, reducing the system's efficacy. Similarly, at a very low pH, $[Fe(H_2O)_6]^{2+}$ is present, which reacts with H_2O_2 and reduces the Fe (II) contents (Navarro, Ichikawa, & Tatsumi, 2010).

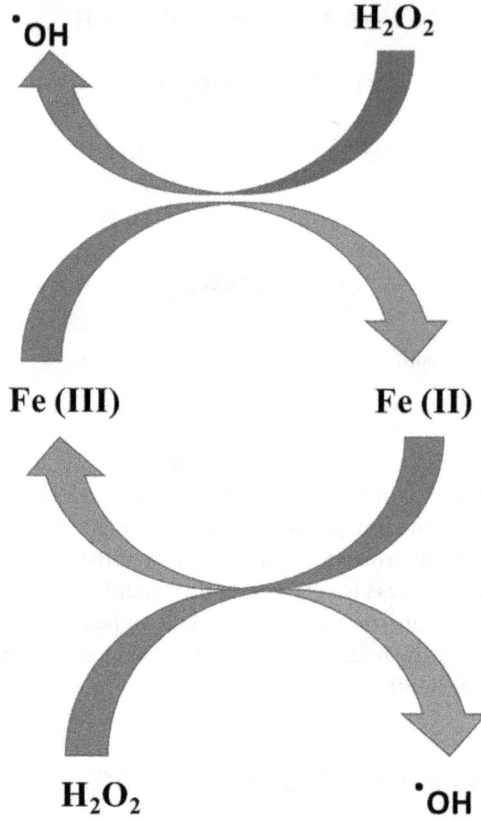

FIGURE 2.6 Interconversion of Fe (III) to Fe(II) along with the formation of •OH free radical (**Divyapriya, Nambi, & Senthilnathan, 2016**).

2.4.2 CONCENTRATIONS OF OXIDIZING SPECIES

Earlier studies have exposed that breakdown efficacy rises with H_2O_2 concentration up to a certain value, after which the degradation ability is reduced due to the H_2O_2 inhibition of •OH at a higher concentration. This is appropriate for any arrangements that comprise H_2O_2 as an oxidizing agent, such as UV/H_2O_2, O_3/H_2O_2, Fenton, and photo-Fenton (Yu, Wu, Ho, & Hong, 2010). At higher concentrations, H_2O_2 is treated with •OH and produces hydroperoxyl radicals, which are not more reactive compared to •OH (Meriç, Selçuk, & Belgiorno, 2005). In count, the recombination of •OH also decreases the degradation proficiency, as rarer radicals exist for oxidation (Neyens & Baeyens, 2003). Hydrogen peroxide can be a source of an upsurge in chemical oxygen demand (COD) and is also expensive. Therefore, there is a need to improve the concentration of H_2O_2 to remove its harmful effect and minimize the price of the operations.

2.4.3 Catalyst Concentration

Photocatalytic activity has to rise when a catalyst is loaded to a specific dosage, and an additional rise results in a decrease in efficiency (Wu, 2008). However, the active sites and beam energy decrease when catalyst loading rises. Further rise in the catalyst concentration marks the nonuniform distribution of light strength and sources the reaction rate (Chen, Du, Zhu, & Zhou, 2013). Therefore, an extreme quantity of unexploited iron salts for homogeneous systems is central to sludge formation and post-treatment (Hammami, Oturan, Oturan, Bellakhal, & Dachraoui, 2012).

2.4.4 Pollutants' Concentration

The early concentration of contaminants has an important effect on the decay process of AOP action systems (Mohammadi & Sabbaghi, 2014). The increase in the impurity concentration results in the lowered path length of photons inflowing the solution, which reduces the number of photons absorbed on the catalyst surface and drops the degree of degradation (Bahrami & Nezamzadeh-Ejhieh, 2015). In addition, a rise in dye concentration in the ozonation process lessens the total organic carbon (TOC) and color elimination at a steady state, which considered that O_3 consumption possibly rises in this state (Soares, Órfão, Portela, Vieira, & Pereira, 2006).

2.4.5 Effects of Inorganic Species and Salts

The existence of inorganic species improved the concentrations of energetic radicals and their abilities to effectively oxidize carbon-based pollutants in the effluent (Merouani, Hamdaoui, Saoudi, Chiha, & Pétrier, 2010). The radical anions designed by this process also act as radical scavengers, resulting in extended dye degradation. They react with OH, producing a high amount of •OH radicals for degradation. The typical inorganic anions present in the effluent are NaCl, KNO_3, $NaNO_3$, Na_3PO_4, Na_2SO_4, Na_2CO_3, and $NaHCO_3$ (Devi, Raju, Kumar, & Rajashekhar, 2011).

2.4.6 Intensity of Radiation

UV irradiation can significantly affect the direct creation of •OH and the photoreduction rate of Fe^{3+} to Fe^{2+} (Papić, Vujević, Koprivanac, & Šinko, 2009). The rate of reaction is directly related to the strength of radiation, but the reaction rate declines beyond a certain extent of intensity. The application of radioactivity in AOPs helps to raise the decay and decolorization process. The decolorization ability of the system is also due to both photosensitivity and the dye's reactivity. Most dyes are light-resistant; UV radiation must be combined with effective oxidants such as H_2O_2 to improve the photolysis process. It was concluded that shorter wavelengths have higher efficiency in removing pollutants (Dükkancı, Vinatoru, & Mason, 2014).

2.5   ADVANTAGES AND LIMITATIONS OF PRIOR METHODS TO NEW HOMOGENEOUS AOPs

AOPs might be well-thought-out as an active substitute and have transformed into the most extensively used techniques for biological effluents. It mainly produces highly reactive species, i.e., hydroxyl free radical, to convert the carbon-based matter to carbon dioxide and water (Trabelsi-Souissi, Oturan, Bellakhal, & Oturan, 2011). Sunlight as an irradiation source decreases procedural expenses and makes it more reasonable for pollutant treatment practices (Babuponnusami & Muthukumar, 2012). These methods can be applied earlier or later on by using old practiced methods, depending on the effluents and distinctive requirements of the sample solution after treatment (Gotvajn, Zagorc-Končan, & Cotman, 2011).

Several authors have reported the advantages and disadvantages of diverse oxidation processes, i.e., UV/H$_2$O$_2$, photo-Fenton ozone-based, and UV/TiO$_2$ processes (Barreto-Rodrigues, Silva, & Paiva, 2009a, 2009b; Barreto-Rodrigues, Souza, Silva, Silva, & Paiva, 2009). Overall, the following positive aspects have been highlighted (Gabardo Filho, 2005; Pignatello, Oliveros, & MacKay, 2006; Pillai et al., 2017; Teixeira & Jardim, 2004):

1. Unlike conventional technologies, these use powerfully oxidant species under specific conditions (HAOPs).
2. Used for the decay of refractory materials resistant to other reactions, as the biological processes.
3. Permitted the conversion of refractory material to biodegradable substances.
4. It might be used in coupling with other processes before or after treatment.
5. Having a strong oxidizing effect with high reaction rates.
6. Best to decline the concentration of the compound designed by another pretreatment, such as decontamination.
7. The generation of by-products can be diminished if adjusted amounts of reactants are used.
8. In several courses, these treatments drive less energy in comparison to the incineration process of liquid wastewater.
9. Permit *in situ* treatment.

AOPs cannot be non-discriminately applied to random residue processing; certain conditions are known to limit their operation, as pointed out by some workers (Domènech, Jardim, & Litter, 2001; Morais, 2005):

1. Not all procedures are readily scaled up to marketable or industrial wants
2. Prices can be high due to the harnessing of electrical energy by radiation-emitting sources, i.e., photo-oxidative degradation processes
3. Limitations are also associated with higher concentrations of pollutants
4. In a few cases, the control of oxidizing agent concentration and pH correction is required for the operation and post-processing of effluent treatment (i.e., Fenton and photo-Fenton)

2.6 APPLICATION IN REMOVAL OF MICROPOLLUTANTS

Micropollutants (MPs) are mainly considered natural and anthropogenic substances present in the water in small quantities, typically in the range of ng/L to µg/L. However, MPs encompass an enormous range of substances, compounds of the manufacturing industry, drugs, pesticides, beauty products, steroid hormones, etc. (Ribeiro, Nunes, Pereira, & Silva, 2015). In addition, several pollutants are known as contaminants of emerging concern (CECs), a name presently employed in the outline of ecological disciplines for novel synthetic substances, unsafe substances, and formerly unidentified compounds that exist in the atmosphere that individually has been sensed owing to developments in the analytical process (Barbosa, Moreira, Ribeiro, Pereira, & Silva, 2016).

2.6.1 IN URBAN WASTEWATER

The bactericidal role of sunlight and solar/H_2O_2-mediated microbial deactivation are significant baseline procedures though their contribution is low and unimportant in consideration of the ideal values for wastewater treatment. As a consequence, it is essential to start these borderline conditions. Initially, solar light deactivates the intracellular bacterial protection mechanisms in contradiction to the natural and photo-generated reactive oxidizing species (ROS, i.e., $HO^•$, H_2O_2, and $O^{2•-}$). On the other side, light persuades a better intracellular Fe release due to the super oxide-mediated potential in its iron-containing components (de la Obra Jimenez et al., 2020). The effluents removed by activated sludge, moving bed bioreactor, and coagulation-flocculation were further treated with innumerable oxidation processes (UV, UV/H_2O_2, Fenton, solar irradiation, solar photo-Fenton) at a research laboratory. The organic micropollutants (Carbamazepine, Diclofenac, Clarithromycin, Benzotriazole, Metoprolol, Mecoprop) promise the new ecological legislation requirements in Switzerland, which were estimated along with the removal of the organic load (Giannakis et al., 2015).

2.6.2 TREATMENT OF LEATHER AND TANNERY WASTES

Homogeneous as well as heterogeneous oxidation processes were used for the treatment of tannery waste. The chemical oxygen demand (COD) value of tannery wastewater was reduced from 3,400 mg/L to 140 mg/L in wastewater treated by combining both processes at ideal pH value of 7, using Fe^{2+} dosage of 0.5 g/L, TiO_2 dosage of 0.2 g/L, and H_2O_2 dosage of 1.8 g/L. The biodegradability of wastewater amplified from 0.4 to 0.7 at an action time of 1 hour (Selvabharathi et al., 2016). After re-tanning, the degradation and decontamination of the synthetic tannin and degreasing agent in the tannery industry were treated using Fenton and photo-Fenton oxidation processes in batch reactors at pH 3 ± 0.2 for 30 minutes. For synthetic tannin (ST) exclusions, photo-Fenton (PF) process provided noticeably high chemical oxygen demand (COD) (>80%) and UV_{254} and UV_{280} (>90%). The highest removal of COD is 57% for Fenton oxidation. It was also supported that degreasing waste water treatment led to 72% COD removal, however, it also led to 86% and 77% of UV_{254} and UV_{280}

eliminations, respectively (Lofrano, Meric, Inglese, Nikolau, & Belgiorno, 2010). The textile water treatment by UV/H$_2$O$_2$/O$_3$ was in 30-minute intervals. However, the time duration for UV/H$_2$O$_2$ was 10 minutes, which was the less effective method (Pourgholi, Jahandizi, Miranzadeh, Beigi, & Dehghan, 2018).

2.6.3 IN PHARMACEUTICAL INDUSTRY

In treating medicinal industry wastewater comprising paracetamol and chloramphenicol from pharmaceutical industry wastewater, the Fenton process stated almost 95% COD removal with a primary COD concentration (12,000 mg/L). Broad mineralization of penicillin was detected in 40 minutes via Fenton/UV radiation method (Gadipelly et al., 2014). The ozonation of antibiotics covering wastewater, specifically quinolones, macrolides, lincosamides, tetracyclines, and sulfonamides, attained extra than 76% degradation. Moreover, the elimination of total organic carbon was small. The entire organic carbon exclusion via the ozonation process was near the ground, and the degraded product showed less harmfulness than its parental molecule (Homem & Santos, 2011).

A complete degradation sequence of diclofenac was explained. The study stated the 72% elimination of diclofenac over BDD electrode at 30 mg/L in 4 hours. The diclofenac intermediates designed throughout the reaction were benzoic acid, 2,6-dichlorobenzenamine, and dihydroxy benzoyl alcohol oxidized into small acidic compounds after covering the reaction time (X. Zhao, Hou, Liu, Qiang, & Qu, 2009). The biodegradation of two cytostatic medicines, cyclophosphamide and ifosfamide, from the clinic and hospital wastewater is improved by employing the O$_3$/H$_2$O$_2$/UV process, as reported by Česen et al. (2015). Likewise, the glutaraldehyde elimination from secondary effluent samples from a local hospital in southwestern Brazil via the O$_3$/UV process was reported by Kist, Rosa, Machado, Camargo, and Moro (2013).

2.6.4 IN LANDFILL LEACHATE

Conventional treatment techniques were not enough to meet the criteria for completely removing micropollutants in landfill leachate. The Fenton oxidation process of the leachate was accomplished with certain conditions, i.e., reaction time at 20–90 minutes, a dose of Fe(II): 0.51–2.55 g/L, H$_2$O$_2$: 5.1–25.5 g/L, and pH ranges between 3 and 5 while for ozonation reaction time being 10–130 minutes and pH values in the range 4–10 (Table 2.1). Reaction time was considered the most important condition for both processes to remove 4-nonylphenol, di-(2-ethylhexyl) phthalate, and 4-tert-octylphenol. However, the degradation efficiency of micropollutants was found in between 5 and 100%. The Fenton reaction was superior due to the easy removal of chemical oxygen demand (COD) and phthalates as micropollutants (Ateş & Argun, 2021). AOPs of hydroxyl and sulfate radicals could be a good method for removing or converting micropollutants to non-hazardous materials. The micropollutants in landfill leachate are detected and eliminated through AOPs (Pisharody, Gopinath, Malhotra, Nidheesh, & Kumar, 2021). Electrochemical AOPs had been an actual substitute for eliminating refractory and recalcitrant organic matter present mainly in urban solid landfill leachates. The electro-Fenton process with a Pt-based anode and a carbon-sensed cathode and the anodic oxidation with a boron-doped diamond

TABLE 2.1

Homogeneous Advanced Oxidation Process for the Removal of Micropollutants.

Sr. No.	Source of Pollutant	Micropollutants	Method	Reactors/Cell	Reactor Conditions	Outcomes	Factors Impacting	References
1)	Waste waters from the dyeing manufacturing industry	Azo dyes	Photo-Fenton AOPs	Photo-reactor in a mirrored box (dimensions = 60 cm × 40 cm × 55 cm)	UV treatment (310–390 nm) by 150 W bulb lamp, flux of incident UV lamp about 2.6 × 10^{-5}, Fe^{3+} catalyst, pH = 7.0	Removal of color (98%), COD (78%), and TOC (59%).	Eliminating COD 83% using 1.5 mM Fe^{3+}, a rise of Fe^{3+} concentration lowered the COD removals due to scavenging of the free radical effect of extreme Fe^{3+}/Fe^{2+} concentrations. COD eliminations also function of H_2O_2 concentration and initial effluent COD. an upsurge in H_2O_2 led to the least COD removal.	(Arslan-Alaton, Tureli, & Olmez-Hanci, 2009)
2)	Landfill leachates	PAHs, VOCs, PCBs, and OCPs	Electrochemical AOPs	An open, undivided, cylindrical electrochemical cell	Pt and BDD (anode), carbon-based (cathode), in situ H_2O_2 by O_2 reduction, 0.2 mM $FeSO_4.7H_2O$ added in Fe^{2+} (0.2 mM) as a catalyst, current (500 or 1000 mA), pH = 8.05 and temperature (23 ± 2 °C).	PAHs (87.4 to 100%), VOCs (78.1 to 100%), PCBs and OCPs (86.9 to 100%).	TOC removal (86.8%) using BDD anode was high, and percentage decreased using Pt anode (71.0%). In AO, initial TOC content is higher than EF treatment due to the pH of the bulk solution not being modified in the earlier case, whereas it is mandatory for EF processes.	(Oturan et al., 2015)

(Continued)

TABLE 2.1 (*Continued*)
Homogeneous Advanced Oxidation Process for the Removal of Micropollutants.

Sr. No.	Source of Pollutant	Micropollutants	Method	Reactors/Cell	Reactor Conditions	Outcomes	Factors Impacting	References
3)	Pesticides comprising industrial wastewater	Organophosphorus comprising substrates as: fenitrothion, diazinon, and profenofos.	Fenton, photo-Fenton, and UV/H_2O_2	Quartz Photo reactor	pH 2–5, Hg-lamp 150 W, UV emitter as Heraeus TQ-150, pH = 3, COD:H_2O_2 = 4.4:1, H_2O_2:Fe^{2+} ratio = 100:1 by the oxidation by the Fenton treatment; and pH 3, COD: H_2O_2 = 2.2:1, H_2O_2:Fe^{2+} ratio = 50:1 by the oxidation by the photo-Fenton treatment.	TOC decay by Fenton and photo-Fenton, respectively, fenitrothion: 54.2%, 86.9% diazinon: 12.9%, 56.7% and profenofos: 46.2%, 89.7% at pH 3.	Optimum pH for Fenton and photo-Fenton process = 3; as pH >3, the decay percentage declined. In the photo-Fenton reaction, more ·OH via photocatalysis (Fe^{3+} complex ions and H_2O_2) by UV irradiation.	(Badawy, Ghaly, & Gad-Allah, 2006)
4)	Polluted concrete	Naphthalene, fluorene, phenanthrene, fluoranthene, and pyrene	Novel photochemical, FHO_3 AOP	Photo-reactor along ozonizer	UV lamps (300 W, 280–400 nm), ozonating ambient air by high voltage and producing 2 g of ozone $hour^{-1}$. H_2O_2 (10%) with added anionic and cationic surfactant (0.2%).	Naphthalene (32.24 ppm), fluorene (0.904 ppm), phenanthrene (1.016 ppm), fluoranthene (1.137 ppm), and pyrene (1.255 ppm)	—	(Nikolić et al., 2017)

5)	Hospital wastewaters (HWW)	Saturated hydrocarbons, aromatic hydrocarbons, fatty acids, and phenols	UV/H_2O_2/O_3	Pyrex glass photo-reactor	The rate of supplied air was 2000 mL/min, and aeration was 400 mL/ minute of O_2. Hg-lamp PUV-1022, Heraeus emission range (200 to 460 nm), 0.02 ml of H_2O_2.	Pollutants (73%) were removed in 20 min.	AOP by UV/H_2O_2/O_3 reduces 50% more chemical compounds than the aeration process. Polycyclic aromatics are not easily removed because C–Cl bonds are not hydrolyzed and resist biotic and photolytic decay.	(Mejía-Morales et al., 2020)
6)	Coal-fired flue gas	NO and SO_2	UV/H_2O_2	Photochemical reactor (constructed by a PMMA bubble column reactor)	In quartz case (UV lamp), sand chip-based gas distributor, water bath temperature maintenance (DCW-1015, ±0.1 °C), 800 mL of $KMnO_4$ and NaOH solution.	Elimination of SO_2 (100%), NO in UV/ H_2O (11.5%) although in UV/H_2O_2 (72.0%)	As H_2O_2 conc. = 0 mol/L very slight NO eliminates while H_2O_2 = 1.5 mol/L elimination of NO (0 to 69%). Elimination percentage of NO 10.8% absence of UV light. As UV power 0–36 increased, NO removal increased from 10.8% to 72.5%.	(Liu, Zhang, Sheng, Zhang, & Zhao, 2010)
7)	Wastewater	Oil removal gas oil has paraffins and a higher relative concentration of C9, C11, C12, C13, C14, and C15.	Photo-Fenton (UV/H_2O_2/ Fe^{+2}) system	Batch mode laboratory-scale reactor,	Fenton-based reagents were added to the solution by $FeCl_3$ (10–100 mg/L) and then H_2O_2 (100–800 mg/L). The optimum temperature was 30°C, and the pH was about 3.	72% of oil from wastewater volume (1,000 mg/L).	Oil degradation increased H_2O_2 (100–400 mg/L) and 54% removal at 400 mg/L after irradiation (150 min). The degradation potential of waste material decreased at pH (<2.5) due to $[Fe(H_2O)_6]^{2+}$ complex formation and the least production of $HO^\bullet$ production.	(Mustafa, Alwared, & Ebrahim, 2013)

(Continued)

TABLE 2.1 (Continued)
Homogeneous Advanced Oxidation Process for the Removal of Micropollutants.

Sr. No.	Source of Pollutant	Micropollutants	Method	Reactors/Cell	Reactor Conditions	Outcomes	Factors Impacting	References
							The removal efficiency decreases linearly from 79% to 50% as the concentration of gas oil increases from 500 to 2,000 mg/L, respectively.	
8)	Bio recalcitrant pesticides	Alachlor, atrazine, chlorfenvinfos, diuron, isoproturon, and pentachlorophenol.	(PhFO) and TiO_2^- photocatalysis/ ozone coupled systems	Pyrex glass cell with thermostatic jacket $(25.0 \pm 0.1\ °C)$	Ozone (1.6 g hour⁻¹) in solution as determined by iodometric titration. A UV source (6 W black light lamp) was employed.	First-order rate constants, kPhFO $(10^3/s^{-1})$ for pesticide at pH = 3.0, Alachlor (2.2) Atrazine (4.0), Chlorfenvinfos (4.2), Diuron (6.2), Isoproturon (6.2).	—	(Farré et al., 2005)

| 9) | Biologically treated wastewater (BTWW) | Carbamazepine, ciprofloxacin, clarithromycin, diclofenac, metoprolol, sitagliptin and sulfamethoxazole | UV/O_3 (pilot-scale study). | Photo-reactor in a pilot container and ozone analyzer (2B Technologies, Boulder, Colorado, USA). Hg lamp (Light Tech Ltd). The UV output of 3 W and 16 W at 185 nm and 254 nm, respectively. Temperature (25.5 ± 1°C–18 ± 2°C) | CARB, CIPR, CLAR, DICL, and SULF were 80% removed. In WWTP-1 METO and SITA removed 66% and 71% while for WWTP-2, 43% and 46%, respectively. | VUV has lesser contact time, penetration power, and ozone supply; therefore, the elimination of micropollutants mainly depends on HO• radicals. On the ozone reaction side, products such as NO_2, NO_3, and NH_4 were also produced; NO_3 concentration up to 50 mg/L was acceptable. | (Krakkó et al., 2021) |

Abbreviations: advanced oxidation process (AOPs), total organic carbon (TOC), volatile organic compounds (VOCs), organochlorine pesticides (OCPs), wastewater treatment plants (WWTPs), carbamazepine (CARB), ciprofloxacin (CIPR), clarithromycin (CLAR), diclofenac (DICL), metoprolol (METO), sitagliptin (SITA), sulfamethoxazole (SULF)

(BDD) anode were also used. Based on the initial procedure, the replacement of the conventional Pt anode with BDD results in improved functionality by effectively preventing the dissolution of total organic carbon (TOC). The elimination of the end products of polycyclic hydrocarbons, alkylphenols, volatile compounds, organochlorine pesticides, polychlorobiphenyls, and polybrominated diphenyl ethers was too inspected (Oturan et al., 2015).

2.7 FUTURE PROSPECTIVE AND CHALLENGES

The effectiveness of each method results from the complex relationship of inhibitory and promoting effects initiated by the water background constituents. The effluent constituents also influence the elimination of MPs consisting of natural organic matter (NOM) (generation of by-products, scavenging effects, light attenuation adsorption to catalyst in photocatalytic processes) and inorganic anions. Promoting effects can result from NOM making reactive oxygen species (ROS), iron ions as the extra basis of catalyst in Fenton methods, and phenolic or other substances having OH and COOH groups, which can reproduce the catalyst. However, it is difficult to understand and discriminate the role of every single component.

It is well recognized that the elimination competencies of MPs rest on their chemical nature. One more account on the degradation of medications exhibited an identical trend for UV-C radiation, Fenton, and photo-Fenton systems (phenacetin < metoprolol < naproxen < amoxicillin). However, metoprolol was the furthermost recalcitrant in the ozone-based processes (Benitez, Acero, Real, Roldan, & Casas, 2011). A new study on UV/chlorine action presented that the degradation of MPs having electron-withdrawing groups (e.g., F^-, Cl^-, NO_2^-) was principally ascribed to HO. However, the degradation of those encompassing electron-donating groups (e.g., alkyl-/alkoxy aromatics, aniline, phenol, and olefins) was attributed to other reactive species such as ClO, Cl, and Cl_2. The alkaline components were recognized to scavenge Cl and HO at an advanced level. Therefore, the MPs were degraded chiefly by these radicals, whose removal is most reduced by alkalinity (Guo et al., 2017).

2.7.1 TOXICITY ESTIMATION

The ecological influence of the treatment techniques concerning toxicity or harmfulness is the utmost important feature that must be considered to assess HAOPs. The treated waste should not be as harmful or poisonous as contrasted to the earlier sewage. This is because some products designed throughout the treatment procedure might enlarge the damaging upshot of a pollutant; for example, long-lasting reaction intermediates may be more toxic than the starting dye (Buthiyappan, Aziz, & Daud, 2016).

2.7.2 DESIGN OF REACTOR

In the homogeneous AOP, designing a well-organized extensive reactor faces various challenges due to the constant interaction of UV light with sample solution and mass transference difficulties in AOPs, especially in ozone presence. Photodegradation of carbamazepine and diclofenac was done in a well-designed compound triangular collector (CTC) reactor under direct UV irradiation. The photo-reactor comprises eight acrylic glass tubes that are present in an aluminum collector and have a good overall volume (10.24 L). The solar unit was recirculated by fitting a Pyrex vessel and a peristaltic pump together (Kowalska et al., 2020). The tannery sewer water was treated over assembled field scale reactor to remove 96% chemical oxygen demand (COD). The average annual UV light intensity for the reported site was 23 ± 2 W/m^2, and the everyday solar light usage was 6 hours (Selvabharathi et al., 2016).

2.7.3   EXISTENCE OF IONIC MOIETIES

The existence of ionic species changes the removal mechanism of micropollutants due to their interaction with $^\bullet$OH and UV light. Anionic groups, i.e., bicarbonate, carbonate, chloride, phosphate nitrate, and sulfate ions, which substantially affect the treatment of toxic materials, are also instances of ionic species that exist in sewerage water. Hence, further research is required to evaluate the effects of both anionic and cationic moieties. According to the literature, the presence of Cl$^-$ retards the degradation of impurities. Chloride ion was supposed to be an effective scavenger to compete with hydroxyl radical. In the degradation process of trichloroethylene, the Cl^{-1} present at 0 to 0.2 mM showed no substantial change in the value of rate constants. Though, once Cl^{-1} ion concentration surpassed 0.2 mM, degradation was stopped (Liang, Wang, & Mohanty, 2006). Hydroxyl radicals can react easily with phosphate ions to form phosphate radicals (PO$_4^{\bullet 2-}$) and are sensed by electron paramagnetic resonance (EPR) and transient spectroscopy. Hydroxyl radicals can also react with hydrogen phosphate and dihydrogen phosphate ions to yield hydrogen phosphate radicals and dihydrogen phosphate radicals, respectively (Black & Hayon, 1970).

2.7.4   CATALYST AS AN INSTANCE OF LIMITATION

Considering a catalyst with higher action and a lesser price is essential since costly catalytic agents are not applied for manufacturing applications. Catalytic inactivation may happen due to changes in its definite area, the intoxication of the catalyst by intermediates, surface deposition, and a strong polymeric carbon layer adsorption (Jafarinejad, 2017). The homogeneous catalytic ozonation method was also employed to efficiently remove organic compounds from aqueous solutions. Still, water treatment charges can be greater due to the formation of ions, resulting in secondary pollution (Kasprzyk-Hordern, Ziółek, & Nawrocki, 2003).

2.7.5   COST OF HYDROGEN PEROXIDE

The major disadvantage of HAOPs is the usage of expensive reagents, particularly H$_2$O$_2$ (Oller, Malato, & Sánchez-Pérez, 2011). The consumption of H$_2$O$_2$ can be lowered via nonstop feed during the reaction and good optimization of AOPs. Although H$_2$O$_2$ is a relatively expensive reagent, the operational expenses are offset by the lower fixed capital cost associated with wet air oxidation and ozonation (Centi, Perathoner, Torre, & Verduna, 2000).

2.7.6   BY-PRODUCTS IN REACTION PROCESS

These reactions may produce various side products with higher polarization and H_2O solubility than the parent substances. These side products were found to be noxious for human and aquatic life. The challenging task in AOPs is the identification of intermediates and by-products formed during the degradation process (Fatta-Kassinos, Vasquez, & Kümmerer, 2011). Although, when the Fenton process occurred in the dark period, acids, i.e., acetic acid, glyoxylic acid, and maleic acid, formed as side products, which made complexes with Fe^{3+} and a source of least elimination of dissolved organic carbon (DOC) (Pignatello et al., 2006). In a homogeneous oxidation reaction, the Fenton catalyst caused a side reaction, forming the least reactive ferryl ion (FeO^{2+}) (Yang et al., 2019).

2.8   CONCLUSION

HAOPs are a promising pretreatment process for decaying and treating industrial pollutants in numerous types of wastewater. The most extensively studied HAOPs included UV-radiation-based, O_3-based processes, Fenton, photo-Fenton and electro-Fenton, and electrochemical oxidation processes. It was concluded that Fenton-based processes are found to be healthier methods for water treatment in different industries. They have various advantages, such as they are readily available, stored and handled very safely, and without environmental impact. Moreover, electrochemical oxidation is also considered one of the emergent methods for wastewater treatment. The different parameters such as pH, the concentration of oxidants, catalyst, pollutants, and light irradiation may influence each process. This chapter provides an analysis of the benefits and drawbacks associated with each process, as well as an examination of significant obstacles that must be overcome in the future. Furthermore, a thorough investigation and assessment of all factors that impact degradation efficiency progress are presented.

REFERENCES

Alfonso-Muniozguren, P., Cotillas, S., Boaventura, R. A., Moreira, F. C., Lee, J., & Vilar, V. J. (2020). Single and combined electrochemical oxidation driven processes for the treatment of slaughterhouse wastewater. *Journal of Cleaner Production*, *270*, 121858.

Arslan-Alaton, I., Tureli, G., & Olmez-Hanci, T. (2009). Treatment of azo dye production wastewaters using Photo-Fenton-like advanced oxidation processes: Optimization by response surface methodology. *Journal of Photochemistry and photobiology A: Chemistry*, *202*(2–3), 142–153.

Ateş, H., & Argun, M. E. (2021). Advanced oxidation of landfill leachate: Removal of micropollutants and identification of by-products. *Journal of Hazardous Materials*, *413*, 125326.

Babuponnusami, A., & Muthukumar, K. (2012). Removal of phenol by heterogenous photo electro Fenton-like process using nano-zero valent iron. *Separation and Purification Technology*, *98*, 130–135.

Babuponnusami, A., & Muthukumar, K. (2013). Treatment of phenol-containing wastewater by photoelectro-Fenton method using supported nanoscale zero-valent iron. *Environmental Science and Pollution Research*, *20*(3), 1596–1605.

Badawy, M. I., Ghaly, M. Y., & Gad-Allah, T. A. (2006). Advanced oxidation processes for the removal of organophosphorus pesticides from wastewater. *Desalination*, *194*(1–3), 166–175.

Bagal, M. V., & Gogate, P. R. (2012). Sonochemical degradation of alachlor in the presence of process intensifying additives. *Separation and Purification Technology*, *90*, 92–100.

Bahrami, M., & Nezamzadeh-Ejhieh, A. (2015). Effect of the supported ZnO on clinoptilolite nano-particles in the photodecolorization of semi-real sample bromothymol blue aqueous solution. *Materials Science in Semiconductor Processing*, *30*, 275–284.

Barbosa, M. O., Moreira, N. F., Ribeiro, A. R., Pereira, M. F., & Silva, A. M. (2016). Occurrence and removal of organic micropollutants: An overview of the watch list of EU Decision 2015/495. *Water Research*, *94*, 257–279.

Barreto-Rodrigues, M., Silva, F. T., & Paiva, T. C. (2009a). Combined zero-valent iron and fenton processes for the treatment of Brazilian TNT industry wastewater. *Journal of Hazardous Materials*, *165*(1–3), 1224–1228.

Barreto-Rodrigues, M., Silva, F. T., & Paiva, T. C. (2009b). Optimization of Brazilian TNT industry wastewater treatment using combined zero-valent iron and fenton processes. *Journal of Hazardous Materials*, *168*(2–3), 1065–1069.

Barreto-Rodrigues, M., Souza, J. V., Silva, E. S., Silva, F. T., & Paiva, T. C. (2009). Combined photocatalytic and fungal processes for the treatment of nitrocellulose industry wastewater. *Journal of Hazardous Materials*, *161*(2–3), 1569–1573.

Beltrán, F., Masa, F., & Pocostales, J. (2009). A comparison between catalytic ozonation and activated carbon adsorption/ozone-regeneration processes for wastewater treatment. *Applied Catalysis B: Environmental*, *92*(3–4), 393–400.

Benitez, F. J., Acero, J. L., Real, F. J., Roldan, G., & Casas, F. (2011). Comparison of different chemical oxidation treatments for the removal of selected pharmaceuticals in water matrices. *Chemical Engineering Journal*, *168*(3), 1149–1156.

Black, E., & Hayon, E. (1970). Pulse radiolysis of phosphate anions $H_2PO_4^-$, HPO_4^{2-}, PO_4^{3-} and $P_2O_7^{4-}$ in aqueous solutions. *The Journal of Physical Chemistry*, *74*(17), 3199–3203.

Buffle, M.-O., Schumacher, J., Meylan, S., Jekel, M., & Von Gunten, U. (2006). Ozonation and advanced oxidation of wastewater: Effect of O_3 dose, pH, DOM and HO•-scavengers on ozone decomposition and HO• generation. *Ozone: Science and Engineering*, *28*(4), 247–259.

Buthiyappan, A., Aziz, A. R. A., & Daud, W. M. A. W. (2016). Recent advances and prospects of catalytic advanced oxidation process in treating textile effluents. *Reviews in Chemical Engineering*, *32*(1), 1–47.

Calderara, V., Jekel, M., & Zaror, C. (2002). Ozonation of 1-naphthalene, 1, 5-naphthalene, and 3-nitrobenzene sulphonic acids in aqueous solutions. *Environmental Technology*, *23*(4), 373–380.

Centi, G., Perathoner, S., Torre, T., & Verduna, M. G. (2000). Catalytic wet oxidation with H_2O_2 of carboxylic acids on homogeneous and heterogeneous Fenton-type catalysts. *Catalysis Today*, *55*(1–2), 61–69.

Česen, M., Kosjek, T., Laimou-Geraniou, M., Kompare, B., Širok, B., Lambropolou, D., & Heath, E. (2015). Occurrence of cyclophosphamide and ifosfamide in aqueous environment and their removal by biological and abiotic wastewater treatment processes. *Science of the Total Environment*, *527*, 465–473.

Chen, D., Du, G., Zhu, Q., & Zhou, F. (2013). Synthesis and characterization of TiO_2 pillared montmorillonites: Application for methylene blue degradation. *Journal of Colloid and Interface Science*, *409*, 151–157.

de la Obra Jimenez, I., Giannakis, S., Grandjean, D., Breider, F., Grunauer, G., López, J. L. C., . . . Pulgarin, C. (2020). Unfolding the action mode of light and homogeneous vs. heterogeneous photo-Fenton in bacteria disinfection and concurrent elimination of micropollutants in urban wastewater, mediated by iron oxides in raceway pond reactors. *Applied Catalysis B: Environmental*, *263*, 118158.

Devi, L. G., Raju, K. A., Kumar, S. G., & Rajashekhar, K. E. (2011). Photodegradation of di azo dye Bismarck Brown by advanced photo-Fenton process: Influence of inorganic anions and evaluation of recycling efficiency of iron powder. *Journal of the Taiwan Institute of Chemical Engineers*, *42*(2), 341–349.

Divyapriya, G., Nambi, I. M., & Senthilnathan, J. (2016). Nanocatalysts in Fenton based advanced oxidation process for water and wastewater treatment. *Journal of Bionanoscience*, *10*(5), 356–368.

Domènech, X., Jardim, W. F., & Litter, M. I. (2001). Procesos avanzados de oxidación para la eliminación de contaminantes. *Eliminación de contaminantes por fotocatálisis heterogénea*, *2016*, 3–26.

Du, J., Zhang, B., Li, J., & Lai, B. (2020). Decontamination of heavy metal complexes by advanced oxidation processes: A review. *Chinese Chemical Letters*, *31*(10), 2575–2582.

Dükkancı, M., Vinatoru, M., & Mason, T. J. (2014). The sonochemical decolourization of textile azo dye Orange II: Effects of Fenton type reagents and UV light. *Ultrasonics Sonochemistry*, *21*(2), 846–853.

Farré, M. J., Franch, M. I., Malato, S., Ayllón, J. A., Peral, J., & Doménech, X. (2005). Degradation of some biorecalcitrant pesticides by homogeneous and heterogeneous photocatalytic ozonation. *Chemosphere*, *58*(8), 1127–1133.

Fatta-Kassinos, D., Dionysiou, D. D., & Kümmerer, K. (2016). *Advanced treatment technologies for urban wastewater reuse*. Springer.

Fatta-Kassinos, D., Vasquez, M. I., & Kümmerer, K. (2011). Transformation products of pharmaceuticals in surface waters and wastewater formed during photolysis and advanced oxidation processes–degradation, elucidation of by-products and assessment of their biological potency. *Chemosphere*, *85*(5), 693–709.

Feng, Y., & Li, X. Y. (2003). Electro-catalytic oxidation of phenol on several metal-oxide electrodes in aqueous solution. *Water Research*, *37*(10), 2399–2407.

Fischbacher, A., von Sonntag, J., von Sonntag, C., & Schmidt, T. C. (2013). The• OH radical yield in the $H_2O_2+ O_3$ (peroxone) reaction. *Environmental Science & Technology*, *47*(17), 9959–9964.

Gabardo Filho, H. (2005). *Estudo e projeto de reatores fotoquimicos para tratamento de efluentes liquidos*. Thesis. https://doi.org/10.47749/T/UNICAMP.2005.358773

Gadipelly, C., Pérez-González, A., Yadav, G. D., Ortiz, I., Ibáñez, R., Rathod, V. K., & Marathe, K. V. (2014). Pharmaceutical industry wastewater: Review of the technologies for water treatment and reuse. *Industrial & Engineering Chemistry Research*, *53*(29), 11571–11592.

Gan, Y., Wang, X., Zhang, L., Wu, B., Zhang, G., & Zhang, S. (2019). Coagulation removal of fluoride by zirconium tetrachloride: Performance evaluation and mechanism analysis. *Chemosphere*, *218*, 860–868.

Ghanbari, F., & Moradi, M. (2017). Application of peroxymonosulfate and its activation methods for degradation of environmental organic pollutants. *Chemical Engineering Journal*, *310*, 41–62.

Giannakis, S., Vives, F. A. G., Grandjean, D., Magnet, A., De Alencastro, L. F., & Pulgarin, C. (2015). Effect of advanced oxidation processes on the micropollutants and the effluent organic matter contained in municipal wastewater previously treated by three different secondary methods. *Water Research*, *84*, 295–306.

Giroto, J., Teixeira, A., Nascimento, C., & Guardani, R. (2008). Photo-Fenton removal of water-soluble polymers. *Chemical Engineering and Processing: Process Intensification*, *47*(12), 2361–2369.

Glaze, W. H., Kang, J.-W., & Chapin, D. H. (1987). The chemistry of water treatment processes involving ozone, hydrogen peroxide and ultraviolet radiation. *Ozone: Science and Engineering*, *4*, 335–352.

Gotvajn, A. Ž., Zagorc-Končan, J., & Cotman, M. (2011). Fenton's oxidative treatment of municipal landfill leachate as an alternative to biological process. *Desalination, 275*(1–3), 269–275.

Gültekin, I., & Ince, N. (2006). Degradation of aryl-azo-naphthol dyes by ultrasound, ozone and their combination: Effect of α-substituents. *Ultrasonics Sonochemistry, 13*(3), 208–214.

Guo, K., Wu, Z., Shang, C., Yao, B., Hou, S., Yang, X., . . . Fang, J. (2017). Radical chemistry and structural relationships of PPCP degradation by UV/chlorine treatment in simulated drinking water. *Environmental Science & Technology, 51*(18), 10431–10439.

Hammami, S., Oturan, M. A., Oturan, N., Bellakhal, N., & Dachraoui, M. (2012). Comparative mineralization of textile dye indigo by photo-Fenton process and anodic oxidation using boron-doped diamond anode. *Desalination and Water Treatment, 45*(1–3), 297–304.

He, W., Liu, Y., Ye, J., & Wang, G. (2018). Electrochemical degradation of azo dye methyl orange by anodic oxidation on Ti_4O_7 electrodes. *Journal of Materials Science: Materials in Electronics, 29*(16), 14065–14072.

Homem, V., & Santos, L. (2011). Degradation and removal methods of antibiotics from aqueous matrices–a review. *Journal of Environmental Management, 92*(10), 2304–2347.

Hübner, U., Zucker, I., & Jekel, M. (2015). Options and limitations of hydrogen peroxide addition to enhance radical formation during ozonation of secondary effluents. *Journal of Water Reuse and Desalination, 5*(1), 8–16.

Ikehata, K., & El-Din, M. G. (2004). Degradation of recalcitrant surfactants in wastewater by ozonation and advanced oxidation processes: A review. *Ozone: Science & Engineering, 26*(4), 327–343.

Ikehata, K., & Li, Y. (2018). Ozone-based processes. In *Advanced oxidation processes for waste water treatment* (pp. 115–134). Elsevier.

Ince, N. H. (2018). Ultrasound-assisted advanced oxidation processes for water decontamination. *Ultrasonics Sonochemistry, 40*, 97–103.

Iram, M., Guo, C., Guan, Y., Ishfaq, A., & Liu, H. (2010). Adsorption and magnetic removal of neutral red dye from aqueous solution using Fe_3O_4 hollow nanospheres. *Journal of Hazardous Materials, 181*(1–3), 1039–1050.

Jafarinejad, S. (2017). Cost-effective catalytic materials for AOP treatment units. In *Applications of advanced oxidation processes (AOPs) in drinking water treatment* (pp. 309–343). Springer.

Kasprzyk-Hordern, B., Ziółek, M., & Nawrocki, J. (2003). Catalytic ozonation and methods of enhancing molecular ozone reactions in water treatment. *Applied Catalysis B: Environmental, 46*(4), 639–669.

Katsoyiannis, I. A., Canonica, S., & von Gunten, U. (2011). Efficiency and energy requirements for the transformation of organic micropollutants by ozone, O_3/H_2O_2 and UV/H_2O_2. *Water Research, 45*(13), 3811–3822.

Kist, L. T., Rosa, E. C., Machado, Ê. L., Camargo, M. E., & Moro, C. C. (2013). Glutaraldehyde degradation in hospital wastewater by photoozonation. *Environmental Technology, 34*(18), 2579–2586.

Kortangsakul, S., & Hunsom, M. (2009). The optimization of the photo-oxidation parameters to remediate wastewater from the textile dyeing industry in a continuous stirred tank reactor. *Korean Journal of Chemical Engineering, 26*(6), 1637–1644.

Kowalska, K., Maniakova, G., Carotenuto, M., Sacco, O., Vaiano, V., Lofrano, G., & Rizzo, L. (2020). Removal of carbamazepine, diclofenac and trimethoprim by solar driven advanced oxidation processes in a compound triangular collector based reactor: A comparison between homogeneous and heterogeneous processes. *Chemosphere, 238*, 124665.

Krakkó, D., Illés, Á., Licul-Kucera, V., Dávid, B., Dobosy, P., Pogonyi, A., . . . Záray, G. (2021). Application of (V) UV/O_3 technology for post-treatment of biologically treated wastewater: A pilot-scale study. *Chemosphere, 275*, 130080.

Kusic, H., Koprivanac, N., & Bozic, A. L. (2006). Minimization of organic pollutant content in aqueous solution by means of AOPs: UV-and ozone-based technologies. *Chemical Engineering Journal*, *123*(3), 127–137.

Kusic, H., Koprivanac, N., & Srsan, L. (2006). Azo dye degradation using Fenton type processes assisted by UV irradiation: A kinetic study. *Journal of Photochemistry and Photobiology A: Chemistry*, *181*(2–3), 195–202.

Lai, B., Zhou, Y., & Yang, P. (2012). Treatment of wastewater from acrylonitrile–butadiene–styrene (ABS) resin manufacturing by Fe0/GAC–ABFB. *Chemical Engineering Journal*, *200*, 10–17.

Liang, C., Wang, Z.-S., & Mohanty, N. (2006). Influences of carbonate and chloride ions on persulfate oxidation of trichloroethylene at 20 C. *Science of the Total Environment*, *370*(2–3), 271–277.

Lin, H., Wu, J., & Zhang, H. (2014). Degradation of clofibric acid in aqueous solution by an EC/Fe^{3+}/PMS process. *Chemical Engineering Journal*, *244*, 514–521.

Liu, Y., Zhang, J., Sheng, C., Zhang, Y., & Zhao, L. (2010). Simultaneous removal of NO and SO$_2$ from coal-fired flue gas by UV/H$_2$O$_2$ advanced oxidation process. *Chemical Engineering Journal*, *162*(3), 1006–1011.

Lofrano, G., Meric, S., Inglese, M., Nikolau, A., & Belgiorno, V. (2010). Fenton oxidation treatment of tannery wastewater and tanning agents: Synthetic tannin and nonylphenol ethoxylate based degreasing agent. *Desalination and Water Treatment*, *23*(1–3), 173–180.

Loures, C. C., Alcântara, M. A., Izário Filho, H. J., Teixeira, A., Silva, F. T., Paiva, T. C., & Samanamud, G. R. (2013). Advanced oxidative degradation processes: Fundamentals and applications. *International Review of Chemical Engineering*, *5*(2), 102–120.

Mahamuni, N. N., & Adewuyi, Y. G. (2010). Advanced oxidation processes (AOPs) involving ultrasound for waste water treatment: A review with emphasis on cost estimation. *Ultrasonics Sonochemistry*, *17*(6), 990–1003.

Martinez-Huitle, C. A., & Ferro, S. (2006). Electrochemical oxidation of organic pollutants for the wastewater treatment: Direct and indirect processes. *Chemical Society Reviews*, *35*(12), 1324–1340.

Matzek, L. W., & Carter, K. E. (2016). Activated persulfate for organic chemical degradation: A review. *Chemosphere*, *151*, 178–188.

Mejía-Morales, C., Hernández-Aldana, F., Cortés-Hernández, D. M., Rivera-Tapia, J. A., Castañeda-Antonio, D., & Bonilla, N. (2020). Assessment of biological and persistent organic compounds in hospital wastewater after advanced oxidation process UV/H$_2$O$_2$/O$_3$. *Water, Air, & Soil Pollution*, *231*(2), 1–10.

Merenyi, G., Lind, J., Naumov, S., & Sonntag, C. V. (2010). Reaction of ozone with hydrogen peroxide (peroxone process): A revision of current mechanistic concepts based on thermokinetic and quantum-chemical considerations. *Environmental Science & Technology*, *44*(9), 3505–3507.

Meriç, S., Selçuk, H., & Belgiorno, V. (2005). Acute toxicity removal in textile finishing wastewater by Fenton's oxidation, ozone and coagulation–flocculation processes. *Water Research*, *39*(6), 1147–1153.

Merouani, S., Hamdaoui, O., Saoudi, F., Chiha, M., & Pétrier, C. (2010). Influence of bicarbonate and carbonate ions on sonochemical degradation of Rhodamine B in aqueous phase. *Journal of Hazardous Materials*, *175*(1–3), 593–599.

Mierzwa, J. C., Rodrigues, R., & Teixeira, A. C. (2018). UV-hydrogen peroxide processes. In *Advanced oxidation processes for waste water treatment* (pp. 13–48). Elsevier.

Mishra, N. S., Reddy, R., Kuila, A., Rani, A., Mukherjee, P., Nawaz, A., & Pichiah, S. (2017). A review on advanced oxidation processes for effective water treatment. *Current World Environment*, *12*(3), 470.

Mohammadi, M., & Sabbaghi, S. (2014). Photocatalytic degradation of 2, 4-DCP wastewater using MWCNT/TiO$_2$ nano-composite activated by UV and solar light. *Environmental Nanotechnology, Monitoring & Management, 1*, 24–29.

Morais, J. L. D. (2005). *Estudo da potencialidade de processos oxidativos avançados, isolados e integrados com processos biológicos tradicionais, para tratamento de chorume de aterro sanitário*. Ph.D Thesis.

Mustafa, Y. A., Alwared, A. I., & Ebrahim, M. (2013). Removal of oil from wastewater by advanced oxidation process/homogenous process. *Journal of Engineering, 19*(6), 686–694.

Navarro, R. R., Ichikawa, H., & Tatsumi, K. (2010). Ferrite formation from photo-Fenton treated wastewater. *Chemosphere, 80*(4), 404–409.

Neyens, E., & Baeyens, J. (2003). A review of classic Fenton's peroxidation as an advanced oxidation technique. *Journal of Hazardous Materials, 98*(1–3), 33–50.

Nikolić, V. M., Karić, S. D., Nikolić, Ž. M., Tošić, M. S., Tasić, G. S., Milovanovic, D. M., & Kaninski, M. P. M. (2017). Novel photochemical advanced oxidation process for the removal of polycyclic aromatic hydrocarbons from polluted concrete. *Chemical Engineering Journal, 312*, 99–105.

Nöthe, T., Fahlenkamp, H., & Sonntag, C. V. (2009). Ozonation of wastewater: Rate of ozone consumption and hydroxyl radical yield. *Environmental Science & Technology, 43*(15), 5990–5995.

Oller, I., Malato, S., & Sánchez-Pérez, J. (2011). Combination of advanced oxidation processes and biological treatments for wastewater decontamination—a review. *Science of the Total Environment, 409*(20), 4141–4166.

Oturan, N., Van Hullebusch, E. D., Zhang, H., Mazeas, L., Budzinski, H., Le Menach, K., & Oturan, M. A. (2015). Occurrence and removal of organic micropollutants in landfill leachates treated by electrochemical advanced oxidation processes. *Environmental Science & Technology, 49*(20), 12187–12196.

Panizza, M., & Cerisola, G. (2009). Direct and mediated anodic oxidation of organic pollutants. *Chemical Reviews, 109*(12), 6541–6569.

Papić, S., Vujević, D., Koprivanac, N., & Šinko, D. (2009). Decolourization and mineralization of commercial reactive dyes by using homogeneous and heterogeneous Fenton and UV/Fenton processes. *Journal of Hazardous Materials, 164*(2–3), 1137–1145.

Peternel, I., Koprivanac, N., & Kusic, H. (2006). UV-based processes for reactive azo dye mineralization. *Water Research, 40*(3), 525–532.

Pignatello, J. J., Oliveros, E., & MacKay, A. (2006). Advanced oxidation processes for organic contaminant destruction based on the Fenton reaction and related chemistry. *Critical Reviews in Environmental Science and Technology, 36*(1), 1–84.

Pillai, S. C., McGuinness, N. B., Byrne, C., Han, C., Lalley, J., Nadagouda, M., . . . OShea, K. (2017). Photocatalysis as an effective advanced oxidation process. *Advanced Oxidation Processes for Water Treatment: Fundamentals and Applications*, 333–381.

Pisharody, L., Gopinath, A., Malhotra, M., Nidheesh, P., & Kumar, M. S. (2021). Occurrence of organic micropollutants in municipal landfill leachate and its effective treatment by advanced oxidation processes. *Chemosphere*, 132216.

Pourgholi, M., Jahandizi, R. M., Miranzadeh, M., Beigi, O. H., & Dehghan, S. (2018). Removal of Dye and COD from textile wastewater using AOP (UV/O$_3$, UV/H$_2$O$_2$, O$_3$/H$_2$O$_2$ and UV/H$_2$O$_2$/O$_3$). *Journal of Environmental Health and Sustainable Development*, 621–629.

Qin, Y., Sun, M., Liu, H., & Qu, J. (2015). AuPd/Fe$_3$O$_4$-based three-dimensional electrochemical system for efficiently catalytic degradation of 1-butyl-3-methylimidazolium hexafluorophosphate. *Electrochimica Acta, 186*, 328–336.

Qu, C., Lu, S., Liang, D., Chen, S., Xiang, Y., & Zhang, S. (2019). Simultaneous electro-oxidation and in situ electro-peroxone process for the degradation of refractory organics in wastewater. *Journal of Hazardous Materials, 364*, 468–474.

Ribeiro, A. R., Nunes, O. C., Pereira, M. F., & Silva, A. M. (2015). An overview on the advanced oxidation processes applied for the treatment of water pollutants defined in the recently launched Directive 2013/39/EU. *Environment International, 75*, 33–51.

Rizzo, L., Malato, S., Antakyali, D., Beretsou, V. G., Đolić, M. B., Gernjak, W., . . . Ribeiro, A. R. L. (2019). Consolidated vs new advanced treatment methods for the removal of contaminants of emerging concern from urban wastewater. *Science of the Total Environment, 655*, 986–1008.

Selvabharathi, G., Adishkumar, S., Jenefa, S., Ginni, G., Rajesh Banu, J., & Tae Yeom, I. (2016). Combined homogeneous and heterogeneous advanced oxidation process for the treatment of tannery wastewaters. *Journal of Water Reuse and Desalination, 6*(1), 59–71.

Silveira, J. E., Cardoso, T. O., Barreto-Rodrigues, M., Zazo, J. A., & Casas, J. A. (2018). Electro activation of persulfate using iron sheet as low-cost electrode: The role of the operating conditions. *Environmental Technology, 39*(9), 1208–1216.

Soares, O. S. G., Órfão, J. J., Portela, D., Vieira, A., & Pereira, M. F. R. (2006). Ozonation of textile effluents and dye solutions under continuous operation: Influence of operating parameters. *Journal of Hazardous Materials, 137*(3), 1664–1673.

Teixeira, C., & Jardim, W. D. F. (2004). Processos oxidativos avançados: Conceitos teóricos. *Caderno temático, 3*, 83.

Tony, M. A., Purcell, P. J., & Zhao, Y. (2012). Oil refinery wastewater treatment using physicochemical, Fenton and Photo-Fenton oxidation processes. *Journal of Environmental Science and Health, Part A, 47*(3), 435–440.

Trabelsi-Souissi, S., Oturan, N., Bellakhal, N., & Oturan, M. A. (2011). Application of the photo-Fenton process to the mineralization of phthalic anhydride in aqueous medium. *Desalination and Water Treatment, 25*(1–3), 210–215.

Vujević, D., Papić, S., Koprivanac, N., & Božić, A. L. (2010). Decolorization and mineralization of reactive dye by UV/Fenton process. *Separation Science and Technology, 45*(11), 1637–1643.

Walling, C. (1975). Fenton's reagent revisited. *Accounts of Chemical Research, 8*(4), 125–131.

Wang, H., Yuan, S., Zhan, J., Wang, Y., Yu, G., Deng, S., . . . Wang, B. (2015). Mechanisms of enhanced total organic carbon elimination from oxalic acid solutions by electro-peroxone process. *Water Research, 80*, 20–29.

Wang, J., Chen, X., Yao, J., & Huang, G. (2015). Decomplexation of electroplating wastewater in a higee electrochemical reactor with rotating mesh-disc electrodes. *International Journal of Electrochemical Science, 10*(7), 5726–5736.

Wang, J., Wang, Z., Vieira, C. L., Wolfson, J. M., Pingtian, G., & Huang, S. (2019). Review on the treatment of organic pollutants in water by ultrasonic technology. *Ultrasonics Sonochemistry, 55*, 273–278.

Wang, Y., & Chu, W. (2013). Photo-assisted degradation of 2,4,5-trichlorophenol by Electro-Fe (II)/Oxone® process using a sacrificial iron anode: Performance optimization and reaction mechanism. *Chemical Engineering Journal, 215*, 643–650.

Wang, Y., Yu, G., Deng, S., Huang, J., & Wang, B. (2018). The electro-peroxone process for the abatement of emerging contaminants: Mechanisms, recent advances, and prospects. *Chemosphere, 208*, 640–654.

Wen, S., Niu, Z., Zhang, Z., Li, L., & Chen, Y. (2018). In-situ synthesis of 3D GA on titanium wire as a binder-free electrode for electro-Fenton removing of EDTA-Ni. *Journal of Hazardous Materials, 341*, 128–137.

Wu, C.-H. (2008). Effects of operational parameters on the decolorization of CI Reactive Red 198 in UV/TiO$_2$-based systems. *Dyes and Pigments, 77*(1), 31–38.

Yang, Q., Ma, Y., Chen, F., Yao, F., Sun, J., Wang, S., . . . Wang, D. (2019). Recent advances in photo-activated sulfate radical-advanced oxidation process (SR-AOP) for refractory organic pollutants removal in water. *Chemical Engineering Journal, 378*, 122149.

Yu, C.-H., Wu, C.-H., Ho, T.-H., & Hong, P. A. (2010). Decolorization of CI reactive Black 5 in UV/TiO$_2$, UV/oxidant and UV/TiO$_2$/oxidant systems: A comparative study. *Chemical Engineering Journal, 158*(3), 578–583.

Zhang, H., Duan, L., & Zhang, D. (2007). Absorption kinetics of ozone in water with ultrasonic radiation. *Ultrasonics Sonochemistry, 14*(5), 552–556.

Zhang, H., Wang, Z., Liu, C., Guo, Y., Shan, N., Meng, C., & Sun, L. (2014). Removal of COD from landfill leachate by an electro/Fe^{2+}/peroxydisulfate process. *Chemical Engineering Journal, 250*, 76–82.

Zhang, Y., Zuo, S., Zhang, Y., Ren, G., Pan, Y., Zhang, Q., & Zhou, M. (2019). Simultaneous removal of tetracycline and disinfection by a flow-through electro-peroxone process for reclamation from municipal secondary effluent. *Journal of Hazardous Materials, 368*, 771–777.

Zhao, X., Hou, Y., Liu, H., Qiang, Z., & Qu, J. (2009). Electro-oxidation of diclofenac at boron doped diamond: Kinetics and mechanism. *Electrochimica Acta, 54*(17), 4172–4179.

Zhao, Z., Dong, W., Wang, H., Chen, G., Tang, J., & Wu, Y. (2018). Simultaneous decomplexation in blended Cu (II)/Ni (II)-EDTA systems by electro-Fenton process using iron sacrificing electrodes. *Journal of Hazardous Materials, 350*, 128–135.

3 Heterogeneous Advanced Oxidation Process

Abdul Sattar Jatoi, Zubair Hashmi,
Nabisab Mujawar Mubarak and Rama Rao Karri

CONTENTS

3.1 Introduction 47
3.2 Advanced Oxidation Process.. 49
3.3 Fenton Process 49
    3.3.1 Heterogeneous Fenton Process 51
        3.3.1.1 Unsupported Heterogeneous Fenton Catalysts 51
        3.3.1.2 Supported Heterogeneous Fenton Catalysts 52
3.4 Photo-Fenton.. 55
3.5 Conclusion 59
References.. 59

3.1 INTRODUCTION

The uncontrolled usage of feedstock, along with the industry itself, has resulted in a slew of difficulties in terms of soil and water contamination in recent years (Fdez-Sanromán et al. 2021; Li 2014). Surface waters, river run-offs, effluent treatment facilities (WWTPs), groundwater, sewage sludge, and even potable water have all been shown to contain various dangerous chemicals classified as emerging contaminants (Cheng et al. 2016; Luo et al. 2020). Pesticides used in agriculture, relentless natural pollutants, foodstuff additives, medications, and FMCG items are among them. Traditional biological treatments used in wastewater treatment facilities fail to eliminate these contaminants from water, resulting in their persistent presence in the ecosystem, even at low levels. This fact offers a significant long-term threat to aquatic and/or terrestrial creatures' health and development, as well as global water supply, exacerbating many countries' drinking water crises (Gosset et al. 2016; Kumar et al. 2019b; Pi et al. 2018). According to the Food and Agriculture Organization, by 2050, the global demand for drinking water will have increased by more than 40%. As a result, the scientific community faces an incredible task in developing innovative methods for wastewater decontamination because conventional treatment plants are ineffective in eliminating such a wide variety of natural contaminants (Casado 2019; Wang & Wang 2016). Biological, physical, and physicochemical processes are

DOI: 10.1201/9781003247913-3

today's most popular technologies employed in water treatment. Biological procedures like biological treatment, biological filtration, and cure with activated sludge are commonly used to eliminate developing pollutants. Despite their cheap cost, some persistent organic pollutants are resistant to immediate biologic cure; hence, physical and chemical approaches are preferred for removing heavy metals and recalcitrant pollutants (Du & Zhou 2021). Adsorption, reverse osmosis, sedimentation, and membrane-filtration-based processes are examples of physical processes, while physicochemical approaches include electrolysis, Fenton-based processes, photolysis, ozonation, and sonolysis (Escudero-Curiel et al. 2021; Luo et al. 2021). Advanced oxidation processes (AOPs) are the common name for these widely used technologies. Due to their high effectiveness, minimalism, and repeatability, the aforementioned AOPs are considered very promising technologies. As a result, they have gotten a lot of consideration in the quest for water treatment solutions (Babuponnusami & Muthukumar 2014; Kumar et al. 2019a; Poza-Nogueiras et al. 2018).

These practices are based on the in-situ production of extremely responsive free radicals like hydroxyl radicals (-OH), sulphate radicals (SO_4), and superoxide radicals (O_2) by the activation of the antecedent oxidants (hydrogen peroxide, persulfate/peroxodisulfate, peroxymonosulfate and sodium percarbonate via oxidation/reduction reactions (Arellano et al. 2019; Brillas et al. 2009; Kang et al. 2016; Ribeiro et al. 2015; Sablas et al. 2020). Organic pollutants are degraded and partially/completely mineralized to CO_2, H_2O, and metallic ions due to the radicals generated in the main solution (Ribeiro et al. 2015; Wang & Wang 2018). Chemical (Fenton or Fenton-like), photochemical (photo-Fenton or photocatalysis), electrochemical (anodic oxidation, electro-Fenton, electrophotocatalysis), and sonochemical (sono-Fenton) methods and technique groupings (photo-electro-Fenton, sono-electro-Fenton) oxidation processes are among the oxidation processes (Giannakis et al. 2015; Puga et al. 2020; Wang & Shih 2015). UV radiation or visible mild, ultrasound, and/or alkalinity of the liquid medium are all used to activate the aforementioned precursor compounds. Other strategies encompass the usage of transition-steel-based catalysts (Fe, Mn, Co, Cu, V, Ru, Mo, Cr, Ce), heat, UV radiation, ultrasound, and/or alkalinity of the liquid medium. The narrow working pH variety of AOPs primarily based on the Fenton technique, as well as the complexity of real water matrices that favour the precipitation of the transition metals present within the catalyst, resulting in the formation of sludge and undesirable by way of merchandise, make them incorrect for huge-scale application. Some drawbacks related to Fenton and photocatalytic reactions can be mitigated using heterogeneous catalysts. Despite the efficiency of homogeneous structures (liquid section), eliminating soluble iron salts from the environment necessitates extraordinarily pricey techniques.

As a result, growing heterogeneous catalysts (stable-liquid levels) has been a vital step closer to higher catalysis. Even though numerous response routes can explain the effectiveness of heterogeneous catalysts, pollutants are generally adsorbed on the catalyst surface during the procedure, followed by chain reactions. This promotes bond cleavage, leading to the formation of consecutive intermediate species that subsequently undergo desorption. On the other hand, these heterogeneous catalysts have some benefits over homogeneous catalysts, inclusive of (i) catalyst regeneration and reuse; (ii) a broader software scope because of some conversion reactions going on at the heterogeneous catalyst; (iii) the use of a wider running pH variety, even on the natural pH value of the water matrix; (iv) clean catalyst recuperation without the formation of metallic sludge; and (v) lengthy-term cloth and energy savings. This chapter specializes in compiling the maximum significant papers on using heterogeneous catalysts in AOPs to remove growing contaminants from water resources from 2015 to the present. It must be mentioned that, due to the latest opinions in the field of heterogeneous ozonation, this trouble isn't always considered in this overview. To begin, AOP investigations based totally on heterogeneous Fenton and Fenton-like reactions and photocatalysts, by and large on the laboratory scale, are mentioned in conjunction with a dialogue of features pertinent to improving method efficacy. Similarly, several realistic suggestions for destiny research and packages inside the area of heterogeneous catalysis for wastewater treatment are made.

3.2 ADVANCED OXIDATION PROCESS

AOPs are generally known as effective water treatment techniques capable of oxidizing a great branch of hard degradation pollutants from water. Nonetheless, AOPs are costly processes solely owing to their electrical costs, but this limitation can be resolved or diminished by applying photovoltaic energy to power the AOP in a trusty and autonomous way. AOPs include using oxidants to oxidize pollutants (Jatoi et al. 2021; Pham et al. 2020), and the most powerful oxidizing radical employed is the hydroxyl radical (OH^{-1}). Sulphate ($SO4^{-2}$) has also been extensively examined to remove organic pollutants in water. AOPs are ecologically gracious chemical technology which can break down C-based pollutants into innocuous outcomes. The pollutants are transferred from one phase to another without producing a large amount of sludge. In addition, this technology has many advantages, including faster reaction speed compared to other conventional processing technologies and shortened retention time. Furthermore, it does not require a sizable space to process the throughput required by the scheme. The AOP classification is presented in Figure 3.1.

Nevertheless, numerous shortcomings can also be emphasized, involving excessive operational and sustentation overheads. It also has a chemical method customized to particular pollutants that require eligible recruits to invent the scheme. AOP is generally apportioned into Ozone Treatment, Ultraviolet (UV) Treatment, Advanced Electrochemical Oxidation Process (eAOP), Advanced Catalytic Oxidation Process (cAOP), and Advanced Photo Oxidation Process (pAOP). The following subsections deal with the most common PDOs used in treating pesticides (Guo et al. 2022; Jiang et al. 2022; Kanakaraju et al. 2018).

3.3 FENTON PROCESS

The Fenton response has been regarded as a reliable method for degrading organic contaminants. It is based on the catalytic breakdown of hydrogen peroxide with the aid of the response between iron salts (ferrous ions) to shape hydroxyl radicals that

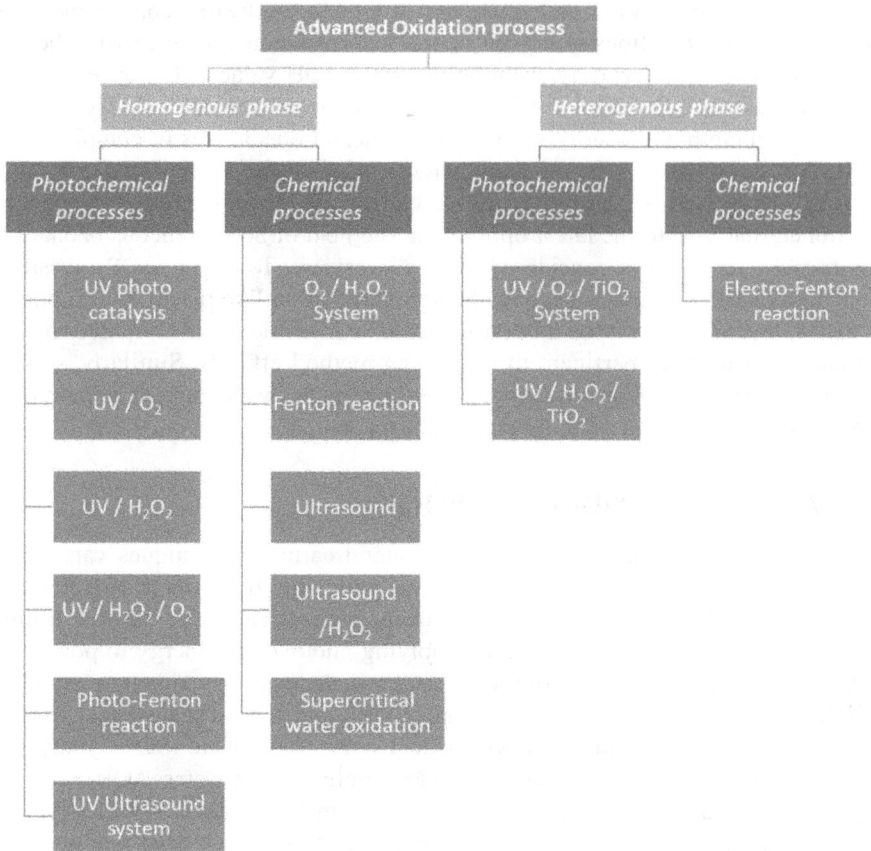

FIGURE 3.1   Advanced oxidation process classification.

can be chargeable for pollutant degradation. Irrespective of the fact that numerous researches on the use of pharmaceuticals and other refractory compounds as goal pollutants have shown high degradation and mineralization probabilities, the homogeneous Fenton manner has some operational limitations, which include a slender running pH, excessive chemical consumption, the manufacturing of iron sludge that is hard to remove, and awkward catalyst reusability. The homogeneous technique has been employed as a suitable pretreatment to enhance the biodegradability of refractory pharmaceutical, chemical substances and/or medical institution effluents.

Notwithstanding those drawbacks, the homogeneous approach has been efficaciously hired as a pretreatment to improve the biodegradability of refractory pharmaceutical, chemical substances, and/or hospital effluents. For that reason, resolving the Fenton method's disadvantages is a modern-day wastewater remedy. Therefore, the medical network has been targeting locating a heterogeneous catalyst for use inside the Fenton technique and combined techniques, including electrochemical remedy (electro-Fenton), photolysis (photo-Fenton), sonolysis (sono-Fenton), and their mixtures (image-electro-Fenton, sono-electro-Fenton) to conquer the aforementioned limitations.

3.3.1 HETEROGENEOUS FENTON PROCESS

The reaction between hydrogen peroxide and the lively websites of the catalyst specific with the aid of AS is the cautioned mechanism for the heterogeneous Fenton manner (Figure 3.2). Low iron concentrations may also be present inside the bulk solution because of the solid heterogeneous catalyst's leaching technique. Because of the standard Fenton reaction, hydroxyl radicals are produced, contributing relatively to the pollutant degradation rate. Many papers published within the last few years have endorsed using alternative types of heterogeneous catalysts with progressed attributes, including activity over a bigger pH range, decreased iron leaching, and high balance, permitting them to be reused for a couple of cycles. In addition, those catalysts permit homogeneous strategies to gain reactivity abilities similar to homogeneous techniques. The two primary catalysts are unsupported catalysts (zero-valent iron, metal minerals, iron oxides/hydroxides, or multimetallic catalysts) and catalysts loaded on substances (polymers, clays, zeolites).

3.3.1.1 Unsupported Heterogeneous Fenton Catalysts

Due to their ease of recuperation and lesser toxicity, the most widespread elements within the Earth's crust have been tested as heterogeneous catalysts, making them a possible alternative to the soluble salts utilized in the homogeneous technique. The common heterogeneous catalysts are the zero-valent iron molecules (ZVI or FeO), iron/copper oxides, hydroxides, oxyhydroxides, and insoluble iron and copper minerals. Under the presence of hydrogen peroxide, the copper-based complexes

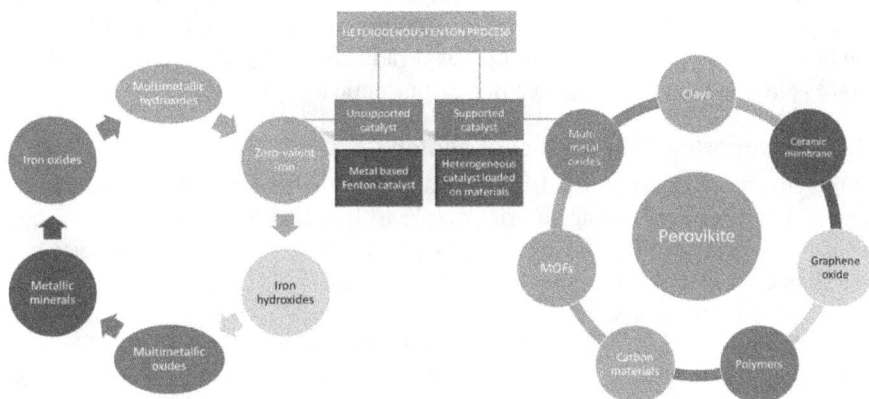

FIGURE 3.2 Heterogeneous Fenton process based on supported and unsupported catalyst.

(Cu^{2+}/Cu^+) behave further like the Fe^{3+}/Fe^{2+} pair, even throughout a bigger pH variety. As a result, using copper and its aggregate with different transition metals in Fenton-like reactions has gotten plenty of attention. Zhang et al. (2020) used a hydrothermal approach to develop a Cu/V bimetallic catalyst for fluconazole degradation, revealing that the organized material has a stronger catalytic pastime over a much broader pH range and higher reusability than monometallic copper compounds. Insoluble minerals such as pyrite (FeS_2), chalcocite (Cu_2S), bornite (Cu_5FeS_4), goethite (-FeOOH), magnetite (Fe_2O_3), hematite (-Fe_2O_3), ferrihydrite (Fh), and wüstite (FeO) had been used to get rid of natural contaminants satisfactorily (Usman et al. 2018). Trichloroethylene, diclofenac, pyrene, and toluene have all been found to be degraded using pyrite as a heterogeneous catalyst under the Fenton technique (Ammar et al. 2015). Kantar et al. (2019) utilized pyrite as a catalyst packed in a column reactor to purify real pharmaceutical effluent below dynamic situations. They confirmed that the pyrite-Fenton machine drastically decreased wastewater toxicity and that including citric acid in the tainted effluent improved the technique's efficacy and the pyrite column's beneficial lifespan.

Further, Muñoz et al. (2018) investigated herbal magnetite as a heterogeneous catalyst for sulfamethoxazole degradation, achieving general elimination and more than 50% mineralization of the target antibiotic below ideal situations (1 g/L catalyst attention, 25 mg/L oxidant awareness, 25 °C, pH value of 5). Despite magnetite's low iron content material, its use for sulfamethoxazole breakdown in health centre effluents necessitated longer response instances to perform complete elimination. Hassani et al. (2018) used an excessive-electricity planetary ball mill to make magnetite nanoparticles for ciprofloxacin degradation. After 120 minutes of remedy, six h ball-milled magnetite accelerated antibiotic clearance via roughly 89%, in step with the information. They postulated a removal mechanism based on adsorption and oxidation phases, with hydroxyl radicals gambling a considerable position.

Furthermore, Nie et al. (2020) used the solvothermal technique to produce magnetite nanospheres for use in tetracycline elimination, accomplishing greater than 80% clearance after 110 minutes. The catalyst's magnetic characteristics made separation from the aqueous solution easy. In addition, they showed the significance of each reactive oxygen species, with hydroxyl radicals produced on the nanospheres' surfaces playing the most vital role in pharmaceutical degradation.

3.3.1.2 Supported Heterogeneous Fenton Catalysts

For the advent of the supported Fenton catalyst, a heterogeneous catalyst may be integrated into several strong matrices (parent 2). Clays, ceramic filtration membranes, polymers, multimetal oxides (Al, Ti, Si, Zr), steel–natural frameworks, perovskites, zeolites, and carbon substances (graphene oxide, carbon nanotubes, biochar, activated carbon, g-C_3N_4 composites) are examples of these support substances (Scaria et al. 2021). Some restrictions are solved through the good assembly of the catalyst into the helping substances. Because typical separation technology (sedimentation and filtering) can't separate nanosize debris, this belonging makes them greater without reusing and separable from the final effluent (Plakas et al. 2019). Because of their widespread abundance, low toxicity, and low fee, clay materials were broadly exploited as catalysts to assist in heterogeneous procedures (Molina et al. 2020). These inorganic materials have better resistance to natural solvents and better thermal balance and inhibit nanoparticle agglomeration. Several methods for incorporating catalysts into clays were proposed in the literature. Khankhasaeva et al. (2017) created Fe/Cu/Al-pillared clay for heterogeneous Fenton oxidative degradation of sulfanilamide as the target pollutant. Inside the presence of hydrogen peroxide, the produced catalyst became active, significantly enhancing the oxidation charge. Xu et al. (2019b) used chemical coprecipitation to make sepiolite-supported magnetite (Fe_3O_4-Sep), resulting in 100% bisphenol. After 15 minutes, there might be a removal. Sétifi et al. (2019) used iron oxyhydroxide debris on a K10 montmorillonite-clay surface for naproxen degradation. At a pH value of 3, nearly complete naproxen elimination and over 90% mineralization were performed after 60 and 300 minutes, respectively. After four cycles, the authors confirmed the catalyst's reusability with a minimum efficiency decrease.

Iron-based composites have numerous capabilities for the Fenton manner (Iglesias et al. 2014). Titouhi and Belgaied (2016) cautioned against using Fe-containing sodium alginate beads as a heterogeneous catalyst for ofloxacin degradation via Fenton oxidation. After three sequential oxidation techniques, almost complete elimination of the antibiotic changed into acquired with low iron leaching and suitable stability. Further, Lyu et al. (2015) used a hydrothermal method to make copper-doped mesoporous silica microspheres. The manufactured microspheres executed admirably to remove ibuprofen, phenytoin, and diphenhydramine. Under environmental standards, nearly 100% pharmaceutical elimination becomes received at neutral pH through copper leaching. Zhang et al. (2019), who created FeO/CeO$_2$ catalyst for 90% tetracycline removal with the aid of the heterogeneous Fenton technique, additionally endorsed using multi-oxides, especially cerium oxides, as a catalytic guide. The catalyst is as-prepared proven software in a pH variety of 3 to 7, with an excessive reusability (as much as five cycles) and minimum iron leaching. Hussain et al. (2020) designed heterogeneous Fe and Cu catalysts supported on Zr (ZrFe and ZrCu), respectively, to assess the removal of ibuprofen via Fenton and Fenton-like reactions. After 120 minutes, the ZrCu catalyst furnished 98% degradation and 50% mineralization beneath the best instances. Ling et al. (2020) used a Fenton-like reaction with an activated-alumina-supported metal-oxide-based catalyst CoMnAl to catalyze the cleansing of pharmaceutical effluent.

They concluded that transition metals at the composite surface had synergistic consequences on the generation of hydroxyl radicals, which had been the primary purpose of pharmaceutical wastewater detoxification.

Because of their crystalline shape of aluminosilicates, which ensures a uniform isostructural distribution of the catalyst, zeolites have been broadly exploited as aid substances in numerous fields in current years (Martinez-Macias et al. 2015). Adityosulindro et al. (2018) tested the heterogeneous Fenton method in simulated and actual wastewater containing ibuprofen, using Fe supported on zeolite (Fe-Zeolite-ZSM5) as a catalyst. After 180 minutes of reaction at herbal pH, the ibuprofen elimination and TOC decay were around 88% and 27%, respectively. Because of the complicated matrix, they discovered that ibuprofen degradation became slower in wastewater effluent than in simulated water, as predicted. Velichkova et al. (2017) used Fe supported on zeolite as a heterogeneous catalyst (Fe/MFI-Zeolite-ZSM5) for the Fenton technique to oxidize paracetamol. Even operating under a continuous process combining oxidation and membrane filtration, paracetamol was eliminated after 300 minutes of treatment, reaching as much as 60% mineralization with minimal leaching. Due to their superior digital, magnetic, and electrochemical properties, using perovskite oxides as heterogeneous catalysts has lately piqued the medical community's hobby. The considerable range of ions inside the crystal shape, the speedy mobility of oxygen molecules, and the stabilization of precise oxidation states of the metals present in their structure all contribute to those capabilities. del Álamo et al. (2020) synthesized LaCu0.5Mn0.5O$_3$ perovskites in an up-waft packed-bed reactor to check their catalytic activity in drug removal from hospital effluents. This on-site pretreatment was a hit in casting off micropollutants in the micrograms consistent with the litre range. Similarly, Nie et al. (2015) used the sol-gel method to make nanoscale LaFeO$_3$ perovskite for sulfamethoxazole removal at impartial pH. The Fenton-like effectiveness becomes more desirable by forming a complicated surface between the hydrogen peroxide and the perovskite, which improves the Fe^{3+}/Fe^{2+} cycle and manufactures loose oxygen radicals. Perovskites doped with other transition metals, which include copper (LaAl$_{1-x}$Cu$_x$O$_3$), have also proven excessive degradation and mineralization percentages for persistent organic pollutants like bisphenol A, phenol, and phthalates through Fenton-like reactions (Wang et al. 2018). Apart from perovskites, membrane-supported Fenton catalysts are a promising technique for eliminating a diffusion of complex pollution from the aqueous matrix while additionally facilitating the separation of decontaminated fractions (Zhang et al. 2021). Plakas et al. (2019) used a catalytic membrane reactor with tubular porous alumina (Al$_2$O$_3$) membrane as a support material for embedded iron oxide nanoparticles to do away with diclofenac at especially low concentrations. Under perfect situations, slight degradation and mineralization had been executed, but the authors advised changing the membrane to enhance adsorption capability and further oxidation inside the catalytic membrane. In addition, Zhang et al. (2021) investigated an ultrafiltration ceramic membrane to help heterogeneous iron oxychloride catalysts, achieving a 100% removal of p-chlorobenzoic acid and bisphenol A with the simplest retention time. Due to their catalytic pastime, stability, and reusability, graphene derivatives have been efficaciously employed for catalyst immobilization on carbonaceous substrates (carbon nanotubes, biochar, g-C$_3$N$_4$, activated carbon, and graphene). Graphene is a two-dimensional carbon allotrope with a large specific floor location (2,360 m^2/g), splendid thermal/electrical conductivity, proper mechanical and chemical balance, high electron transfer capacity, and sincere chemical production from graphite (Liu et al. 2018). Even though various scientists have documented its agglomeration and restacking to produce photo-fenton due to strong interplanar contacts, these operational regulations can be circumvented by using other molecules and nanomaterials to functionalize it externally (Chowdhury & Balasubramanian 2014). As a result, these structures are promising for carrying out nanotechnology-based processes. Yang et al. (2019) advanced nanoscale zero-valent iron enclosed in a three-dimensional graphene network (3D-GN@nZVI) to eliminate sulfadiazine, an antibiotic. In addition, Xu et al. (2019a) created a heterogeneous catalyst using three-dimensional macroporous zero-valent copper nanoparticles self-assembled in graphene (three-D-GN@CuO). Their findings revealed that metronidazole elimination became excessive during an extensive pH range. Jiang et al. (2018) designed FeO catalysts hooked up on mesoporous carbon (FeO/MC) for persulfate activation to do away with tetracycline from aqueous effluent. Below perfect situations, these catalysts had been capable of reducing the aggregation fashion related to zero-valent iron nanoparticles, ensuing in 97% tetracycline removal. Qin et al. (2020) recently defined the growth due to the production of manganese ferrite nanoparticles. For the antimicrobial degradation of ofloxacin, amoxicillin, and tetracycline, they created magnetic centre-shell $MnFe_2O_4$@C-NH_2. The notable activity of the magnetic catalyst was attributed to the carbon shell due to the high specific surface area and negligible steel leaching. Furthermore, the addition of –NH_2 multiplied the carbon shell's electron density. Those one-of-a-kind characteristics help grow metallic regeneration cycles, which improve the efficiency of the process. Metallic-organic frameworks (MOFs) are made of multiple metallic ions related to polyfunctional organic ligands via moderately strong bonds, giving them qualities like excessive porosity, well-described periodic network structure, considerable lively sites, and a wide surface region. Moreover, their predesignate synthesis allows greater particular manipulation of structural and chemical houses with tailor-made functionalities for a positive goal characteristic (Jiang et al. 2018). Researchers are interested in those particular capabilities because they promise proper performance as heterogeneous catalysts for Fenton or Fenton-like reactions (Howarth et al. 2016). Regardless, adding a second energetic steel centre increases MOF performance because of a possible synergistic effect (Cheng et al. 2018).

Most studies reported in the literature investigated the use of MOFs as support for metallic nanoparticles (Bavykina et al. 2020). Tang and Wang (2020) developed bimetallic Fe and Cu MOFs (Fe_xCu_{1-x}(BDC)) as heterogeneous Fenton catalysts for sulfamethoxazole degradation. Complete antibiotic breakdown took 120 minutes at natural answer pH. However, the removal performance changed considerably higher than monometallic copper and iron catalysts, implying a synergistic effect among the two transition metals in the sulfamethoxazole breakdown technique. Tang and Wang (2019) additionally created a three-dimensional "flower-like" MOF with the usage of Fe–Cu bimetallic nanoparticles embedded in a mesoporous carbon layer (FeCu@C). After 90 and 240 minutes, this MOF became utilized as a heterogeneous Fenton catalyst for the degradation of sulfamethazine, obtaining 100% degradation and 73% mineralization, respectively. Moreover, the catalyst's magnetic characteristics allowed for facile recuperation and reuse in successive experimental runs without the want for laundry.

3.4 PHOTO-FENTON

UV–Vis radiation (580 nm) complements the Fenton response by producing extra hydroxyl radicals through (i) photoreduction of ferric ions to ferrous ions leading to the formation of iron complexes underneath acidic conditions and (ii) hydrogen peroxide photolysis.

Underneath perfect running situations, this oxidation-based generation produces biodegradable intermediates and almost general mineralization. However, due to its multiplied electricity intake, several solutions have been proposed to lessen operating costs and improve photograph-Fenton performance, consisting of heterogeneous catalysts and/or chelating retailers.

In addition, instead of simulated UV irradiation, photo-Fenton strategies may be driven by way of visible mild, sun power, or alternative light (LED, fibre optic). Because of their low cost, low toxicity, sturdy semiconducting characteristics, and simplicity of recovery, many iron-based oxides have been widely used as heterogeneous picture-Fenton catalysts. The semiconductors can be activated to yield photogenerated electrons by light irradiation, a likely semiconducting mechanism for iron oxides can be described by way of the subsequent equations: Ferrites (with the everyday molecular formula $M_xFe_3O_4$, wherein M is a bivalent transition metallic ion) have piqued attention among iron oxides employed as heterogeneous picture-Fenton catalysts because of their high-quality balance and narrow bandgap, taking into consideration energetic catalysis below seen mild. Hematite (slender bandgap around 2.2 eV) can absorb mild till 560 nm and capture around 40% of sun spectrum energy, making it an appropriate picture-Fenton catalyst. Due to its decreased kinetic constant price, the regeneration step of ferrous ions is the most critical restricting component in a Fenton-primarily based system. As a result, several research studies have suggested using electron-rich materials in combination with Fenton catalysts to expedite the synthesis of ferrous ions through the electrons in these substances. Copper has long been utilized as a heterogeneous catalyst due to its catalytic competencies, which might be similar to iron ones. Copper ferrites were used as a heterogeneous picture-Fenton catalyst for sulfamethoxazole degradation at circumneutral pH. After 120 minutes of treatment, nearly the entire deterioration and kind of 32% mineralization had been finished under perfect conditions. Further, Emidio et al. (2020) looked into the position of copper within the magnetite lattice ($Fe_{3x}Cu_xO_4$) in the image-Fenton degradation of 5-fluorouracil and cyclophosphamide once they had been exposed to UV mild. Modified magnetite outperformed natural magnetite in terms of catalytic pastime and drug degradation price, reaching 100% removal after 150 minutes on four consecutive runs. In addition, Xu et al. (2016) created "nanoflowers" of goethite doped with copper ((Fe-Cu) OOH) for diclofenac removal and underneath mild irradiation (>420 nm). After five cycles, the nanoflowers had been solid, with a pollutant degradation rate of approximately 95% using 3% Cu-doped (Fe-Cu)OOH in comparison to 75% of the usage of pure -FeOOH. The catalytic activity became attributed to the process within the produced catalyst that covered Fenton-like photocatalysis, and synergistic steel activation using commercial metallurgical waste as a heterogeneous catalyst has been proposed as a viable alternative for capturing these risky wastes. Slag is characterized by diverse proportions of ferrous metallic oxides, including magnetite, goethite, and hematite, coupled with silicates, aluminium, calcium, magnesium, copper, and titanium to a lesser or better extent relying on the authentic industry.

Two copper and metal business wastes were hired as Fenton-type photocatalysts for diclofenac elimination beneath simulated sunshine irradiation. Arzate-Salgado et al. (2016), while using metal, copper-based total catalysts (2.2 eV), had a higher photocatalytic interest. At near-impartial pH, general elimination and roughly 87% mineralization of the medicine were received after 90 and 300 minutes, respectively. A few microorganisms can oxidize steel ions in the presence of various enzymes.

Nontoxic chemical substances are used in those microbial techniques, making the process far greater ecologically benign and price-powerful. Du et al. (2020) used biogenic Fe–Mn oxides generated by *Pseudomonas* sp. F2, which is a heterogeneous catalyst below UV irradiation (420 nm), within the photograph-Fenton manner for ofloxacin degradation. Maghemite and blended oxidation states of manganese were the maximum commonplace oxides generated by the bacteria. In line with their findings, biogenic Fe–Mn oxide has twice the catalytic activity of chemically generated Fe–Mn oxides, casting off more than 95% of ofloxacin after 120 minutes beneath best situations.

Because of the inclusion of oxygenated-metallic cations into the clay shape, which results in more advantageous porosity and floor region, pillared clays have excessive photocatalytic abilities. Hurtado et al. (2019) investigated the utilization of Cu/Fe pillared clay as a catalyst in photograph-Fenton paracetamol mineralization underneath circumneutral pH situations. Beneath ideal situations, almost 80% of TOC reduction is achieved, with iron leaching of approximately 3% following remedy and negligible activity loss in the reusability cycles. Sétifi et al. (2019) used K10 montmorillonite clay to manufacture goethite debris for naproxen photograph-Fenton degradation. UV light radiation increased the performance of the procedure around six times, resulting in a 100% naproxen breakdown after 10 mins. Furthermore, combining the picture-Fenton gadget and photocatalysis with incorporating a few semiconductor substances on clay-based catalysts considerably improves iron regeneration and the formation of hydroxyl radicals, ensuing in a growth inside the degradation performance of natural pollution. Heidari et al. (2022) used business bentonite clay to synthesize supported Fe_2O_3–TiO_2 heterostructures, which have been used as image-Fenton catalysts to remove acetaminophen and antipyrine. Acetaminophen and antipyrine were completely degraded in less than 60 minutes, with TOC conversions of approximately 40% and 25%, respectively. Iron leaching turned into less than 5% in all the studies. Bansal and Verma (2018) also evolved heterogeneous catalysts based totally on Fe–TiO_2 composites that were supported with the aid of a clay/foundry sand/ash combination. These components are excessive-iron-content waste merchandise from the metallurgical region. Underneath sun irradiation, those heterogeneous catalysts were examined at the picture-Fenton process for the degradation of real pharmaceutical effluents. After 120 mins, the contaminated effluent DQO decreased by 80%, and the catalyst's reusability (>70 cycles) proved its potential for usage on a huge scale. Aside from clays, membranes have also been studied as a heterogeneous Fenton-kind catalytic support. Catalá et al. (2015) used a photo-Fenton system below UV–visible irradiation to degrade illicit drugs of abuse (benzodiazepines, cannabinoids, opioids, LSD, etc.) in a complicated herbal aqueous effluent using an iron-based mesoporous heterogeneous catalyst (crystalline iron oxides in hematite form) supported on mesostructured silica (Fe after 6 hours of treatment, removal efficiency become proven to be greater than 95% in all cases beneath best situations (catalyst = 0.6 g/L; pH = 3; H_2O_2 = 0.5 mg/L; T = 22 °C; 500 rpm)). Following that, phytotoxicity assays using Polystichum setiferum spores have been used to evaluate the efficacy of the Fenton remedy inside the purification of the effluent. In addition, Liu et al. (2020) investigated the activation and degradation of nitrobenzene (NB) under UV irradiation using a ceramic membrane lined with iron oxychloride (FeOCl-M). After 7 mins of treatment, the NB is completely removed, demonstrating the efficacy, recyclability, and stability of this shape of a supported heterogeneous catalyst.

Biochar (produced by biomass combustion below minimum oxygen conditions) has additionally been proposed as a thrilling assist fabric for image-Fenton-type catalysts due to its physicochemical properties, excessive catalytic ability, excessive availability, and coffee value while as compared to different carbonaceous substances. Lai et al. (2019) created a biochar-supported $MnFe_2O_4$ compound employed as a heterogeneous catalyst in a photo-Fenton method to break tetracycline. Below ideal conditions (tetracycline = 40 mg/L; H_2O_2 = 100 mmol/L; pH = 5.5), a 95% degradation turned into finished with low $MnFe_2O_4$ (0.2 mg/L). Bocos et al. (2016) hired activated carbon with Fe as a heterogeneous catalyst in a photo-Fenton method to degrade diatrizoic acid (DIA) in the presence of UV irradiation. The results showed that the TOC was removed by approximately 67% in 4 hours with no iron leaching into the medium.

On the other hand, Guo et al. (2019) proposed novel components for forming a heterogeneous catalyst of the Fe_2O_3@g-C_3N_4 heterostructure: They started by calcining MIL-53 (Fe) and melamine, as well as attaching that debris in g-C_3N_4. This novel catalyst became utilized in a photo-Fenton method to degrade TC (92% in 60 mins), resulting in degradation efficiencies that were 7–14 times better than those done with Fe_2O_3/MIL-53, g-C_3N_4, and MIL-53 alone. Furthermore, these novel catalysts also operated nicely for the duration of a wide pH range and remained solid for five cycles in a row.

In recent years, there has been little research into polymers as a support for heterogeneous Fenton-type catalysts. Sodium alginate is the polymer that has been explored the maximum under the photograph-Fenton remedy. Cuervo Lumbaque et al. (2019) employed three distinctive substances as catalysts in a photograph-Fenton process: Mining waste (an aggregate of goethite, magnetite, and hematite), Fe^{3+}/mining waste, and Fe^{2+}/mining waste included in sodium alginate spheres, all of which contributed to this knowledge. To break down medicines, the three catalysts were dosed into three awesome water matrices (distilled water, simulated wastewater, and health facility wastewater). The results confirmed that the eight drugs were degraded in the diverse matrices after 116 mins under the examined instances (catalyst = 3 g alginate spheres, H_2O_2 = 25 mg/L; pH = 5, simulated sunlight). However, their success revealed that the release of iron into the solution enhanced performance in the removal of pharmaceuticals. Compared to free systems, their findings found that iron-alginate substances were homo/heterogeneous systems that delivered iron dose with stepped-forward overall performance within the removal of pharmaceuticals. In addition, Cruz et al. (2017) produced a Fe^{3+} catalyst embedded in alginate spheres for the image-Fenton degradation of sulfamethoxazole in distinctive water matrices, namely deionized water and bottled drinking water, and at distinctive pH stages. At pH values of 2 and 3, sulfamethoxazole was removed after half an hour. However, at a pH value of 5 or above, the simplest 20% of the sulfamethoxazole changed to be destroyed in 30 minutes. Underneath the circumstances studied (catalyst = 5 g/L; SMX = 20 mg/L; UV = eight W, 365 nm; 700 rpm, T = 25 °C), scarcely any degradation change was noticed at pH value of 11. Their findings also found that large quantities of iron are launched into the response media under the tested situations, converting the manner to be homogeneous with a minimal heterogeneous contribution.

3.5 CONCLUSION

This chapter examined current developments in the usage of heterogeneous AOPs for managing aquatic ecosystems, focusing on the important elements of each option. These sophisticated oxidation techniques have several limitations. With increased energy usage (ultraviolet-light irradiation) and operational expenses, they are not currently considered commercially feasible procedures on a big scale. Consequently, solar radiation and other renewable energy provide a potential growth path of efficient and economical technology. It is necessary to evaluate the efficacy of these advanced wastewater treatment systems. Other types of pollutants exist in the organic waste and continuous-flow systems for genuine industrial effluent. Despite their promising bench results in terms of deterioration and mineralization percentage, the primary disadvantages of these technologies are their relatively high energy consumption and the need to improve their performance and stability to accomplish environmentally friendly and cost-effective wastewater treatment in pilot and large pilot stages, and the resulting increased operational expenses.

REFERENCES

Adityosulindro S, Julcour C, Barthe L (2018): Heterogeneous Fenton oxidation using Fe-ZSM5 catalyst for removal of ibuprofen in wastewater. Journal of Environmental Chemical Engineering 6, 5920–5928.

Ammar S, Oturan MA, Labiadh L, Guersalli A, Abdelhedi R, Oturan N, Brillas E (2015): Degradation of tyrosol by a novel electro-Fenton process using pyrite as heterogeneous source of iron catalyst. Water Research 74, 77–87.

Arellano M, Sanromán MA, Pazos M (2019): Electro-assisted activation of peroxymonosulfate by iron-based minerals for the degradation of 1-butyl-1-methylpyrrolidinium chloride. Separation and Purification Technology 208, 34–41.

Arzate-Salgado SY, Morales-Pérez AA, Solís-López M, Ramírez-Zamora RM (2016): Evaluation of metallurgical slag as a Fenton-type photocatalyst for the degradation of an emerging pollutant: Diclofenac. Catalysis Today 266, 126–135.

Babuponnusami A, Muthukumar K (2014): A review on Fenton and improvements to the Fenton process for wastewater treatment. Journal of Environmental Chemical Engineering 2, 557–572.

Bansal P, Verma A (2018): In-situ dual effect studies using novel Fe-TiO$_2$ composite for the pilot-plant degradation of pentoxifylline. Chemical Engineering Journal 332, 682–694.

Bavykina A, Kolobov N, Khan IS, Bau JA, Ramirez A, Gascon J (2020): Metal-organic frameworks in heterogeneous catalysis: Recent progress, new trends, and future perspectives. Chemical Reviews 120, 8468–8535.

Bocos E, Oturan N, Pazos M, Sanromán MÁ, Oturan MA (2016): Elimination of radiocontrast agent diatrizoic acid by photo-Fenton process and enhanced treatment by coupling with electro-Fenton process. Environmental Science and Pollution Research 23, 19134–19144.

Brillas E, Sirés I, Oturan MA (2009): Electro-Fenton process and related electrochemical technologies based on Fenton's reaction chemistry. Chemical Reviews 109, 6570–6631.

Casado J (2019): Towards industrial implementation of electro-Fenton and derived technologies for wastewater treatment: A review. Journal of Environmental Chemical Engineering 7, 102823.

Catalá M, Domínguez-Morueco N, Migens A, Molina R, Martínez F, Valcárcel Y, Mastroianni N, de Alda ML, Barceló D, Segura Y (2015): Elimination of drugs of abuse and their toxicity from natural waters by photo-Fenton treatment. Science of the Total Environment 520, 198–205.

Cheng M, Lai C, Liu Y, Zeng G, Huang D, Zhang C, Qin L, Hu L, Zhou C, Xiong W (2018): Metal-organic frameworks for highly efficient heterogeneous Fenton-like catalysis. Coordination Chemistry Reviews 368, 80–92.

Cheng M, Zeng G, Huang D, Lai C, Xu P, Zhang C, Liu Y (2016): Hydroxyl radicals based advanced oxidation processes (AOPs) for remediation of soils contaminated with organic compounds: A review. Chemical Engineering Journal 284, 582–598.

Chowdhury S, Balasubramanian R (2014): Recent advances in the use of graphene-family nanoadsorbents for removal of toxic pollutants from wastewater. Advances in Colloid and Interface Science 204, 35–56.

Cruz A, Couto L, Esplugas S, Sans C (2017): Study of the contribution of homogeneous catalysis on heterogeneous Fe (III)/alginate mediated photo-Fenton process. Chemical Engineering Journal 318, 272–280.

Cuervo Lumbaque E, Wielens Becker R, Salmoria Araújo D, Dallegrave A, Ost Fracari T, Lavayen V, Sirtori C (2019): Degradation of pharmaceuticals in different water matrices by a solar homo/heterogeneous photo-Fenton process over modified alginate spheres. Environmental Science and Pollution Research 26, 6532–6544.

del Álamo AC, González C, Pariente MI, Molina R, Martínez F (2020): Fenton-like catalyst based on a reticulated porous perovskite material: Activity and stability for the on-site removal of pharmaceutical micropollutans in a hospital wastewater. Chemical Engineering Journal 401, 126113.

Du X, Zhou M (2021): Strategies to enhance catalytic performance of metal–organic frameworks in sulfate radical-based advanced oxidation processes for organic pollutants removal. Chemical Engineering Journal 403, 126346.

Du Z, Li K, Zhou S, Liu X, Yu Y, Zhang Y, He Y, Zhang, Y (2020): Degradation of ofloxacin with heterogeneous photo-Fenton catalyzed by biogenic Fe-Mn oxides. Chemical Engineering Journal 380, 122427.

Emídio ES, Hammer P, Nogueira RFP (2020): Simultaneous degradation of the anticancer drugs 5-fluorouracil and cyclophosphamide using a heterogeneous photo-Fenton process based on copper-containing magnetites ($Fe_{3-x}Cu_xO_4$). Chemosphere 241, 124990.

Escudero-Curiel S, Penelas U, Sanromán MÁ, Pazos M (2021): An approach towards Zero-Waste wastewater technology: Fluoxetine adsorption on biochar and removal by the sulfate radical. Chemosphere 268, 129318.

Fdez-Sanromán A, Pazos M, Rosales E, Sanromán MÁ (2021): Prospects on integrated electrokinetic systems for decontamination of soil polluted with organic contaminants. Current Opinion in Electrochemistry 27, 100692.

Giannakis S, Papoutsakis S, Darakas E, EscAlas-Cañellas A, Pétrier C, Pulgarin C (2015): Ultrasound enhancement of near-neutral photo-Fenton for effective E. coli inactivation in wastewater. Ultrasonics Sonochemistry 22, 515–526.

Gosset A, Ferro Y, Durrieu C (2016): Methods for evaluating the pollution impact of urban wet weather discharges on biocenosis: A review. Water Research 89, 330–354.

Guo K, Wu Z, Chen C, Fang J (2022): UV/Chlorine process: An efficient advanced oxidation process with multiple radicals and functions in water treatment. Accounts of Chemical Research 55, 286–297.

Guo T, Wang K, Zhang G, Wu X (2019): A novel α-Fe_2O_3@ g-C_3N_4 catalyst: Synthesis derived from Fe-based MOF and its superior photo-Fenton performance. Applied Surface Science 469, 331–339.

Hassani A, Karaca M, Karaca S, Khataee A, Açışlı Ö, Yılmaz B (2018): Preparation of magnetite nanoparticles by high-energy planetary ball mill and its application for ciprofloxacin degradation through heterogeneous Fenton process. Journal of Environmental Management 211, 53–62.

Heidari A, Shahbazi A, Aminabhavi TM, Barceló D, Rtimi S (2022): A systematic review of clay-based photocatalysts for emergent micropollutants removal and microbial inactivation from aqueous media: Status and limitations. Journal of Environmental Chemical Engineering 108813.

Howarth AJ, Liu Y, Li P, Li Z, Wang TC, Hupp JT, Farha OK (2016): Chemical, thermal and mechanical stabilities of metal-organic frameworks. Nature Reviews Materials 1, 15018.

Hurtado L, Romero R, Mendoza A, Brewer S, Donkor K, Gómez-Espinosa RM, Natividad R (2019): Paracetamol mineralization by photo Fenton process catalyzed by a Cu/Fe-PILC under circumneutral pH conditions. Journal of Photochemistry and Photobiology A: Chemistry 373, 162–170.

Hussain S, Aneggi E, Briguglio S, Mattiussi M, Gelao V, Cabras I, Zorzenon L, Trovarelli A, Goi D (2020): Enhanced ibuprofen removal by heterogeneous-Fenton process over Cu/ZrO_2 and Fe/ZrO_2 catalysts. Journal of Environmental Chemical Engineering 8, 103586.

Iglesias O, Gómez J, Pazos M, Sanromán MÁ (2014): Electro-Fenton oxidation of imidacloprid by Fe alginate gel beads. Applied Catalysis B: Environmental 144, 416–424.

Jatoi AS, Hashmi Z, Adriyani R, Yuniarto A, Mazari SA, Akhter F, Mubarak NM (2021): Recent trends and future challenges of pesticide removal techniques—A comprehensive review. Journal of Environmental Chemical Engineering 9, 105571.

Jiang D, Xu P, Wang H, Zeng G, Huang D, Chen M, Lai C, Zhang C, Wan J, Xue W (2018): Strategies to improve metal organic frameworks photocatalyst's performance for degradation of organic pollutants. Coordination Chemistry Reviews 376, 449–466.

Jiang S, Zhao Z, Chen J, Yang Y, Ding C, Yang Y, Wang Y, Liu N, Wang L, Zhang X (2022): Recent research progress and challenges of MIL-88 (Fe) from synthesis to advanced oxidation process. Surfaces and Interfaces 101843.

Kanakaraju D, Glass BD, Oelgemöller M (2018): Advanced oxidation process-mediated removal of pharmaceuticals from water: A review. Journal of Environmental Management 219, 189–207.

Kang J, Duan X, Zhou L, Sun H, Tadé MO, Wang S (2016): Carbocatalytic activation of persulfate for removal of antibiotics in water solutions. Chemical Engineering Journal 288, 399–405.

Kantar C, Oral O, Oz NA (2019): Ligand enhanced pharmaceutical wastewater treatment with Fenton process using pyrite as the catalyst: Column experiments. Chemosphere 237, 124440.

Khankhasaeva ST, Dashinamzhilova ET, Dambueva DV (2017): Oxidative degradation of sulfanilamide catalyzed by Fe/Cu/Al-pillared clays. Applied Clay Science 146, 92–99.

Kumar A, Rana A, Sharma G, Naushad M, Dhiman P, Kumari A, Stadler FJ (2019a): Recent advances in nano-Fenton catalytic degradation of emerging pharmaceutical contaminants. Journal of Molecular Liquids 290, 111177.

Kumar R, Sarmah AK, Padhye LP (2019b): Fate of pharmaceuticals and personal care products in a wastewater treatment plant with parallel secondary wastewater treatment train. Journal of Environmental Management 233, 649–659.

Lai C, Huang F, Zeng G, Huang D, Qin L, Cheng M, Zhang C, Li B, Yi H, Liu S, Chen L (2019): Fabrication of novel magnetic $MnFe_2O_4$/bio-char composite and heterogeneous photo-Fenton degradation of tetracycline in near neutral pH. Chemosphere 224, 910–921.

Li WC (2014): Occurrence, sources, and fate of pharmaceuticals in aquatic environment and soil. Environmental Pollution 187, 193–201.

Ling L, Liu Y, Pan D, Lyu W, Xu X, Xiang X, Lyu M, Zhu L (2020): Catalytic detoxification of pharmaceutical wastewater by Fenton-like reaction with activated alumina supported CoMnAl composite metal oxides catalyst. Chemical Engineering Journal 381, 122607.

Liu F, Yao H, Sun S, Tao W, Wei T, Sun P (2020): Photo-Fenton activation mechanism and antifouling performance of an FeOCl-coated ceramic membrane. Chemical Engineering Journal 402, 125477.

Liu Y, Huang H, Gan D, Guo L, Liu M, Chen J, Deng F, Zhou N, Zhang X, Wei Y (2018): A facile strategy for preparation of magnetic graphene oxide composites and their potential for environmental adsorption. Ceramics International 44, 18571–18577.

Luo H, Zeng Y, Cheng Y, He D, Pan X (2020): Recent advances in municipal landfill leachate: A review focusing on its characteristics, treatment, and toxicity assessment. Science of the Total Environment 703, 135468.

Luo H, Zeng Y, He D, Pan X (2021): Application of iron-based materials in heterogeneous advanced oxidation processes for wastewater treatment: A review. Chemical Engineering Journal 407, 127191.

Lyu L, Zhang L, Hu C (2015): Enhanced Fenton-like degradation of pharmaceuticals over framework copper species in copper-doped mesoporous silica microspheres. Chemical Engineering Journal 274, 298–306.

Martinez-Macias C, Serna P, Gates BC (2015): Isostructural zeolite-supported rhodium and iridium complexes: Tuning catalytic activity and selectivity by ligand modification. ACS Catalysis 5, 5647–5656.

Molina CB, Sanz-Santos E, Boukhemkhem A, Bedia J, Belver C, Rodriguez JJ (2020): Removal of emerging pollutants in aqueous phase by heterogeneous Fenton and photo-Fenton with Fe_2O_3-TiO_2-clay heterostructures. Environmental Science and Pollution Research 27, 38434–38445.

Muñoz M, Conde J, de Pedro ZM, Casas JA (2018): Antibiotics abatement in synthetic and real aqueous matrices by H_2O_2/natural magnetite. Catalysis Today 313, 142–147.

Nie M, Li Y, He J, Xie C, Wu Z, Sun B, Zhang K, Kong L, Liu J (2020): Degradation of tetracycline in water using Fe_3O_4 nanospheres as Fenton-like catalysts: Kinetics, mechanisms and pathways. New Journal of Chemistry 44, 2847–2857.

Nie Y, Zhang L, Li YY, Hu C (2015): Enhanced Fenton-like degradation of refractory organic compounds by surface complex formation of $LaFeO_3$ and H_2O_2. Journal of Hazardous Materials 294, 195–200.

Pham TH, Bui HM, Bui TX (2020): Chapter 13 —Advanced oxidation processes for the removal of pesticides. In: Varjani S, Pandey A, Tyagi RD, Ngo HH, Larroche C (Editors), Current Developments in Biotechnology and Bioengineering. Elsevier, pp. 309–330.

Pi Y, Li X, Xia Q, Wu J, Li Y, Xiao J, Li Z (2018): Adsorptive and photocatalytic removal of persistent organic pollutants (POPs) in water by metal-organic frameworks (MOFs). Chemical Engineering Journal 337, 351–371.

Plakas KV, Mantza A, Sklari SD, Zaspalis VT, Karabelas AJ (2019): Heterogeneous Fenton-like oxidation of pharmaceutical diclofenac by a catalytic iron-oxide ceramic microfiltration membrane. Chemical Engineering Journal 373, 700–708.

Poza-Nogueiras V, Rosales E, Pazos M, Sanroman MA (2018): Current advances and trends in electro-Fenton process using heterogeneous catalysts–a review. Chemosphere 201, 399–416.

Puga A, Rosales E, Pazos M, Sanromán M (2020): Prompt removal of antibiotic by adsorption/electro-Fenton degradation using an iron-doped perlite as heterogeneous catalyst. Process Safety and Environmental Protection 144, 100–110.

Qin H, Cheng H, Li H, Wang Y (2020): Degradation of ofloxacin, amoxicillin and tetracycline antibiotics using magnetic core–shell $MnFe_2O_4$@C-NH_2 as a heterogeneous Fenton catalyst. Chemical Engineering Journal 396, 125304.

Ribeiro AR, Nunes OC, Pereira MF, Silva AM (2015): An overview on the advanced oxidation processes applied for the treatment of water pollutants defined in the recently launched Directive 2013/39/EU. Environment International 75, 33–51.

Sablas MM, de Luna MDG, Garcia-Segura S, Chen C-W, Chen C-F, Dong C-D (2020): Percarbonate mediated advanced oxidation completely degrades recalcitrant pesticide imidacloprid: Role of reactive oxygen species and transformation products. Separation and Purification Technology 250, 117269.

Scaria J, Gopinath A, Nidheesh PV (2021): A versatile strategy to eliminate emerging contaminants from the aqueous environment: Heterogeneous Fenton process. Journal of Cleaner Production 278, 124014.

Sétifi N, Debbache N, Sehili T, Halimi O (2019): Heterogeneous Fenton-like oxidation of naproxen using synthesized goethite-montmorillonite nanocomposite. Journal of Photochemistry and Photobiology A: Chemistry 370, 67–74.

Tang J, Wang J (2019): MOF-derived three-dimensional flower-like FeCu@C composite as an efficient Fenton-like catalyst for sulfamethazine degradation. Chemical Engineering Journal 375, 122007.

Tang J, Wang J (2020): Iron-copper bimetallic metal-organic frameworks for efficient Fenton-like degradation of sulfamethoxazole under mild conditions. Chemosphere 241, 125002.

Titouhi H, Belgaied JE (2016): Heterogeneous Fenton oxidation of ofloxacin drug by iron alginate support. Environmental Technology 37, 2003–2015.

Usman M, Byrne JM, Chaudhary A, Orsetti S, Hanna K, Ruby C, Kappler A, Haderlein SB (2018): Magnetite and green rust: Synthesis, properties, and environmental applications of mixed-valent iron minerals. Chemical Reviews 118, 3251–3304.

Velichkova F, Delmas H, Julcour C, Koumanova B (2017): Heterogeneous fenton and photo-fenton oxidation for paracetamol removal using iron containing ZSM-5 zeolite as catalyst. AlChE Journal 63, 669–679.

Wang C, Shih Y (2015): Degradation and detoxification of diazinon by sono-Fenton and sono-Fenton-like processes. Separation and Purification Technology 140, 6–12.

Wang H, Zhang L, Hu C, Wang X, Lyu L, Sheng G (2018): Enhanced degradation of organic pollutants over Cu-doped LaAlO$_3$ perovskite through heterogeneous Fenton-like reactions. Chemical Engineering Journal 332, 572–581.

Wang J, Wang S (2016): Removal of pharmaceuticals and personal care products (PPCPs) from wastewater: A review. Journal of Environmental Management 182, 620–640.

Wang J, Wang S (2018): Activation of persulfate (PS) and peroxymonosulfate (PMS) and application for the degradation of emerging contaminants. Chemical Engineering Journal 334, 1502–1517.

Xu J, Li Y, Yuan B, Shen C, Fu M, Cui H, Sun W (2016): Large scale preparation of Cu-doped α-FeOOH nanoflowers and their photo-Fenton-like catalytic degradation of diclofenac sodium. Chemical Engineering Journal 291, 174–183.

Xu L, Yang Y, Li W, Tao Y, Sui Z, Song S, Yang J (2019a): Three-dimensional macroporous graphene-wrapped zero-valent copper nanoparticles as efficient micro-electrolysis-promoted Fenton-like catalysts for metronidazole removal. Science of the Total Environment 658, 219–233.

Xu X, Chen W, Zong S, Ren X, Liu D (2019b): Magnetic clay as catalyst applied to organics degradation in a combined adsorption and Fenton-like process. Chemical Engineering Journal 373, 140–149.

Yang Y, Xu L, Li W, Fan W, Song S, Yang J (2019): Adsorption and degradation of sulfadiazine over nanoscale zero-valent iron encapsulated in three-dimensional graphene network through oxygen-driven heterogeneous Fenton-like reactions. Applied Catalysis B: Environmental 259, 118057.

Zhang N, Chen J, Fang Z, Tsang EP (2019): Ceria accelerated nanoscale zerovalent iron assisted heterogenous Fenton oxidation of tetracycline. Chemical Engineering Journal 369, 588–599.

Zhang N, Xue C, Wang K, Fang Z (2020): Efficient oxidative degradation of fluconazole by a heterogeneous Fenton process with Cu-V bimetallic catalysts. Chemical Engineering Journal 380, 122516.

Zhang S, Hedtke T, Zhu Q, Sun M, Weon S, Zhao Y, Stavitski E, Elimelech M, Kim JH (2021): Membrane-confined iron oxychloride nanocatalysts for highly efficient heterogeneous Fenton water treatment. Environmental Science & Technology 55, 9266–9275.

4 Photochemical and Sonochemical Strategies in Advanced Oxidation Processes for Micropollutants' Treatments

4 Photochemical
and Sonochemical
Strategies in Advanced
Oxidation Processes
for Micropollutants'
Treatments

Franklin Vargas, Edmanuel Lucena-Mendoza,
Tamara Zoltan, Yesica Torres, Miguel León,
Beatriz Angulo and Gabriel Ibrahin Tovar

CONTENTS

4.1 Introduction 65
4.2 Photolysis 66
    4.2.1 Direct Photolysis 66
    4.2.2 Photochemical Processes in Surface Waters 68
    4.2.3 Photolysis Combined with Oxidants (H_2O_2/O_3) 70
4.3 Photocatalysis 72
4.4 Nano-photocatalysts 77
    4.4.1 Photoelectrocatalysis 78
    4.4.2 Photocatalytic Degradation of Microplastics 80
4.5 Sonolysis 81
    4.5.1 Sonochemical Process 82
4.6 Concluding Remarks 88
References 89

4.1 INTRODUCTION

Recently, small concentrations of inorganic, organic, and mineral compounds in the aquatic environment have increased noticeably, mostly by human activities such as excessive and rapid industrialization, urban encroachment, and improved agricultural operations. One feasible option for eliminating organic pollutants from wastewater is the application of the Advanced Oxidation Processes (AOPs) or their combinations with other treatments. These methods have been commonly recognized to be

DOI: 10.1201/9781003247913-4

highly capable of removing recalcitrant contaminants or being used as pretreatment to transform contaminants into shorter-chain compounds that traditional biological processes can treat.

AOPs such as photocatalysis, direct photolysis, photo-Fenton, and sonolysis have been proposed as tertiary treatments for urban effluents due to their ability to detoxify wastewater streams containing persistent contaminants. The treatment of industrial wastewater effluents is a very complex challenge due to the broad array of substances and high concentrations that they can contain. Treatment by activated sludge is more efficient and less expensive for removing high concentrations of organic compounds. Nevertheless, there are some circumstances where AOPs may offer some advantages. AOPs typically have a small footprint and seamless integration capability with other treatment methods. They serve as a viable option for eliminating non-biodegradable pollutants that persist even after undergoing biological treatment. Industrial wastewater often contains toxic substances or compounds that exhibit a high level of bio-recalcitrance, rendering them unsuitable for activated sludge treatment due to the presence of a considerable amount of dissolved organic carbon. However, AOPs have been demonstrated to effectively degrade these toxic compounds, resulting in effluents that are more readily biodegradable prior to biological treatment. It is important to highlight that one AOP or a combination must be appropriately selected for the remediation of a specific industrial or urban wastewater, considering factors such as wastewater characteristics, technical applicability, regulatory requirements, economical aspects, and long-term environmental impacts. The proposal of the use of solar AOPs as tertiary treatments for contaminants of emerging concern and removal due to the use of solar energy diminishes costs, resulting in wastewater treatments that are simple, sturdy, and inexpensive.

4.2 PHOTOLYSIS

4.2.1 DIRECT PHOTOLYSIS

Photochemistry plays a fundamental role during the transformation of organic pollutants in natural systems using solar energy; therefore, its application frontiers in this area are limited by those spaces to which sunlight penetrates. For its part, environmental geochemistry is the discipline of earth sciences dedicated to the study of the source, transformation, transport, and destination of chemical pollutants on the planet, specifically in surface systems such as terrestrial, aquatic, biotic, and atmospheric, in most of which sunlight is present and has a fundamental role in the degradation of some pollutants. This is how sunlight links photochemistry with environmental geochemistry. On the one hand, photochemistry studies the photodegradation reactions of organic pollutants, while, on the other, environmental geochemistry provides information of interest on the natural factors that may be involved, impacting, either promoting or limiting, the photochemical reactions that transform organic pollutants. In this sense, some authors have used the term photogeochemistry, referring to the study of chemical reactions that occur in the terrestrial surface system due to the action of sunlight on minerals, plant residues, dissolved

$$P\bullet + X\bullet \quad \text{Homolysis}$$

$$PX\bullet \longrightarrow (PX)^* \quad \begin{array}{c} P^+ + X^- \\ \text{or} \\ P^- + X^+ \end{array} \quad \text{Heterolysis}$$

$$(PX)\bullet^+ + e_{ac}^- \quad \text{Photoionization}$$

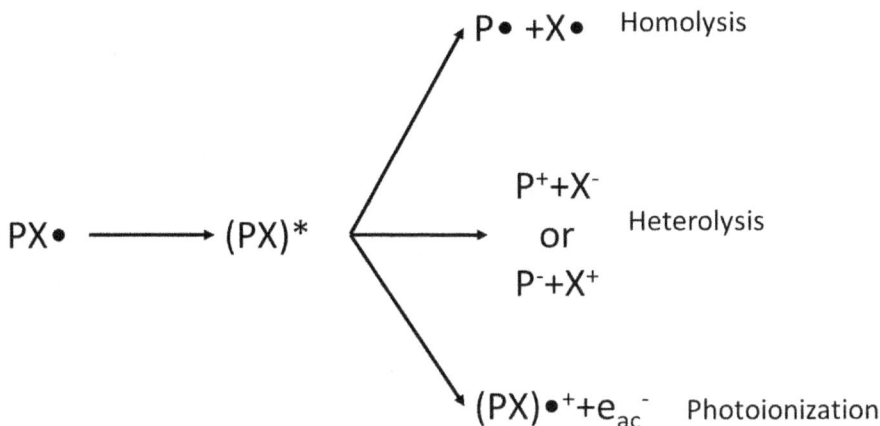

FIGURE 4.1 Possible chemical events that can take place under direct photolysis.

organic matter, and organic and inorganic aerosols and gases, associated with terrestrial, aquatic, or tropospheric systems.[1,2]

A brief review of the synergy between both disciplines and its impact on understanding the transformation phenomena of organic pollutants in the natural surface environment is described here. Information on the link between surface water photochemistry and climate is currently scarce, as only a few studies have been devoted to the subject. Based on the limited knowledge currently available, the present inferences can be made as follows:

(1) Global warming can cause greater leaching (or solid-liquid extraction) of ionic solutes from soils and sediments to surface waters, including, in addition to cations, major anions such as bicarbonate, a product of the chemical weathering of siliciclastic minerals utilizing hydrolysis reaction with carbonic acid, as well as biologically unstable species such as sulphate. Although the main source of bicarbonate supply is chemical weathering, the preferential biodegradation of sulphate followed by its microbiological elimination can also increase alkalinity, favouring the generation of the carbonate radical, $CO_3^{\bullet-}$. However, these phenomena would be easily compensated by fluctuations in dissolved organic carbon (DOC), which is inversely correlated with $CO_3^{\bullet-}$. Therefore, gaining insight into the evolution of DOC is a key topic in understanding the link between photochemistry and climate.

(2) Climate change could exacerbate water scarcity in the dry season for some regions. Fluctuations in the water column could profoundly alter the photochemistry, generally favoured in shallow waters. However, this negative water imbalance would strongly affect the predominant photo-induced processes. The decrease in the volume of run-off water without important changes in the concentration of solutes would mainly favour the reactions induced by hydroxyl and carbonate radicals ($^\bullet OH$ and $CO_3^{\bullet-}$). On the

contrary, concentration by evaporation would promote reactions mediated by singlet oxygen ($^1O_2$) and by the triple states of chromophoric dissolved organic matter ($^3MODC^*$).

(3) In a warmer climate, the stratification period of the lakes would last longer, increasing the photochemical reactions in the epilimnion while keeping the hypolimnion's water in the dark for long periods.

Photochemical reactions depend highly on solar radiation, water chemistry, and depth. The chemistry of water and the irradiation produced by incident sunlight affect the concentration values of photo-reactants (MODC, nitrate, and nitrite) and of transient reactive species scavengers (mainly ˙OH and $CO_3{^{˙-}}$). Finally, depth is important because deep areas of a body of water are poorly lit by sunlight. Therefore, photochemical reactions are favoured in shallow water bodies. Climate change can affect the chemistry and depth of water in several ways. For example, a change in run-off and the rate of weathering can change the concentration of cationic and anionic species in receiving waters. The chemical weathering of minerals and the biological ability of anions such as sulphate would cause an increase in alkalinity, which would alter the concentration values of bicarbonate and carbonate and modify the rate of formation of CO_3 after carbonate oxidation and bicarbonate by OH.[3,4]

4.2.2 PHOTOCHEMICAL PROCESSES IN SURFACE WATERS

The absorption of radiation triggers sensitized photolysis by photoactive compounds called photosensitizers; the main ones in surface waters are nitrate, nitrite, and, most notably, MODC. Dissolved organic matter (MOD) consists of organic compounds

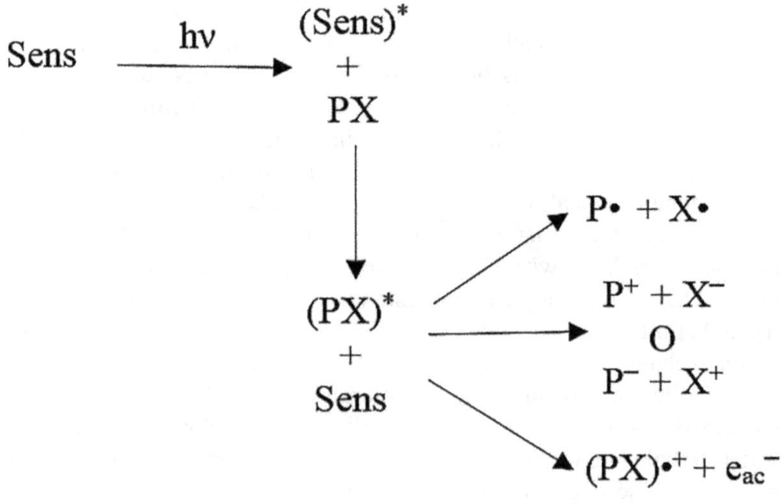

FIGURE 4.2 Chemical events that occur under indirect photolysis (photosensitized photolysis involving energy transfer).

dissolved in water that can be derived from the microbiological transformation of animal and plant remains. MOD comprises small primary molecular fractions (100–200 Da) that are organized into supramolecular structures. These aggregates are composed of aromatic and aliphatic hydrocarbon structures with various functional groups (e.g. $CONH_2$, -COOH, -OH, -CO).[5] The aromatic/quinone fractions play a very important role as they can absorb sunlight. They are contained in humic and fulvic acids, a product of the biodegradation of lignin or the oligomerization/supramolecular association of smaller compounds caused by photo-oxidation.[6] The fraction of MOD that can absorb sunlight is called MODC. The absorption of radiation by MODC causes the formation of excited singlet states (^{1}MODC*) that are transformed by inter-system crossover (ISC) into excited triplet states (MODC*). The reactivity of dissolved compounds with ^{3}MODC* is greater than with ^{1}MODC* due to the longer useful life of the former.[7,8] The •OH radical is produced by irradiation of nitrate, nitrite, and MODC. In the latter case, the process details are not yet clear, although the significant but not exclusive role of H_2O_2 is known, which could, for example, be involved in Fenton reactions.[9,10]

On the other hand, there is ample evidence of an independent H_2O_2 pathway from OH generated by irradiated MODC,[10] which could possibly be counted among the processes by triplet-sensitized oxidation of OH- and/or H_2O.[11] The radical CO_3— is produced after the oxidation of CO_3^{2-} and $H—O_3^-$ by OH, as well as by the oxidation of carbonate by ^{3}MODC*. The ^{3}MODC* triplet state can degrade pollutants on its own but can also react with O_2 to form the reactive transient $^1O_2$, which is also involved in pollutant breakdown. The reactions that can take place in natural water systems (apart from the not-yet-clear formation of OH from irradiated MODC) are illustrated here:[7]

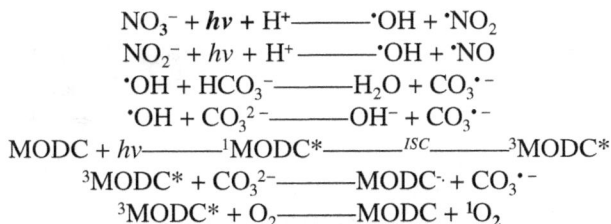

$$NO_3^- + h\nu + H^+ \longrightarrow {}^•OH + {}^•NO_2$$
$$NO_2^- + h\nu + H^+ \longrightarrow {}^•OH + {}^•NO$$
$${}^•OH + HCO_3^- \longrightarrow H_2O + CO_3^{•-}$$
$${}^•OH + CO_3^{2-} \longrightarrow OH^- + CO_3^{•-}$$
$$MODC + h\nu \longrightarrow {}^1MODC^* \overset{ISC}{\longrightarrow} {}^3MODC^*$$
$${}^3MODC^* + CO_3^{2-} \longrightarrow MODC^{•-} + CO_3^{•-}$$
$${}^3MODC^* + O_2 \longrightarrow MODC + {}^1O_2$$

The transient species •OH, $CO_3^{•-}$, $^1O_2$, and ^{3}MODC* can induce the degradation of xenobiotics, but with important differences. In particular, the •OH radical is primarily involved in decontamination with a generally limited formation of harmful intermediates. Compared with •OH, the probability of forming harmful intermediates is considerably higher in the case of $^1O_2$, $CO_3^{•-}$, and ^{3}MODC*.

Photochemical reactions involving •OH, $CO_3^{•-}$, $^1O_2$, and ^{3}MODC* are highly dependent on the chemistry and depth of the water. The depth effect is mainly affected by absorbent substances dissolved in water and, most notably, by MODC. The latter is the main absorber of radiation in surface waters between 300 and 500 nm, which is the most significant wavelength range for photoinduced processes. The absorption of sunlight by MODC plays an important role in reducing the intensity of solar radiation in the water column. Therefore, the lower region of a body of water will be less illuminated than the surface, which also promotes faster photochemical processes in shallow water bodies compared to deep ones. In the latter, the high photoactivity in the superficial zone is compensated by the lack of photoactivity in depth. Such an effect reduces the importance of photochemical reactions as the water column's depth increases and protects aquatic life from exposure to harmful UV radiation.[12] The absorption of MODC shows an exponential decrease with the increase of the wavelength. Also, the absorption in the different spectral ranges decreases in the order UVB > UVA > visible. Consequently, penetration of the column by radiation is in the order UVB < UVA < visible. Nitrate primarily absorbs UVB radiation, and its photochemistry is highly inhibited with depth. A lower degree of inhibition is observed with nitrite that absorbs UVA rays, while MODC, which also absorbs visible rays, is less affected by depth. Nitrite and nitrate are direct sources of $\cdot$OH and indirect sources of $CO_3^{\cdot-}$, while ODC produces only $^1O_2$ and $^3MODC^*$. Therefore, the relative importance of $^1O_2$ and $^3MODC^*$ versus $\cdot$OH and $CO_3^{\cdot-}$ increases as depth increases.[13,14]

The photochemical reactivity of surface waters has been the subject of specific studies in some environments, but so far, no long-term campaigns have been carried out. Consequently, there has been no information on photochemical behaviour in defined environments in the last 20–30 years, and data for the next few decades is not yet available. For this reason, the relationship between climate change and surface water photochemistry is largely unknown and almost non-existent as a research topic.

4.2.3 Photolysis Combined with Oxidants (H_2O_2/O_3)

The efficiency of photolysis, if combined with chemical oxidants such as hydrogen peroxide or ozone, can be improved to increase the rate of pesticide degradation further. In this sense, in the synergistic process between UV/H_2O_2,[15] the photolysis of hydrogen peroxide leads to the formation of HO $\cdot$ radicals with a quantum efficiency of around 0.536. These radicals, by reaction with other H_2O_2 molecules, lead to the formation of $HO_2 \cdot$ (hydroperoxide radical) and O_2, (superoxide anion).[16] The reactivity of hydroperoxide and superoxide radicals with organic matter is comparable in order of magnitude that HO $\cdot$ radicals are generated in the primary process.[17]

Regarding the mechanisms involved in this type of reaction, the $\cdot$OH radicals formed, due to their high reactivity, initiated the oxidative degradation of aqueous pollutants mainly through addition reactions to aromatic substrates to form cyclohexadienyl radicals or through hydrogen abstraction to aliphatic substrates to form alkyl radicals. Subsequently, the generated cyclohexadienyl or alkyl radicals add molecular oxygen and form peroxide radicals that initiate thermal reactions that finally lead to the formation of carbon dioxide, water, and inorganic acids (HNO_3 and HCl). This series of processes can be represented by the following reaction scheme[17–24,24–27]:

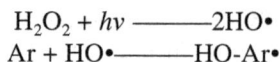

$$H_2O_2 + h\nu \longrightarrow 2HO\bullet$$
$$Ar + HO\bullet \longrightarrow HO\text{-}Ar\bullet$$

$$HO\text{-}Ar\bullet + O_2 \text{——————} INT + RH$$
$$RH + HO\bullet \text{——————} R\bullet + H_2O$$
$$R\bullet + O_2 \text{——————} RO_2\bullet$$
$$RO_2\bullet + INT \text{——————} CO_2 + H_2O$$

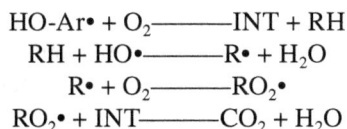

where INT represents the set of reaction intermediates, while Ar and RH correspond to aromatic and aliphatic compounds, respectively.

In this sense, Chelme-Ayala et al.[28] carried out an experiment in which the relative performance of UV/H_2O_2 photolysis was evaluated during the degradation of two pesticides (bromoxynil and trifluralin). The degradation efficiency of both pesticides was double using H_2O_2 compared to direct photolysis, showing the improvement using oxidants. In another work, the photochemical degradation of thiacloprid was studied in the UV/H_2O_2 system under solution conditions with different initial concentrations of H_2O_2 and pH.[29] The results showed 97% degradation of thiacloprid in approximately 2 hours with the H_2O_2/thiacloprid molar ratio of 220 and pH of 2.8. Thiacloprid has been degraded by forming several ionic by-products (chloride, acetate, and formate) and organic intermediates. In a similar investigation, the degradation of γ-hexachlorocyclohexane (lindane) was also investigated under various solution conditions.[30] At pH 7 and an initial concentration of 1 mM H_2O_2, optimal degradation (~90%) of lindane was achieved in less than 4 min. Likewise, it was reported that after 15 minutes of reaction, all the chlorine atoms were converted into chloride ions. This observation suggests, therefore, that chlorinated organic by-products do not accumulate. In this same sense, Wu et al.31 also reported the powerful function of ions in the UV/H_2O_2 system for the degradation of two organophosphate pesticides, parathion and chlorpyrifos. The presence of carbonate and bicarbonate ions in the aqueous solution contributed to two completely divergent effects.[31] It was found that bicarbonate ions hinder the efficiency of the UV/H_2O_2 system by quenching the hydroxyl radical, while carbonate ions promote degradation. These competitive reactions can be one of the most dramatic mechanisms in natural water bodies that contain carbonates and bicarbonates.

In addition to the synergistic UV/H_2O_2 process, there is another similarly used process, UV/O_3. After UV absorption, ozone undergoes an intermediate structural change and combines with water to form the hydroxyl radical.[32,33]

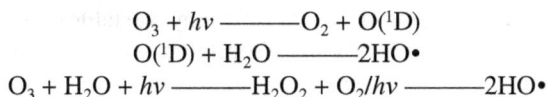

$$O_3 + h\nu \text{———} O_2 + O(^1D)$$
$$O(^1D) + H_2O \text{———} 2HO\bullet$$
$$O_3 + H_2O + h\nu \text{———} H_2O_2 + O_2/h\nu \text{———} 2HO\bullet$$

This method has proven its effectiveness for the oxidation and degradation of toxic organic compounds that are refractory to other treatments and for destroying bacteria and viruses in water. From a photochemical point of view, ozone has a molar extinction coefficient of 254 nm higher than that of H_2O_2; thus, the internal filter effects with aromatic compounds are less problematic than in the UV/H_2O_2 technique. In systems where the UV/O_3 combination has been applied, the observed degradation rates are much higher than those obtained separately with O_3 or UV radiation. In general, this synergism has been used on a pilot and industrial scale to treat pollutants in concentrations of ppm or ppb without generating hazardous waste. However, this technique has two serious drawbacks. First, the low solubility of ozone in water limits the process's speed and makes the design of reactors difficult due to mass transfer problems. The second drawback is related to the high production costs of O_3.

The formation of the • OH radical is the fundamental principle of the two previous advanced oxidation photochemical processes. Many studies have been conducted to understand the importance of oxidants (such as ozone) concerning conventional photolysis. For example Kearney et al.[34] explored the effectiveness of the UV/O_3 system to determine the time required to destroy nine formulated herbicides and two formulated insecticides prepared at three concentration levels (10, 100, and 1,000 ppm). The results showed that the time required to achieve 90% destruction depends on its concentration levels, which increased with the increasing concentration of the pesticide.[34] In the same way, Rao and Chu[35] conducted a comparative study to evaluate the degradation of the pesticide Linuron, between the three processes: photolysis, ozonization, and UV/O_3. The degradation rate of the O_3/UV system is 3.5 and 2 times higher than photolysis and conventional ozone treatment, respectively.

Intrinsically, it is important to address that degradation should also help decrease unfavourable products. For example, under conventional UV photolysis, the Linuron degradation pathway was followed through demethoxylation, photohydrolysis, and N-terminal demethylation. In contrast, the degradation of the same pesticide during ozonation proceeded through N-methoxylation, dechlorination, and hydroxylation at the benzene ring. In this way, the UV/O_3 system has proven to be the best degradation route to obtain the mineralization, dechlorination, and denitrogenation of Linuron. Therefore, it is considered one of the best ways to treat such pesticides with the least harmful by-products. Making a comparison between synergistic processes whose systems involve UV/O_3 and UV/H_2O_2, we can point out that the absorption of radiation by O_3 is immensely pragmatic concerning hydrogen peroxide. Therefore, its operation is easy, even in the low radiation range. In contrast, since H_2O_2 needs higher radiation energy with low absorption capacity, it exhibits high efficiency compared to UV/O_3 due to forming more radicals. As the competing relationships between capital and operating costs can vary widely, the selection between UV/O_3 and UV/H_2O_2 systems depends on many variables, including the type of effluent, types and concentrations of pollutants present, and the degree of removal required.

4.3 PHOTOCATALYSIS

In photocatalytic oxidation processes, pesticides are destroyed in the presence of semiconductor materials known as photocatalysts (e.g. TiO_2, ZnO), an energetic light source, and an oxidizing agent such as oxygen or air. As illustrated in

Figure 4.1, only photons with energies greater than the energy gap (energy band gap, ΔE) can give rise to the excitation of electrons of the valence band (BV) of the material, which then promotes possible reactions with pollutants. The absorption of photons with energy less than ΔE or greater than the wavelengths generally causes energy dissipation in the form of heat. Illumination of the photocatalytic surface with sufficient energy leads to the formation of a positive hole (H^+) in the valence band and an electron ($e-$) in the conduction band. The positive gap either directly oxidizes the contaminant or water to produce the hydroxyl radical $\bullet OH$, while the electron in the conduction band reduces the oxygen adsorbed on the photocatalyst. Activation of TiO_2, for example, by UV light, can be represented by the following steps.

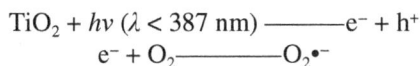

$$TiO_2 + hv \ (\lambda < 387 \ nm) \longrightarrow e^- + h^+$$
$$e^- + O_2 \longrightarrow O_2^{\bullet-}$$

In this reaction, the $h +$ and $e›$ pair are potent oxidizing and reducing agents, respectively. The oxidative and reductive reaction steps can be expressed according to Figure 4.3.

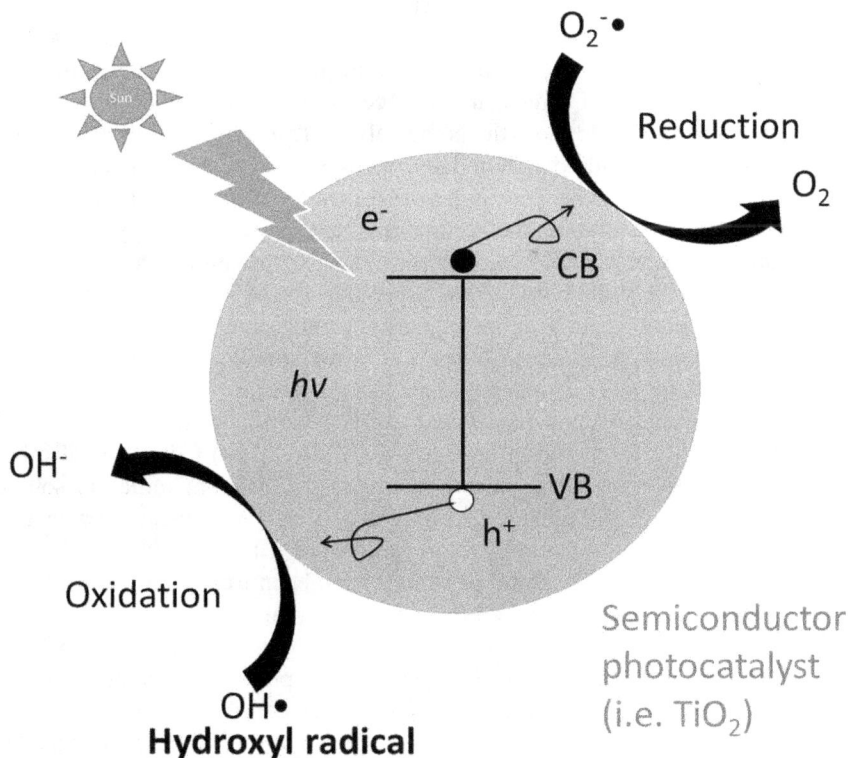

FIGURE 4.3 Principle of photocatalysis with TiO_2.

Oxidative reaction:

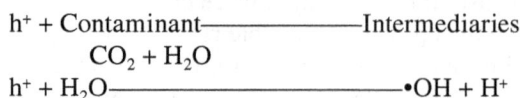

$$h^+ + Contaminant \text{———————} Intermediaries$$
$$CO_2 + H_2O$$
$$h^+ + H_2O \text{———————————} \bullet OH + H^+$$

Reductive reaction:

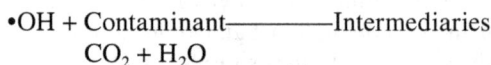

$$\bullet OH + Contaminant \text{———————} Intermediaries$$
$$CO_2 + H_2O$$

The generation of hydroxyl radicals by the photocatalytic oxidation process is shown in the previous steps. In the degradation of organic pollutants, the hydroxyl radical • OH generated from the oxidation of adsorbed water is the primary oxidant, and the presence of oxygen can prevent the recombination of an electron–hole pair. • OH thus attacks organic compounds, for example, chlorinated aromatic compounds, aniline, and nitrophenols, among others, which give rise to different reaction intermediates depending on the nature of the compounds. The resulting intermediates react with more • OH to produce degradation products such as CO_2 and H_2O. In the photocatalytic degradation of pollutants, when the oxygen reduction process and the oxidation of pollutants do not advance simultaneously, there is an accumulation of electrons in the BC, which causes an increase in the recombination rate of e› and h^+.[36,37] Therefore, this is of utmost importance to avoid electron accumulation in effective photocatalytic oxidation. In photocatalysis, TiO_2 (commonly known as titania) is the most widely studied system due to its high activity, desirable physical and chemical properties, low cost, and high availability. Of the three common crystalline forms of TiO_2, the anatase and rutile forms have been extensively investigated as photocatalysts. Anatase has been reported as being the more active phase as a photocatalyst than the rutile phase. Oxidation pathways similar to those of TiO_2 are confirmed in ZnO photocatalysts, including the formation of the • OH radical and direct oxidation by photogenerated holes. ZnO was reported to have a similar and comparable reactivity to TiO_2 under sunlight. The high activity of ZnO is attributed to the fact that the energy band gap of the material is similar to that of TiO_2, that is, 3.2 eV. Other metal oxides, including CeO_2, SnO_2, WO_3, and CdS, have also been examined for degradation of organic pollutants.[38–40] Different radiation sources, such as UV lamps and solar radiation, have been used in the photocatalytic degradation of various pesticides and herbicides in wastewater effluents.[41–43]

Based on these principles, many pesticides have been treated by photocatalytic degradation. In fact, a wide variety of semiconductor materials (ZnO, CuO, TiO_2, and WO_3) are used for photocatalytic purposes. Among them, titanium dioxide has been used most widely due to its favourable catalytic properties, the anatase phase being the most widely used due to the mentioned characteristics.

Organochlorine pesticides (POC) have been treated by the photocatalysis method. In this sense, the photocatalytic degradation of dicofol on TiO_2 nanoparticles (TiO_2-NPs) under irradiation with UV light has been reported. The authors have documented that dicofol has the potential to fully decompose into inorganic chloride ions, indicating a significant level of mineralization efficiency. It's worth noting that titania photocatalysts are currently available in various trade names, such as Degussa P25, Millennium PC 500, etc. Madani et al.[44] conducted a study to compare the activity of the two aforementioned commercial anatase products, using the pesticide diuron as a model. The results obtained by the researchers showed that the Degussa P25 material was more efficient in activity than Milenio PC 500.

Although photocatalysis is shown as an excellent methodology for eliminating pesticides, titania is limited for commercial applications due to its high activity in the UV range concerning natural sunlight. Therefore, many researchers have investigated the possible expansion of its absorption range in the visible region (e.g. through structural modifications and surface doping, among others). On the other hand, the recombination of excited electrons and the gap must be controlled and inhibited because it is another limiting factor in the photocatalysis process. In this sense, other options, such as metal/non-metal doping and other types of surface modifications, have been actively studied to improve the photocatalytic activity of semiconductor materials. Under this premise, Senthilnathan and Philip[45] report using TiO_2 doped with nitrogen-containing compounds (specifically trimethylamine). The authors found that such doping leads to an increase in the degradation of the pesticide lindane. However, it is widely accepted that metal doping is more effective in changing the absorbance range to a visible region relative to non-metal doped titanium oxide. Therefore, a wide variety of metal-doped titanium oxide compounds (i.e. transition metal-doped catalysts) have been prepared and studied to determine their ability to degrade against the pesticide chlorpyrifos. From what has been reported in the literature, it has been found that certain metals (V, Mo, Th, etc.) demonstrated greater degradation efficiency concerning undoped titanium oxide under solar lighting. However, when the same catalysts were subjected to pesticide degradation under UV radiation, the undoped material exhibited superior performance over titania with V, Mo, and Th.[46]

In this way, several dopants were used: Sn,[47] Ag and Pd,[48] Re,[49] Bi^{3+},[50] V^{5+}, Mo^{6+} and Th^{4+},[51-52] and Pt^{6+}[53] were shown to increase the photocatalytic activity of the systems examined substantially. However, the photoactivity of the metal-doped TiO_2 photocatalyst depends largely on the nature of the doping ion and its concentration, as well as the preparation methodology and operating conditions.[54] Both positive and negative results have been recorded in metal ion doping. It is desired that the deposition of metal ions in TiO_2 can modify the photoconductive properties by increasing the efficiency of charge separation between electrons and holes.[55] Increasing charge separation efficiency will improve the formation of hydroxyl radicals and reactive oxygen species.[56] In contrast, the photocatalytic activity with doped metals is clouded by thermal instability and the recombination of electrons and photogenerated holes.[57] As mentioned earlier, in addition to titania, many other semiconductor materials have been used as photocatalysts (ZnO, WO_3, etc.) in treating pesticides.

In another work,[58] the degradation based on ZnO to degrade Diazinion in an aqueous solution was studied under UV irradiation. To improve the photocatalytic activity, ZnO was doped with lanthanum; this doping decreased particle size and increased surface area. As a result, the photocatalytic degradation of monocrotophos species (inhibitors of the endocrine system) was markedly increased (compared to non-doped material).[59] The doping concentrations and a summary of the pesticides degraded using a modified catalyst through doping are shown in Table 4.1.

Similarly, photodegradation using WO_3 to degrade Monuron has also been shown to be effective in the presence of solar irradiation.[66] In a typical photocatalysis, several factors influence the performance of the reaction, for example, catalyst concentration, pH, temperature, irradiation light, and the presence of oxidants, among others. Accordingly, the effects of various factors on the degradation of methamidophos (organophosphate pesticide) were examined using TiO_2 as a photocatalyst.[67] The authors reported that the appropriate catalyst concentration must be optimized by selecting a suitable wavelength. Alkaline media favoured the reaction rate, and the addition of oxidants (such as H_2O_2, KBrO, Fe^{3+}, and Cu^{2+}) accelerated the reaction rate. On the other hand, ions like (Na^+, K^+, Ca^{2+}, and Zn^{2+}) or anions (like $Cl^›$ and $Br^›$) did not affect the degradation of methamidophos.

A great effort has also been made to modify the applicability of conventional photocatalytic techniques. One such modification includes photocatalytic ozonation $(O_3/TiO_2/UV)$. An aqueous solution performed a degradation study of the neonicotinoid insecticides (thiacloprid and imidacloprid). In a preliminary stability test,

TABLE 4.1
Doped Photocatalysts Showing the Optimal Doping Percentage in the Degradation of Different Pollutants.

Pollutant	Light	Material	Doped (%)	% Optimal Doping (%)	Ref.
Diuron	UV	$Pt–TiO_2$	0–2.0	0.2	53
Monocrotophos	Solar	$La–ZnO$	0–1.0	0.8	59
2,4-DCAA	Visible	$CeO_2–TiO_2$	0–10.0	5.0	60
Methyl parathion	UV	$Bi^{3+}–TiO_2$	0–2.0	1.5	50
4-Chlorophenol	Visible	$N–TiO_2$	0.21–0.45	0.45	61
Clopyralid	Visible	$Fe^{3+}–TiO_2$	0.13–1.48	1.27	62
Mecropod, MCPP	Visible	$Fe^{3+}–TiO_2$	0.13–1.48	0.13	63
O-cresol	Visible	$Pt–TiO_2$	0–1.0	0.5	45
Lindano	Visible	$Ag–TiO_2$	0–2.5	1.5	64
Lindano	Visible	$Cr–TiO_2$	0–2.5	2.0	49
Beta-cypermethrin	UV	$RuO_2–TiO_2$	0.1–0.8	0.3	52
Metamidofos	UV	$Re–TiO_2$	2.0–6.0	5.0	65
Orizalina	Solar	$Th–TiO_2$	0–0.1	0.06	47
Acetamiprid	UV	$Ag–TiO_2$	0–0.12	0.75	48
Chlorosulfuron	UV	$Sn–TiO_2$	0–0.2	0.11	58
Phenol	UV	$Ag–ZnO$	0–1.0	0.75	59
Phenol	UV	$Pd–ZnO$	0–1.0	0.5	66

thiacloprid showed higher photostability compared to imidacloprid. A detailed study was conducted to evaluate the suitability of the various treatments for the degradation and mineralization of thiacloprid in water under varied pH and ozone doses.[68] These authors evaluated the study of ozonation (O_3) and three different advanced oxidation photochemical processes, such as ozonation together with UV radiation (O_3/UV), O_3/TiO_2/UV, and O_2/TiO_2/UV. UV-A range radiation was applied in all three processes. Photocatalytic ozonation (O_3/TiO_2/UV) was found to be the most efficient process, regardless of pH conditions. The synergistic effect of ozone and TiO_2 photocatalysis was evident at acidic and neutral pH. Still, this phenomenon was lost at basic pH, probably due to the rapid self-decomposition of ozone under alkaline conditions. At acidic pH, the oxidation of chloride anions to chlorate (V) was also observed, which in turn hampered the degradation efficiency. Thus, photocatalysis is now widely used as a photochemical degradation method.[69] On the other hand, although photocatalysis has certain limitations (e.g. recombination), its main advantages are recognized due to the simple operation, economic feasibility, and high recyclability since the catalyst can be used repeatedly with a very slight decrease in catalytic activity for each use.

Another way to improve the spectral response of semiconductors is through sensitization of the semiconductor with molecules that absorb visible radiation and are capable of, in their excited state, injecting electrons into the conduction band, utilizing a surface charge complex (known as "MLCT", for its acronym in English: metal-to-ligand charge transfer).[70]

This has been a fruitful research field initiated by A. J. Bard's group in the 1980s, with spectral sensitization by phthalocyanines.[71] These studies aim to find the molecule that allows greater efficiency in converting solar energy. In this regard, sensitization in the visible has been achieved using complexes of Zn (II),[72] phosphorous derivatives of proline,[36] complexes of Pt (II),[73] catechol,[74] derivatives of fluorescein and anthracene,[54] thionin and eosin,[75] enediols or salicylic acid.[76,77] In addition, sensitization to TiO_2 in the near IR zone has been achieved through modified phthalocyanines.[78]

4.4 NANO-PHOTOCATALYSTS

Nanophotocatalytic materials have also played a key role in effectively treating pollutants in wastewater. These nanomaterials are relatively inexpensive and lead to a greater active surface area for illumination, resulting in improved photocatalytic activity.[79] In particular, the materials based on transition metal oxide nanoparticles have been widely documented in the literature (TiO_2, CuO, ZnO, among others), with excellent photocatalytic performances reported.[80] However, the application of metal oxide-based nano photocatalysts has the serious disadvantage of requiring UV light for their activation, confining their applications only in the UV range of the solar spectrum. To increase the absorption range in the visible light region, some structural modification strategies have been developed for nano photocatalysts, for example, by doping metal oxides or supporting them on carbonaceous materials, such as graphene.[81]

Graphene and its derivatives (graphene oxide, GO, and reduced graphene oxide, r-GO) are materials that have aroused an increasing interest due to their excellent properties. Their large surface area, excellent electrical, thermal and chemical stability, conjugated π bond system, and increased carrier mobility make them very promising materials for photocatalysis applications. Metal oxide nanocomposites based on graphene have shown improved photocatalytic activity due to inhibiting the recombination of electron–hole pairs[82] by reducing the band gap and increasing the surface area of the catalyst. In particular, graphene has been reported to act as an electron acceptor/transporter for TiO_2 particles; therefore, graphene is expected to substantially improve the lifetime of photoexcited electron–hole pairs[83]. TiO_2–graphene nanoparticles can absorb a wider region of light in the UV and visible light range, in addition to exhibiting faster photocatalytic kinetic, as observed in the catalytic photodegradation of methylene blue with TiO_2 coated with GO using UV and visible radiation.[84]

Cruz et al. studied the degradation of four organic pesticides using $GO–TiO_2$ and bare TiO_2 as catalysts.[85] The photocatalytic activity of $GO–TiO_2$ nanocomposite was higher than that of bare TiO_2. This is possibly due to the combined contribution of a lower energy band-gap and the known extinction of photoluminescence under NIR and visible laser excitation found for $GO–TiO_2$, compared to pristine TiO_2 (i.e. electrons photogenerated in TiO_2 can be transferred to GO by inhibiting charge recombination).

Regarding treating some pharmaceutical contaminants, the degradation of metronidazole was evaluated using $ZnSnO_3$ hollow nanospheres and $ZnSnO_3$/r-GO nanocomposites as photocatalysts. When applying 60 minutes of UV irradiation, the $ZnSnO_3$ hollow nanospheres exhibited excellent photocatalytic activity in the degradation of metronidazole (~100%) compared to the low result obtained in the blank test in the absence of $ZnSnO_3$ (~40%). When applying visible light irradiation for 180 min, the highest photodegradation of metronidazole was achieved using the $ZnSnO_3$/r-GO nanocomposites (73%) compared to the low performance observed for the $ZnSnO_3$ hollow nanospheres (42%).[86] In another study, Tao et al. used graphene–TiO_2 nanotubes to decompose acetaminophen. A photodegradation rate of up to 96% was achieved under UV light irradiation for 3 hours.[64] Similar results are reported in the literature when TiO_2/r-GO nanocomposites are used in the photodegradation of various active pharmaceutical ingredients such as diclofenac,[87] carbamazepine ibuprofen, and sulfamethoxazole.[88]

4.4.1 Photoelectrocatalysis

The application of external voltage to the purely photocatalytic process (then called the photoelectrocatalytic process) converts the photocatalyst into a photo-anode so that the recombination processes of the electrons generated due to UV radiation are delayed, as well as the photo-reaction. The reduction of oxygen itself happens according to the following reaction:

$$e^- + O_2ads \text{———} O_2^-$$

This reaction is minimized because the electron does not act on the semiconductor but is incorporated into the electrical circuit.[89–92]

Modifications have been made to these processes in order to improve their performance. For example, Zhao et al.[91] reported the degradation of 2,4-dichlorophenol (2,4-DCP) using a new integrative oxidation procedure based on the invention of three TiO_2/Ti–Fe–graphite electrodes. The authors compared the photoelectrocatalytic efficiency of TiO_2 assisted by electro-Fenton and H_2O_2. Here, H_2O_2 was produced near a cathode, while Fe^{2+} was generated continuously from the anode of Fe in solution (when current and O_2 were applied). As a result, the degradation ratio of 2,4-DCP was 93% in this integrative oxidation process, while the photoelectrocatalytic efficiency by electro-Fenton and H_2O_2 was only 31% and 46%, respectively. This way, it was shown that the degradation of 2,4-DCP was greatly improved through photoelectric cooperation. This integration process also worked well over a wide pH range.

Similarly, the research seeks to find materials that turn out to be good support for the processes involved since, by the nature of the reactions involved, they share characteristics like those used in electrochemical applications. Degradation by photoelectrocatalysis was used to determine mineralization in the degradation of the insecticide imidacloprid in an aqueous solution. The studies were carried out using electrodes made of TiO_2 P-25 coated with Ti (TiO_2/Ti). The photoelectrocatalytic efficiency of the TiO_2/Ti electrodes concerning imidacloprid oxidation was evaluated in terms of degradation and mineralization under various experimental conditions. The degradation efficiency increased with the increase of the applied potential up to +1.5 V versus Ag/AgCl, which is more favourable in acidic environments than in alkaline ones. In fact, this efficiency exhibited a much higher activity than the typical photocatalysis in the case of imidacloprid.

Depending on the nature of the photosensitizer/photocatalyst, the photodynamic action proceeds through type I or type II mechanisms or both. Type I mechanism involves electron transfer from excited sensitizer to substrate molecule or oxygen yielding free radicals and superoxide ion.

Photocatalysis research has been focused on the depuration of water contaminated mainly with three kinds of recalcitrant organic compounds: dyes, pesticides, and pharmaceuticals, and in this field, TiO_2 has established itself as the benchmark material. However, the fact that it can only be activated with ultraviolet radiation due to its wide bandgap constitutes a drawback. Because of this, more recently, the scientific effort has been directed to broaden the visible light response band of TiO_2 using different approaches such as metal and non-metal doping, coupling with other narrower bandgap semiconductors and dye sensitization.

Activating TiO_2 and several other methods can be used.[93] However, due to its simplicity, the impregnation of TiO_2 has been employed predominantly. For the latter, the dye's solubility must be considered to ensure that only single molecules are adsorbed onto TiO_2 instead of clusters that eventually will quench the photoactivated species during the degradation process. Another important aspect to consider to increase the photocatalytic activity of the hybrid photocatalytic system is the molecular structure of the phthalocyanine sensitizer. In this regard, Sharma et al.[94] suggested that the incorporation of push–pull phthalocyanines shows great potential to improve the electron injection into the semiconductor while simultaneously avoiding the back electron transfer (recombination).

Incorporating the asymmetric zinc phthalocyanine (ZnPc) into the semiconductor oxide revealed apparent changes in its optical and structural properties; particularly, a synergetic enhancement of the photocatalytic activity of the dye-sensitized materials was observed compared to the non-sensitized products. Furthermore, it was found that the TiO/AZnPc composite has a higher photocatalytic activity for the degradation of Rh-B (rhodamine B) under visible light than pure TiO_2. This outstanding result was rationalized in terms of the orientation of the dye (favoured by its asymmetric nature) that improves the electron injection from the photo-activated phthalocyanine to TiO_2.

Applying photodynamic action to the inactivation and decontamination of water certainly seems to be a promising alternative to conventional chemical treatment methods. The numerous selections of photosensitizers render the possibility to explore more into this field. Many semiconductors whose bandwidths lie in the visible region are available. Also, the absorption of TiO_2 in the visible region has been the topic of study in many trials by doping TiO_2 with other metals. Metal phthalocyanines, porphyrins, and transition metal complexes have proven useful in many photoinactivation studies trials but are still limited to lab-scale and pilot plant studies. However, it is desirable to avoid using metal centres for water disinfection purposes. Metal-free porphyrin-related molecules are also gaining much attention in disinfection studies owing to their long-lived triplet excited states and presence in natural systems. The rich chemistry of macrocyclic compounds allows for formulating them into photosensitizers with maximum desired absorption and functional groups that are transformable to attach to the support. Photocatalysis also brings about the irreversible oxidation of any unwanted and harmful pollutants in water. However, to carry this phenomenon to actual treatments of wastewater, industrial water, and drinking water treatment, it is necessary to consider the need for pretreatments to water prior to illumination. Both inorganic and organic photosensitizers are being studied worldwide to improve water treatment processes, but the scope in modelling organic sensitizers seems promising.

4.4.2 Photocatalytic Degradation of Microplastics

AOPs, characterized by the generation of reactive species in situ, mainly hydroxyl radicals (OH), have been recognized as one of the potential technologies for the degradation of plastic waste. Ozonization, the combination of ozone and UV radiation, the combination of peroxide (H_2O_2) and UV radiation, as well as the combination of UV radiation and photocatalysts are the most studied AOPs. The main drawback of all these processes is the high cost of maintenance, which somewhat limits the widespread practical application of these compelling technologies. With the emergence of highly efficient sources of UV radiation, catalysts that absorb visible spectrum radiation, an improved reactor design, and technologies based on UV and solar radiation applications have great potential for widespread use. However, more research is needed to make this possible.

Full mineralization of organic contaminants is generally not economically feasible and is not always necessary. Partial oxidation of the starting compound to fewer stable intermediates is a possible alternative, especially if the resulting intermediates are susceptible to further environmental degradation and do not pose a risk to the aquatic environment and human health. However, the partial oxidation of organic contaminants can sometimes lead to forming more toxic intermediates than the starting compound. The nature and number of degradation products depend on the oxidation process applied, the treatment time, and the chemical characteristics of the aqueous matrix.

Photocatalytic oxidation of LDPE led to the formation of low molecular weight (Mw) compounds such as hydroperoxides, peroxides, carbonyl, and unsaturated groups. Furthermore, the results of their study showed that the degree of oxidation of LDPE was directly proportional to the surface area of the catalyst. Shengying Li et al. (2010) carried out the degradation of polyethylene (PE) plastic under solar irradiation with polypyrrole/TiO_2 nanocomposite (PPy/TiO_2) as a photocatalyst, which was prepared by emulsion and sol-gel polymerization methods. They found that irradiating PE plastic for 240 hours in sunlight reduced its weight to 35.4% and 54.4% Mw, respectively.[95] According to this study, atomic force microscopy (AFM) images showed the formation of cavities on the surface of PE plastic. Fourier transform infrared spectroscopy (FTIR) spectroscopic studies indicated a strong interaction between the PE and PPy/TiO_2 interface, causing degradation of PE. Tian et al. (2019) studied the degradation and mineralization of polystyrene (PS) nanoplastics under ultraviolet (UV) radiation at 254 nm using 14C radioisotope tracing technology.[96] 14C polystyrene (PS) nanoplastics were synthesized from 14C-labeled styrene. To study the role of water during the photodegradation of PS nanoplastics, 14C-PS nanoplastics were exposed to UV radiation in the air or suspended in water. The results of X-ray photoelectron spectroscopy (XPS) showed that after 48 hours of UV irradiation, CO groups were formed on the surface, while no significant change was observed in the FTIR analysis, indicating that the short-term photo-oxidation only occurs on the surface layer of PS nanoplastics. The Mw of PS nanoplastics increased in the air after irradiation, suggesting crosslinking of PS chains, while it did not show significant changes in the presence of water. The mineralization of the PS nanoplastics was higher in water (17.1 ± 0.55%) than in air (6.17 ± 0.1%). A significant amount (11.0 ± 0.1%) of by-products was detected with a small Mw in water during UV irradiation, much higher than that removed from the surface of the nanoplastics exposed to air. The higher photoreactivity in water suggests that the mechanisms underlying the phototransformation of the PS nano plastics in the two matrices could be different. The present study provided the first evidence of photodegradation of PS nanoplastics in aqueous environments.

4.5 SONOLYSIS

In ultrasound-assisted processes, several factors can be optimized to achieve high efficiency in the treatment of water contaminated with organic compounds, among which the operating parameters of the ultrasound equipment (frequency, power density, and calorimetric efficiency) and the properties of the liquid medium (initial concentration of the substrate, pH, water matrix) are the most widely studied.[97–101] Regarding the sonochemical transducer system, Tran et al. reviewed the characteristics of irradiation sources and the types of reactors used in the sonochemical degradation of pharmaceutical compounds. In addition, they discussed the important factors that affect sonochemical oxidation efficiency, including electrical power, frequency, and temperature.[101]

4.5.1 SONOCHEMICAL PROCESS

Another AOP based on the generation of • OH radicals consists of the sonication of water using high-frequency ultrasonic waves (from 16 to 2,000 kHz) that lead to the formation, growth, and collapse of cavities (millions of microscopic bubbles in vapour or gas phase) in the liquid, a process better known as sonolysis or acoustic cavitation. The collapse or implosion of the cavities generates high pressures, temperatures, and highly reactive • OH radicals.[102,103] A schematic representation of the sonolysis process is shown in Figure 4.4.

Experience in homogeneous sonochemistry has shown that for ultrasound-assisted liquids, there are three different sites where chemical reactions occur, as shown in Figure 4.5:[103,104] (i) In the cavitation bubble itself, (ii) in the cavitation bubble–liquid interface, and (iii) in the bulk of the solution. The collapse of the

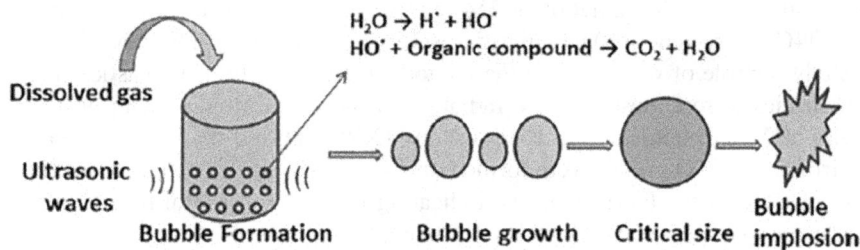

FIGURE 4.4 Schematic representation of sonolysis process. Reprinted with permission of Reference 103.

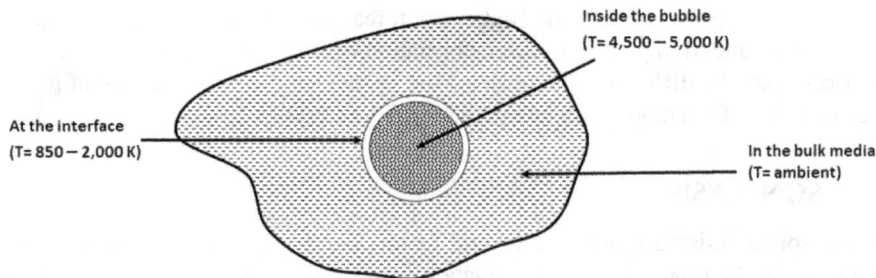

FIGURE 4.5 Possible sites of chemical reactions in homogeneous reaction media. Reprinted with permission of Reference 104.

cavitation bubble produces a large concentration of energy. Some authors have reported that temperatures reaching 4,500–5,200 K occur within the cavitation bubble and 850–2,000 K for the bubble–liquid interface.[104,105] While Pinjari and Pandit (2010) reported that the collapse of the bubbles creates even more extreme conditions of temperature (5,000–10,000 K) and pressure (1,000–2,000 atm) within the hot spots in the aqueous medium.[106] Based on this, in water treatment, organic pollutants can be destroyed in the first two sites due to the combined effects of pyrolytic decomposition and hydroxylation or in the bulk solution by oxidative reactions with •OH and H_2O_2.104

In recent years, the sonolysis process has been extensively used as an AOP for removing many micropollutants in wastewater.[107–110] In sonolysis, various reaction mechanisms can generate highly oxidizing reactive species such as the •OH, HOO•, and O• radicals (Eqs. 4.1–4.4).[111–113] Hydrogen peroxide is also produced during sonolysis (Eqs. 4.5–4.6).[114,115] The reactive oxygen species, mainly radicals •OH, promote the oxidation of pollutants (Eqs. 4.7–4.8). In fact, the concentration of • OH radicals at the bubble/solution interface has been estimated to be as high as 1×10^{-2} M.[115] H_2O_2 has also contributed to the degradation of pollutants in wastewater (Eq. 4.9).[110]

$$H_2O +))) \rightarrow •OH + H• \tag{4.1}$$

$$H_2O_2 +))) \rightarrow •OH + H• \tag{4.2}$$

$$H• + O_2 +))) \rightarrow HOO• \tag{4.3}$$

$$2 •OH \rightarrow •O + H_2O \tag{4.4}$$

$$H_2O +))) \rightarrow ½ H_2 + ½ H_2O_2 \tag{4.5}$$

$$2 •OH \rightarrow H_2O_2 \tag{4.6}$$

$$\text{Pollutants} + •OH +))) \rightarrow \text{Degradation products} \tag{4.7}$$

$$\text{Pollutants} + HOO• +))) \rightarrow \text{Degradation products} \tag{4.8}$$

$$\text{Pollutants} + H_2O_2 +))) \rightarrow \text{Degradation products} \tag{4.9}$$

[*where*))) *refers to ultrasound irradiation*]

The optimal ultrasound frequency for the degradation of an organic compound is also influenced by its properties, such as solubility and volatility. The degradation rate for a low volatility molecule is more efficient at the frequency where the greatest amounts of •OH and H_2O_2 oxidants are generated. Isariebel et al. (2009) evaluated the sonolytic degradation of aqueous solutions of levodopa and paracetamol using various frequencies (574, 860, and 1,134 kHz). For the lowest frequency (574 kHz), both pollutants reached 100% efficiency (32 W, 25 mg L^{-1}, 20 ° C and 8 hours). Experimental tests using 1-butanol (as a radical scavenger) and H_2O_2 (as a promoter) confirmed that the degradation of drugs occurred mainly through radical reactions.[116]

Similar results were obtained by De Bel et al. (2011), who studied the influence of operational parameters in the sonolysis of aqueous solutions of ciprofloxacin (15 mg/L) at 25 °C. The researchers found that of the three frequencies evaluated (544, 801, and 1,081 kHz), the lowest (544 kHz) produced the maximum amount of •OH radicals that led to the highest degradation rate constant (0.0067 min^{-1}) for ciprofloxacin.[117] In a study carried out by Rahmani et al. (2014), the application of frequencies above 120 kHz did not have a notable effect on the degradation of tinidazole by sonolysis in the presence of hydrogen peroxide. They reported 40 and 60% degradation at 40 and 80 kHz, respectively. While at higher frequencies, the change in degradation rate was very low at 74% at 120 kHz and 75% at 180 kHz. [118]

Some authors have shown that the degradation efficiency of pharmaceutical contaminants is highly dependent not only on the ultrasound power but also on the frequency of the applied ultrasonic irradiation.[101,105,109,119] Therefore, to achieve remarkable efficiency in the degradation of diclofenac and carbamazepine, high power density values are required (100–400 W L^{-1}). Similarly, Lan et al. (2012) obtained the complete decomposition of naproxen in just 33 minutes by linearly increasing the ultrasonic power (46% at 200 W; 96% at 300 W and ~100% at 450 W). Also, an increase in power density from 20 to 60 W L^{-1} improved the degradation of the drug piroxicam. Similar results were found in the degradation of ibuprofen and 4-cumylphenol.[120,121] This behaviour is because the high ultrasound power densities generate a more significant number of cavitation bubbles in the liquid, promoting the formation of greater amounts of •OH radicals.

Ghafoori et al. (2015) found that greater mineralization of the TOC (Total Organic Carbon) is achieved by increasing the ultrasound power.[120] They obtained an elimination of 60.22% at 20 W, compared to 75.26% obtained at a higher power of 140 W. The high ultrasound power causes the highest breakdown rate of H_2O_2 molecules in an aqueous solution, which causes an increase in the concentration of hydroxyl radicals that attack organic matter.[101] On the other hand, Sutar and Rathod (2015) also investigated the effect of ultrasonic power irradiation on the degradation efficiency of cetirizine dihydrochloride. A low percentage of degradation (~50%) was obtained for the lowest powers (30 and 50 W) compared to the percentages achieved in the highest irradiation powers, which were 91% at 100 W and 80% at 150 W. The authors noted that for higher irradiation power, the bubbles combine to form larger bubbles, causing a weak implosion that reduces degradation efficiency.[121]

De Bel et al. showed that the degradation of ciprofloxacin was highly dependent on the temperature of the bulk solution.[117] The degradation constant increased markedly with increasing temperature, from 0.0055 min^{-1} at 15°C to 0.0105 min^{-1} at 45°C. Various authors have shown that the degradation rate of organic compounds is directly proportional to the temperature as the organic molecules migrate from the bulk solution to the gas–liquid interface, where the temperature and the •OH concentration are high. [101,122]

On the other hand, Lianou et al. carried out experiments at 20 °C, 40 °C, and 60 °C to study the effect of temperature on the degradation of piroxicam.[97] Interestingly, the results showed that an increase in volume temperature reduces the efficiency of the process. The total degradation of piroxicam was achieved in 10, 20, and 40 minutes of sonochemical treatment at 20 °C, 40 °C, and 60 °C, respectively. The negative effect observed can be explained by the following: the increase in temperature increases the vapour pressure of the solvent. Due to this, a higher water vapour content is inside the cavitation bubbles, making the collapse of cavitation bubbles less violent, known as the "cushioning effect". This causes a reduction in the collapse temperature and thus a reduced production of •OH radicals.[123]

The sonodegradation of ibuprofen in wastewater was also studied by Adityosulindro et al. the authors observed that a decrease in the pH of the solution did not cause an appreciable change in the degradation of the drug; however, adjusting the pH to an alkaline value led to a significant reduction in the degradation rate of ibuprofen.[108] This effect is due to the two possible forms of ibuprofen at a given pH value, as the pKa of the carboxyl group is 4.9, and at pH values of 2.6 and 4.3, ibuprofen is in its molecular form. In contrast, at pH value of 8, it is completely deprotonated. In the ionic form, ibuprofen is less present on the surface of the cavitation bubbles, the place where the -OH radicals responsible for the degradation of the molecule are produced. This behaviour was also observed in the sonolysis of diclofenac, dicloxacillin, paracetamol, and sulfadiazine, which, despite having been performed at high sonication frequency (300–862 kHz), exhibited a considerable reduction in the degradation rate at a pH value higher than the pKa of the molecule.[123,124]

Naddeo et al. obtained promising performances in the sonochemical degradation of 23 organic pollutants (mainly pharmaceuticals) in urban wastewater.[109] The authors found that the degradation rate depended on the chemical structure of each compound. For example, it was observed that for a 180-minute treatment, triclosan was degraded by 95%, while erythromycin was only degraded by 50%. Recently, Serna-Galvis et al. (2019) evaluated the sonochemical treatment of six antibiotics belonging to the fluoroquinolone classes (ciprofloxacin and norfloxacin), penicillins (oxacillin and cloxacillin), and cephalosporins (cephalexin and cefadroxil).[125] The results showed that the sonodegradation efficiency of these antibiotics was highly dependent on the chemical structure. The highest degradation was observed for the most hydrophobic molecule (cloxacillin) and the lowest degradation for the most hydrophilic molecule (cefadroxil).[110]

The degree of sonochemical degradation of aqueous solutions of dicloxacillin was studied by Villegas-Guzman et al. The results indicated that the sonochemical treatment could completely degrade the pollutant in 180 min, transforming it into more oxidized by-products.[123] However, TOC analysis only showed 10% mineralization, indicating that the by-products are recalcitrant to ultrasonic action. In other research, Nejumal et al. (2013) studied the sonolysis of the antihypertensive atenolol. Although sonolysis gave a 100% transformation, only a 62% reduction in TOC was observed during 60 minutes of irradiation. Furthermore, the analysis of the degradation of the product showed the formation of several stable intermediates during the irradiation.[126]

It is well known that many surfactants from households and industries are discharged daily into effluents, resulting in environmental water pollution, damage to aquatic life, and serious risk to human health.[127] Singla et al.[128] studied the sonodegradation of the non-ionic surfactant octaethylene glycol monododecyl ether ($C_{12}E_8$) at different initial concentrations lower and higher than its critical micellar concentration. They observed that the degradation of $C_{12}E_8$ increased with increasing surfactant concentration up to its critical micellar concentration (from 40 to 90 μM). Still, above this concentration, a behaviour similar to saturation was evidenced, suggesting that the surfactant in its monomeric form is better involved in the degradation process. Therefore, the authors proposed that the $C_{12}E_8$ monomer undergoes two possible degradation mechanisms taking place at the bubble/solution interface: hydroxylation/oxidation and pyrolytic fragmentation of the surfactant.

Although cationic surfactants are not as widely used as anionic and non-ionic surfactants, they are known to be 10 to 100 times more toxic than the others. Singla et al. evaluated the sonodegradation of the cationic surfactant, laurylpyridinium chloride (LPC), in water.[129] They observed that the degradation rate of LPC decreased with increasing its initial concentration and did not show a direct relationship with the OH radicals generated ultrasonically. Thus, the mechanism of LPC decomposition was mostly governed by pyrolysis reactions. The authors concluded that sonolysis alone was not an effective technique for the complete degradation of LPC since, at prolonged sonication times (20 h), they only obtained 60% mineralization.[128,129]

Other environmental applications where sonolytic processes have gained increasing popularity have been in the degradation of recalcitrant organic molecules in wastewater, such as phenol,[130] chloroaromatic pesticides (1,4-dichlorobenzene and 1-chloronaphthalene),[131,132] and organic dyes [159]. A summary of several studies on the removal of pollutants in wastewater using sonolysis as an AOP is compiled in Table 4.2. In addition, the experimental and operational conditions of ultrasound are also described, together with the degradation efficiencies of the micropollutants.

TABLE 4.2
Ultrasound Operational Conditions and Degradation Efficiencies of Some Micropollutants in Wastewater Using Sonolysis as an Advanced Oxidation Process.

Micropollutant	Operating Conditions	Degradation Efficiency and Significant Findings	Ref.
Losartan and valsartan	f_{US} = 375, 575 and 990 kHz, PD = 13.4, 37.4 y 88.0 W L⁻¹, T = 20 °C, C_0 = 40 µM, pH = 6.5, V = 300 mL, matrix = distilled water, simulated seawater, and simulated hospital wastewater	At the highest power density (88.0 W L⁻¹) and the lowest frequency (375 kHz), antihypertensives were degraded by ~70 and 60% of losartan and valsartan, respectively. The high potentiality of ultrasound was observed to degrade losartan or valsartan in complex water matrices	125
Ibuprofen	f_{US} = 20, 40, 200, 572 and 1130 kHz; PD = 67, 185, 179, 231, 466 W L⁻¹; C_0 = 50 µM; pH = 3.0; V= 250 mL, matrix = ultrapure water	Complete degradation of ibuprofen was obtained at 572 kHz, 231 W L⁻¹ and just 20 min. Carbon mineralization was generally low, peaking at 577 kHz (10.4%). An increase in ibuprofen degradation was observed with an increase in frequency.	119
Oxacillin	f_{US} = 275 kHz; P = 60 W; t = 120 min; C_0 = 47.23 µM; pH = 5.6, matrix = distilled water	Sonochemical process (275 kHz) efficiently degraded oxacillin over 2 h, but the process was unable to achieve complete mineralization of the pollutant. However, with a 30 hour of additional sonication, oxacillin mineralization (99% TOC) was achieved in a combination of a conventional biological system	123

TABLE 4.2 (*Continued*)
Ultrasound Operational Conditions and Degradation Efficiencies of Some Micropollutants in Wastewater Using Sonolysis as an Advanced Oxidation Process.

Micropollutant	Operating Conditions	Degradation Efficiency and Significant Findings	Ref.
Acetaminophen	f_{US} = 600 kHz; P = 20, 30, 40, 50 y 60 W; C_0 = 33–1,323 µM; pH = 3.0, 5.6, 9.5, 11.0, 12.0; V = 300 mL; T: 20 °C; matrix = deionized and mineral natural water	Sonolysis in acidic medium (pH 3.0–5.6) was higher than in basic aqueous solutions (pH 9.5–12.0). Under acidic conditions, the initial concentration of ACP decreased from 82.5 mM to 45 mM at 60 minutes of treatment (600 kHz, 60W and 20 °C). High ultrasonic powers and low and natural acidic pH values favored the efficiency of the treatment	99
C.I. reactive Orange 107	Two different ultrasonic power generators were used: K8 (f_{US} = 850 kHz; P = 60 W) and MFLG (f_{US} = 378 and 992 kHz; P = 44 and 62 W); t = 4 h; T = 20 and 30 °C; C_0 = 50, 100, and 200 mg L⁻¹; pH = 11; V = 500 mL; matrix = deionized water	The discoloration of the dye solution increased by 99% at 850 kHz (50 mg/L at 20 °C). A further increase in frequency at 992 kHz, the discoloration (69%) decreased considerably. Degradation of the dye Red-107 followed pseudo-first-order kinetics	133
Crystal violet	f_{US} = 800 kHz; P = 20, 40, 60, and 80 W; T = 20 °C; C_0 = 2.45, 12.3, 24.5, 49, 245, 735 and 1225 µM; pH = 3, 5.5, and 9; V = 250 mL; matrix = water containing anions (Cl⁻, SO_4^{2-}, and HCO_3^-) and cations (Fe^{2+})	Crystal violet (49 µM) was completely removed after 360 minutes of irradiation time at 80 W and 800 kHz, in the presence of Fe^{2+} (1 mM) with argon as a saturating gas. Only 12% of the initial TOC was removed, indicating low mineralization of the contaminant.	134
Benzophenone-3	f_{US} = 20 kHz; P = 40.2, 55.9, 73.8 and 80.1 W; T = 25 °C; C_0 = 6.6–19.7 µM; pH = 2, 6.5 y 10; V = 200 mL; matrix = Milli-Q water and anions (Cl⁻, NO_3. and HCO_3^-)	After sonolytic irradiation for 60 min, a decrease in the pollutant concentration was achieved by 91.4 % with the highest power (80.1 W). The experiments were carried out at pH values of 2 and 10 as extreme values, at pH 2 the highest degradation of Benzophenone-3 (3.9 mg/L) was reached by 82% (55.9 W and 60 min)	135

(*Continued*)

TABLE 4.2 (*Continued*)

Ultrasound Operational Conditions and Degradation Efficiencies of Some Micropollutants in Wastewater Using Sonolysis as an Advanced Oxidation Process.

Micropollutant	Operating Conditions	Degradation Efficiency and Significant Findings	Ref.
Pyridine	f_{US} = 40 kHz; P = 360 W; T = 20, 30, 40 °C; C_0 = 10, 20, 60, 80 and 100 mg L^{-1}; pH = 2.4 to 10.5; V = 100 mL; matrix = synthetic wastewater and deionized water	The removal rates of pyridine at 180-minute sonication time decreased from 52% to 15%, increasing the initial concentration from 10 to 100 mg/L (pH 6.7 and 20 °C). The optimal pH was found to be 9.1, which resulted in 25% pyridine removal after sonolysis for 180 min.	100
Furfural	f_{US} = 20 kHz; PD = 600, 1,500, and 3,000 W L^{-1}; C_0 = 1, 2, 5, 8, and 10 mg L^{-1}, pH = 2, 5, 7, 9, and 11, V = 250 mL, matrix = double distilled water	The maximum removal efficiency of furfural was found to be 83% at 2 hours (10 mg L^{-1} and 3000 W L^{-1}). It was observed that sonolysis of furfural decreased with increasing the pH of solution. The sonochemical degradation is a promising clean technique for furfural removal in wastewaters	136
4-Cumylphenol	f_{US} = 300 and 600 kHz, P = 20, 40, 60, 80, and 100 W, T = 20, 30, 40, and 50 °C, C_0 = 0.05, 0.1, 0.5, 1, 5, 10, 20, and 30 mg L^{-1}; pH = 2, 6.5, and 10, V = 300 mL; matrix = pure and natural water	The destruction rate of 4-cumylphenol was observed to occur at the lowest frequency by 92% (300 kHz). The degradation rate increased significantly with increasing power by 37, 92, and 97% at 20, 80, and 100 W, respectively. Complete degradation was achieved at 50 °C in just 30 minutes of ultrasonic treatment (5 mg L^{-1}, 300 kHz, and 80 W)	137

f_{US} = ultrasound frequency, P = electrical power, PD = electrical power density, T = temperature, C_0 = initial concentration, V = volume.

Although the sonolysis process does not require the use of additional chemical compounds, some studies have reported several limitations of the method, such as low efficiency to reach acceptable percentages in the mineralization of organic contaminants, formation of hydrophilic products with low degradation, and high operating costs due to the use of high amounts of energy[113,123,124,128]. These disadvantages confine the individual application of the method to a laboratory-scale since it would not be very efficient on an industrial scale. Therefore, sonolysis must be combined with other conventional AOPs to overcome these challenges.

4.6 CONCLUDING REMARKS

Advances in the field of water treatment have made it possible to improve the oxidative degradation processes of organic compounds with the application of catalytic, photochemical, and sonochemical methods. In each method, some advantages and disadvantages were discussed, but both can be compensated when used in combination as synergistic systems. However, this chapter details the performance aspects and the considerations to consider. Today, the greatest challenge for more widespread use of these methodologies is in the design of reactors since the performance in the degradation of pollutants has been improved with the integration of nanomaterials and nanocomposites. Both the processes derived from photolysis and sonolysis present great advantages for green chemistry in more efficient and sustainable water treatment systems.

REFERENCES

1. Schrauzer, G. N.; Strampach, N.; Hui, L. N.; Palmer, M. R.; Salehi, J. N. Photoreduction on Desert Sands Under Sterile Conditions. *Proc. Natl. Acad. Sci.* **1983**, *80* (12), 38 **73–3876. https://doi.org/10.1073/pnas.80.**12.3873.
2. Canfield, D. E.; Glazer, A. N.; Falkowski, P. G. The Evolution and Future of Earth's Nitrogen Cycle. *Science* **2010**, *330* (6001), **192–196. https://doi.org/10.1126/ science.**1186120.
3. Nesbitt, H. W.; Markovics, G.; Price, R. C. Chemical Processes Affecting Alkalis and Alkaline Earths During Continental Weathering. *Geochim. Cosmochim. Acta* **1980**, *44* (11), 16 **59–1666. https://doi.org/10.1016/0016-7037(80)**90218-5.
4. Delpla, I.; Jung, A.-V.; Baures, E.; Clement, M.; Thomas, O. Impacts of Climate Change on Surface Water Quality in Relation to Drinking Water Production. *Environ. Int.* **2009**, *35* (8), 12 **25–1233. https://doi.org/10.1016/j.envint.20**09.07.001.
5. Leavitt, P. R.; Cumming, B. F.; Smol, J. P.; Reasoner, M.; Pienitz, R.; Hodgson, D. A. Climatic Control of Ultraviolet Radiation Effects on Lakes. *Limnol. Oceanogr.* **2003**, *48* (5), 20 **62–2069. https://doi.org/10.4319/lo.2003.**48.5.2062.
6. Leenheer, J.; Croué, J.-P. Peer Reviewed: Characterizing Aquatic Dissolved Organic Matter. *Environ. Sci. Technol.* **2003**, *37* (1), **18A-26A. https://doi.org/10.1021/**es032333c.
7. Canonica, S.; Jans, U.; Stemmler, K.; Hoigne, J. Transformation Kinetics of Phenols in Water: Photosensitization by Dissolved Natural Organic Material and Aromatic Ketones. https://pubs.acs.org/doi/pdf/10.1021/es00007a020 (accessed 2021–10–30).
8. Halladja, S.; ter Halle, A.; Aguer, J.-P.; Boulkamh, A.; Richard, C. Inhibition of Humic Substances Mediated Photooxygenation of Furfuryl Alcohol by 2,4,6-Trimethylphenol. Evidence for Reactivity of the Phenol with Humic Triplet Excited States. *Environ. Sci. Technol.* **2007**, *41* (17), 60 **66–6073. https://doi.org/10.1021/**es070656t.
9. Page, S. E.; Arnold, W. A.; McNeill, K. Assessing the Contribution of Free Hydroxyl Radical in Organic Matter-Sensitized Photohydroxylation Reactions. *Environ. Sci. Technol.* **2011**, *45* (7), 28 **18–2825. https://doi.org/10.1021/**es2000694.
10. Dong, M. M.; Rosario-Ortiz, F. L. Photochemical Formation of Hydroxyl Radical from Effluent Organic Matter. *Environ. Sci. Technol.* **2012**, *46* (7), 37 **88–3794. https://doi. org/10.1021/**es2043454.
11. Sur, B.; Rolle, M.; Minero, C.; Maurino, V.; Vione, D.; Brigante, M.; Mailhot, G. Formation of Hydroxyl Radicals by Irradiated 1-Nitronaphthalene (1NN): Oxidation of Hydroxyl Ions and Water by the 1NN Triplet State. *Photochem. Photobiol. Sci.* **2011**, *10* (11), 18 **17–1824. https://doi.org/10.1039/C1PP05216K.**
12. Dattilo, A. M.; Decembrini, F.; Bracchini, L.; Focardi, S.; Mazzuoli, S.; Rossi, C. Penetration of Solar Radiation into the Waters of Messina Strait (Italy). *Ann. Chim.* **2005**, *95* (3–4), **177–184. https://doi.org/10.1002/adic.**200590020.
13. De Laurentiis, E.; Maurino, V.; Minero, C.; Vione, D.; Mailhot, G.; Brigante, M. Could Triplet-Sensitised Transformation of Phenolic Compounds Represent a Source of Fulvic-like Substances in Natural Waters? *Chemosphere.* **2013**, *90* (2), **881–884. https://doi.org/10.1016/j.chemosphere.20**12.09.031.

14. Vermilyea, A. W.; Voelker, B. M. Photo-Fenton Reaction at Near Neutral PH. *Environ. Sci. Technol.* **2009**, *43* (18), 69 **27–6933. https://doi.org/10.1021/es900721x.**

15. Elabd, H.; Jury, W. A.; Cliath, M. M. Spatial Variability of Pesticide Adsorption Parameters. https://pubs.acs.org/doi/pdf/10.1021/es00145a006 (accessed 2021–10–30).

16. Lopez, J. L.; Einschlag, F. S. G.; González, M. C.; Capparelli, A. L.; Oliveros, E.; Hashem, T. M.; Braun, A. M. Hydroxyl Radical Initiated Photodegradation of 4-Chloro-3,5-Dinitrobenzoic Acid in Aqueous Solution. *J. Photochem. Photobiol. Chem.* **2000**, *137* (2), **177–184. https://doi.org/10.1016/S1010-6030(00)00357-9.**

17. Jonsson, M.; Lind, J.; Reitberger, T.; Eriksen, T. E.; Merenyi, G. Free radical combination reactions involving phenoxyl radicals. https://pubs.acs.org/doi/pdf/10.1021/j100133a018 (accessed 2021–10–30).

18. Karpel Vel Leitner, N.; Gombert, B.; Ben Abdessalem, R.; Doré, M. Kinetics and Mechanisms of the Photolytic and OH° Radical Induced Oxidation of Fluorinated Aromatic Compounds in Aqueous Solutions. *Chemosphere.* **1996**, *32* (5), **893–906. https://doi.org/10.1016/0045-6535(96)00011-2.**

19. Weir, B. A.; Sundstrom, D. W. Destruction of Trichloroethylene by UV Light-Catalyzed Oxidation with Hydrogen Peroxide. *Chemosphere.* **1993**, *27* (7), 12 **79–1291. https://doi.org/10.1016/0045-6535(93)90175-5.**

20. Sundstrom, D. W.; Klei, H. E.; Nalette, T. A.; Reidy, D. J.; Weir, B. A. Destruction of Halogenated Aliphatics by Ultraviolet Catalyzed Oxidation with Hydrogen Peroxide. *Hazard. Waste Hazard. Mater.* **1986**, *3* (1), **101–110. https://doi.org/10.1089/hwm.1986.3.101.**

21. Guittonneau, S.; de Laat, J.; Dore, M.; Duguet, J. P.; Bonnel, C. Etude Comparative de La Degradation de Quelques Molecules Aromatiques Simples En Solution Aqueuse Par Photolyse UV et Par Photolyse Du Peroxyde d'hydrogene. *Environ. Technol. Lett.* **1988**, *9* (10), 11 **15–1128. https://doi.org/10.1080/09593338809384673.**

22. Zhang, Z.; Xiang, Q.; Glatt, H.; Platt, K. L.; Goldstein, B. D.; Witz, G. Studies on Pathways of Ring Opening of Benzene in a Fenton System. *Free Radic. Biol. Med.* **1995**, *18* (3), **411–419. https://doi.org/10.1016/0891-5849(94)00148-D.**

23. Mohan, H.; Mudaliar, M.; Aravindakumar, C. T.; Rao, B. S. M.; Mittal, J. P. Studies on Structure–Reactivity in the Reaction of OH Radicals with Substituted Halobenzenes in Aqueous Solutions. *J. Chem. Soc. Perkin Trans.* **1991**, *2* (9), 13 **87–1392. https://doi.org/10.1039/P29910001387.**

24. Stefan, M. I.; Mack, J.; Bolton, J. R. Degradation Pathways During the Treatment of Methyl Tert-Butyl Ether by the UV/H_2O_2 Process. *Environ. Sci. Technol.* **2000**, *34* (4), **650–658. https://doi.org/10.1021/es9905748.**

25. Stefan, M. I.; Bolton, J. R. Mechanism of the Degradation of 1,4-Dioxane in Dilute Aqueous Solution Using the UV/Hydrogen Peroxide Process. *Environ. Sci. Technol.* **1998**, *32* (11), 15 **88–1595. https://doi.org/10.1021/es970633m.**

26. Stefan, M. I.; Hoy, A. R.; Bolton, J. R. Kinetics and Mechanism of the Degradation and Mineralization of Acetone in Dilute Aqueous Solution Sensitized by the UV Photolysis of Hydrogen Peroxide. *Environ. Sci. Technol.* **1996**, *30* (7), 23 **82–2390. https://doi.org/10.1021/es950866i.**

27. Stefan, M. I.; Bolton, J. R. Reinvestigation of the Acetone Degradation Mechanism in Dilute Aqueous Solution by the UV/H_2O_2 Process. *Environ. Sci. Technol.* **1999**, *33* (6), **870–873. https://doi.org/10.1021/es9808548.**

28. Chelme-Ayala, P.; El-Din, M. G.; Smith, D. W. Degradation of Bromoxynil and Trifluralin in Natural Water by Direct Photolysis and UV Plus H_2O_2 Advanced Oxidation Process. *Water Res.* **2010**, *44* (7), 22 **21–2228. https://doi.org/10.1016/j.watres.20**09.12.045.

29. Abramović, B. F.; Banić, N. D.; Šojić, D. V. Degradation of Thiacloprid in Aqueous Solution by UV and UV/H₂O₂ Treatments. *Chemosphere.* **2010**, *81* (1), **114–119.** **https://doi.org/10.1016/j.chemosphere.20**10.07.016.

30. Nienow, A. M.; Bezares-Cruz, J. C.; Poyer, I. C.; Hua, I.; Jafvert, C. T. Hydrogen Peroxide-Assisted UV Photodegradation of Lindane. *Chemosphere.* **2008**, *72* (11), 17 **00–1705. https://doi.org/10.1016/j.chemosphere.20**08.04.080.

31. Wu, C.; Linden, K. G. Phototransformation of Selected Organophosphorus Pesticides: Roles of Hydroxyl and Carbonate Radicals. *Water Res.* **2010**, *44* (12), 35 **85–3594.** **https://doi.org/10.1016/j.watres.20**10.04.011.

32. Šojić, D.; Despotović, V.; Orčić, D.; Szabó, E.; Arany, E.; Armaković, S.; Illés, E.; Gajda-Schrantz, K.; Dombi, A.; Alapi, T.; Sajben-Nagy, E.; Palágyi, A.; Vágvölgyi, Cs.; Manczinger, L.; Bjelica, L.; Abramović, B. Degradation of Thiamethoxam and Metoprolol by UV, O₃ and UV/O₃ Hybrid Processes: Kinetics, Degradation Intermediates and Toxicity. *J. Hydrol.* **2012**, *472*, **314–327. https://doi.org/10.1016/j.jhydrol.20**12.09.038.

33. Gurol, M. D.; Vatistas, R. Oxidation of Phenolic Compounds by Ozone and Ozone + u.v. Radiation: A Comparative Study. *Water Res.* **1987**, *21* (8), **895–900. https://doi.** **org/10.1016/S0043-1354(87)**80006-4.

34. Kearney, P. C.; Muldoon, M. T.; Somich, C. J. UV-Ozonation of Eleven Major Pesticides as a Waste Disposal Pretreatment. *Chemosphere.* **1987**, *16* (10), 23 **21–2330. https://** **doi.org/10.1016/0045-6535(87)**90289-X.

35. Rao, Y. F.; Chu, W. A New Approach to Quantify the Degradation Kinetics of Linuron with UV, Ozonation and UV/O₃ Processes. *Chemosphere.* **2009**, *74* (11), 14 **44–1449.** **https://doi.org/10.1016/j.chemosphere.20**08.12.012.

36. Hoffmann, M. R.; Martin, S. T.; Choi, W.; Bahnemann, D. W. Environmental Applications of Semiconductor Photocatalysis. *Chem. Rev.* **1995**, *95* (1)**, 69–96. https://doi.org/** **10.1021/cr**00033a004.

37. Herrmann, J.-M. Heterogeneous Photocatalysis: Fundamentals and Applications to the Removal of Various Types of Aqueous Pollutants. *Catal. Today.* **1999**, *53* (1), **115–129.** **https://doi.org/10.1016/S0920-5861(99)**00107-8.

38. Swarnalatha, B.; Anjaneyulu, Y. Studies on the Heterogeneous Photocatalytic Oxidation of 2,6-Dinitrophenol in Aqueous TiO₂ Suspension. *J. Mol. Catal. Chem.* **2004**, *223* (1), **161–165. https://doi.org/10.1016/j.molcata.20**04.03.058.

39. Ji, P.; Zhang, J.; Chen, F.; Anpo, M. Ordered Mesoporous CeO₂ Synthesized by Nanocasting from Cubic Ia3d Mesoporous MCM-48 Silica: Formation, Characterization and Photocatalytic Activity. *J. Phys. Chem. C.* **2008**, *112* (46), 1780 **9–17813. https://** **doi.org/10.1021/**jp8054087.

40. Lin, C.-F.; Wu, C.-H.; Onn, Z.-N. Degradation of 4-Chlorophenol in TiO₂, WO₃, SnO₂, TiO₂/WO₃ and TiO₂/SnO₂ Systems. *J. Hazard. Mater.* **2008**, *154* (1–3), 10 **33–1039.** **https://doi.org/10.1016/j.jhazmat.20**07.11.010.

41. Bahnemann, W.; Muneer, M.; Haque, M. M. Titanium Dioxide-Mediated Photocatalysed Degradation of Few Selected Organic Pollutants in Aqueous Suspensions. *Catal. Today.* **2007**, *124* (3), **133–148. https://doi.org/10.1016/j.cattod.20**07.03.031.

42. Lachheb, H.; Houas, A.; Herrmann, J.-M. Photocatalytic Degradation of Polynitrophenols on Various Commercial Suspended or Deposited Titania Catalysts Using Artificial and Solar Light. *Int. J. Photoenergy.* **2008**, *2008*, **e497895. https://doi.org/10.1155/** **20**08/497895.

43. Yu, B.; Zeng, J.; Gong, L.; Yang, X.; Zhang, L.; Chen, X. Photocatalytic Degradation Investigation of Dicofol. *Chin. Sci. Bull.* **2008**, *53* (1) **, 27–32. https://doi.org/10.1007/** **s11434-0**07-0482-8.

44. Madani, M. E.; Guillard, C.; Pérol, N.; Chovelon, J. M.; Azzouzi, M. E.; Zrineh, A.; Herrmann, J. M. Photocatalytic Degradation of Diuron in Aqueous Solution in Presence of Two Industrial Titania Catalysts, Either as Suspended Powders or Deposited on

Flexible Industrial Photoresistant Papers. *Appl. Catal. B Environ.* **2006**, *65* (1) **, 70–76. https://doi.org/10.1016/j.apcatb.20**05.12.005.

45. Senthilnathan, J.; Philip, L. Photocatalytic Degradation of Lindane Under UV and Visible Light Using N-Doped TiO_2. *Chem. Eng. J.* **2010**, *161* (1) **, 83–92. https://doi. org/10.1016/j.cej.20**10.04.034.

46. Gomathi Devi, L.; Narasimha Murhty, B.; Girish Kumar, S. Photo Catalytic Degradation of Imidachloprid Under Solar Light Using Metal Ion Doped TiO_2 Nano Particles: Influence of Oxidation State and Electronic Configuration of Dopants. *Catal. Lett.* **2009**, *130* (3), **496–503. https://doi.org/10.1007/s10562-0**09-9938-6.

47. Fresno, F.; Guillard, C.; Coronado, J. M.; Chovelon, J.-M.; Tudela, D.; Soria, J.; Herrmann, J.-M. Photocatalytic Degradation of a Sulfonylurea Herbicide over Pure and Tin-Doped TiO_2 Photocatalysts. *J. Photochem. Photobiol. Chem.* **2005**, *173* (1), **13–20. https://doi.org/10.1016/j.jphotochem.20**04.12.028.

48. Liqiang, J.; Dejun, W.; Baiqi, W.; Shudan, L.; Baifu, X.; Honggang, F.; Jiazhong, S. Effects of Noble Metal Modification on Surface Oxygen Composition, Charge Separation and Photocatalytic Activity of ZnO Nanoparticles. *J. Mol. Catal. Chem.* **2006**, *244* (1), **193–200. https://doi.org/10.1016/j.molcata.20**05.09.020.

49. Zhang, L.; Yan, F.; Su, M.; Han, G.; Kang, P. A Study on the Degradation of Methamidophos in the Presence of Nano-TiO_2 Catalyst Doped with Re. *Russ. J. Inorg. Chem.* **2009**, *54* (8), 12 **10–1216. https://doi.org/10.1134/S0036023**609080075.

50. Rengaraj, S.; Li, X. Z.; Tanner, P. A.; Pan, Z. F.; Pang, G. K. H. Photocatalytic Degradation of Methylparathion—An Endocrine Disruptor by Bi^{3+}-Doped TiO_2. *J. Mol. Catal. Chem.* **2006**, *247* (1–2) **, 36–43. https://doi.org/10.1016/j.molcata.20**05.11.030.

51. Devi, L. G.; Murthy, B. N.; Kumar, S. G. Photocatalytic Activity of V^{5+}, Mo^{6+} and Th^{4+} Doped Polycrystalline TiO_2 for the Degradation of Chlorpyrifos Under UV/ Solar Light. *J. Mol. Catal. Chem.* **2009**, *308* (1), **174–181. https://doi.org/10.1016/j. molcata.20**09.04.007.

52. Devi, L. G.; Murthy, B. N. Structural Characterization of Th-Doped TiO_2 Photocatalyst and Its Extension of Response to Solar Light for Photocatalytic Oxidation of Oryzalin Pesticide: A Comparative Study. *Open Chem.* **2009**, *7* (1), **118–129. https://doi. org/10.2478/s11532-0**08-0101-9.

53. Katsumata, H.; Sada, M.; Nakaoka, Y.; Kaneco, S.; Suzuki, T.; Ohta, K. Photocatalytic Degradation of Diuron in Aqueous Solution by Platinized TiO_2. *J. Hazard. Mater.* **2009**, *171* (1), 10 **81–1087. https://doi.org/10.1016/j.jhazmat.20**09.06.110.

54. Dvoranová, D.; Brezová, V.; Mazúr, M.; Malati, M. A. Investigations of Metal-Doped Titanium Dioxide Photocatalysts. *Appl. Catal. B Environ.* **2002**, *37* (2), **91–105. https://doi.org/10.1016/S0926-3373(01)**00335-6.

55. Sanchez, E.; Lopez, T. Effect of the Preparation Method on the Band Gap of Titania and Platinum-Titania Sol-Gel Materials. *Mater. Lett.* **1995**, *25* (5), **271–275. https:// doi.org/10.1016/0167-577X(95)**00190-5.

56. Kato, S.; Hirano, Y.; Iwata, M.; Sano, T.; Takeuchi, K.; Matsuzawa, S. Photocatalytic Degradation of Gaseous Sulfur Compounds by Silver-Deposited Titanium Dioxide. *Appl. Catal. B Environ.* **2005**, *57* (2), **109–115. https://doi.org/10.1016/j.apcatb. 20**04.10.015.

57. Bouras, P.; Stathatos, E.; Lianos, P. Pure Versus Metal-Ion-Doped Nanocrystalline Titania for Photocatalysis. *Appl. Catal. B Environ.* **2007**, *73* (1) **, 51–59. https://doi. org/10.1016/j.apcatb.20**06.06.007.

58. Daneshvar, N.; Aber, S.; Seyed Dorraji, M. S.; Khataee, A. R.; Rasoulifard, M. H. Photocatalytic Degradation of the Insecticide Diazinon in the Presence of Prepared Nanocrystalline ZnO Powders Under Irradiation of UV-C Light. *Sep. Purif. Technol.* **2007**, *58* (1) **, 91–98. https://doi.org/10.1016/j.seppur.20**07.07.016.

59. Anandan, S.; Vinu, A.; Sheeja Lovely, K. L. P.; Gokulakrishnan, N.; Srinivasu, P.; Mori, T.; Murugesan, V.; Sivamurugan, V.; Ariga, K. Photocatalytic Activity of La-Doped ZnO for the Degradation of Monocrotophos in Aqueous Suspension. *J. Mol. Catal. Chem.* **2007**, *266* (1), **149–157. https://doi.org/10.1016/j.molcata.20**06.11.008.

60. Galindo, F.; Gómez, R.; Aguilar, M. Photodegradation of the Herbicide 2,4-Dichlorophenoxyacetic Acid on Nanocrystalline TiO$_2$–CeO$_2$ Sol–Gel Catalysts. *J. Mol. Catal. Chem.* **2008**, *281* (1), **119–125. https://doi.org/10.1016/j.molcata. 2007.**10.008.

61. Sun, H.; Bai, Y.; Liu, H.; Jin, W.; Xu, N. Photocatalytic Decomposition of 4-Chlorophenol Over an Efficient N-Doped TiO$_2$ Under Sunlight Irradiation. *J. Photochem. Photobiol. Chem.* **2009**, *201* (1) , **15–22. https://doi.org/10.1016/j.jphotochem.20**08.08.021.

62. Sojić, D. V.; Despotović, V. N.; Abazović, N. D.; Comor, M. I.; Abramović, B. F. Photocatalytic Degradation of Selected Herbicides in Aqueous Suspensions of Doped Titania Under Visible Light Irradiation. *J. Hazard. Mater.* **2010**, *179* (1–3)**, 49–56. https://doi.org/10.1016/j.jhazmat.20**10.02.055.

63. Chen, H.-W.; Ku, Y.; Kuo, Y.-L. Effect of Pt/TiO$_2$ Characteristics on Temporal Behavior of o-Cresol Decomposition by Visible Light-Induced Photocatalysis. *Water Res.* **2007**, *41* (10), 20 **69–2078. https://doi.org/10.1016/j.watres.20**07.02.021.

64. Tao, H.; Liang, X.; Zhang, Q.; Chang, C.-T. Enhanced Photoactivity of Graphene/ Titanium Dioxide Nanotubes for Removal of Acetaminophen. *Appl. Surf. Sci.* **2015**, *324*, **258–264. https://doi.org/10.1016/j.apsusc.20**14.10.129.

65. Cao, Y.; Tan, H.; Shi, T.; Tang, T.; Li, J. Preparation of Ag-Doped TiO$_2$ Nanoparticles for Photocatalytic Degradation of Acetamiprid in Water. *J. Chem. Technol. Biotechnol.* **2008**, *83* (4), **546–552. https://doi.org/10.1002/**jctb.1831.

66. Chu, W.; Rao, Y. F. Photocatalytic Oxidation of Monuron in the Suspension of WO$_3$ Under the Irradiation of UV–Visible Light. *Chemosphere.* **2012**, *86* (11), 10 **79–1086. https://doi.org/10.1016/j.chemosphere.20**11.11.062.

67. Zhang, L.; Yan, F.; Wang, Y.; Guo, X.; Zhang, P. Photocatalytic Degradation of Methamidophos by UV Irradiation in the Presence of Nano-TiO$_2$. *Inorg. Mater.* **2006**, *42* (12), 13 **79–1387. https://doi.org/10.1134/S002016**50612017X.

68. Černigoj, U.; Štangar, U. L.; Trebše, P. Degradation of Neonicotinoid Insecticides by Different Advanced Oxidation Processes and Studying the Effect of Ozone on TiO$_2$ Photocatalysis. *Appl. Catal. B Environ.* **2007**, *75* (3), **229–238. https://doi.org/10.1016/ j.apcatb.20**07.04.014.

69. Kabra, K.; Chaudhary, R.; Sawhney, R. L. Treatment of Hazardous Organic and Inorganic Compounds Through Aqueous-Phase Photocatalysis: A Review. *Ind. Eng. Chem. Res.* **2004**, *43* (24), 76 **83–7696. https://doi.org/10.1021/**ie0498551.

70. Wang, H.; Lindgren, T.; He, J.; Hagfeldt, A.; Lindquist, S.-E. Photolelectrochemistry of Nanostructured WO$_3$ Thin Film Electrodes for Water Oxidation: Mechanism of Electron Transport. *J. Phys. Chem. B.* **2000**, *104* (24), 56 **86–5696. https://doi. org/10.1021/**jp0002751.

71. Alnaizy, R.; Akgerman, A. Advanced Oxidation of Phenolic Compounds. *Adv. Environ. Res.* **2000**, *4* (3), **233–244. https://doi.org/10.1016/S1093-0191(00)**00024-1.

72. Lana-Villarreal, T.; Rodes, A.; Pérez, J. M.; Gómez, R. A Spectroscopic and Electrochemical Approach to the Study of the Interactions and Photoinduced Electron Transfer between Catechol and Anatase Nanoparticles in Aqueous Solution. *J. Am. Chem. Soc.* **2005**, *127* (36), 1260 **1–12611. https://doi.org/10.1021/**ja052798y.

73. Santato, C.; Ulmann, M.; Augustynski, J. Enhanced Visible Light Conversion Efficiency Using Nanocrystalline WO$_3$ Films. *Adv. Mater.* **2001**, *13* (7), 511–514. https://doi. org/10.1002/1521-4095(200104)13:7 < 511::AID-ADMA511 > 3.0.CO;2-W.

74. Sivalingam, G.; Priya, M. H.; Madras, G. Kinetics of the Photodegradation of Substituted Phenols by Solution Combustion Synthesized TiO$_2$. *Appl. Catal. B Environ.* **2004**, *51* (1), **67–76. https://doi.org/10.1016/j.apcatb.20**04.02.006.

75. Dobosz, A.; Sobczyński, A. The Influence of Silver Additives on Titania Photoactivity in the Photooxidation of Phenol. *Water Res.* **2003**, *37* (7), 14 **89–1496. https://doi. org/10.1016/S0043-1354(02)**00559-6.

76. Sobczyński, A.; Duczmal, Ł. Photocatalytic Destruction of Catechol on Illuminated Titania. *React. Kinet. Catal. Lett.* **2004**, *82* (2), **213–218. https://doi.org/10.1023/ B:REAC.0000034829.**78255.80.

77. Sobczyński, A.; Duczmal, Ł.; Zmudziński, W. Phenol Destruction by Photocatalysis on TiO$_2$: An Attempt to Solve the Reaction Mechanism. *J. Mol. Catal. Chem.* **2004**, *213* (2), **225–230. https://doi.org/10.1016/j.molcata.20**03.12.006.

78. Zmudziński, W.; Sobczyńska, A.; Sobczyński, A. Oxidation of Phenol and Hexanol in Their Binary Mixtures on Illuminated Titania: Kinetic Studies. *React. Kinet. Catal. Lett.* **2007**, *90* (2), **293–300. https://doi.org/10.1007/s11144-0**07-4984-9.

79. Bagal, M. V.; Raut-Jadhav, S. Chapter 31—The Process for the Removal of Micropollutants Using Nanomaterials. In *Handbook of Nanomaterials for Wastewater Treatment*; Bhanvase, B., Sonawane, S., Pawade, V., Pandit, A., Eds.; Micro and Nano Technologies; Elsevier, 2021; pp. 9 **57–1007. https://doi.org/10.1016/B978-0-12- 821496-**1.00020-9.

80. Ahmed, S.; Rasul, M. G.; Brown, R.; Hashib, M. A. Influence of Parameters on the Heterogeneous Photocatalytic Degradation of Pesticides and Phenolic Contaminants in Wastewater: A Short Review. *J. Environ. Manage.* **2011**, *92* (3), **311–330. https://doi. org/10.1016/j.jenvman.20**10.08.028.

81. Gusain, R.; Gupta, K.; Joshi, P.; Khatri, O. P. Adsorptive Removal and Photocatalytic Degradation of Organic Pollutants Using Metal Oxides and Their Composites: A Comprehensive Review. *Adv. Colloid Interface Sci.* **2019**, *272*, **102009. https://doi. org/10.1016/j.cis.20**19.102009.

82. Gusain, R.; Kumar, P.; Sharma, O. P.; Jain, S. L.; Khatri, O. P. Reduced Graphene Oxide–CuO Nanocomposites for Photocatalytic Conversion of CO$_2$ into Methanol under Visible Light Irradiation. *Appl. Catal. B Environ.* **2016**, *181*, **352–362. https:// doi.org/10.1016/j.apcatb.20**15.08.012.

83. Zhang, Y.; Zhang, N.; Tang, Z.-R.; Xu, Y.-J. Improving the Photocatalytic Performance of Graphene–TiO$_2$ Nanocomposites via a Combined Strategy of Decreasing Defects of Graphene and Increasing Interfacial Contact. *Phys. Chem. Chem. Phys.* **2012**, *14* (25), 91 **67–9175. https://doi.org/10.1039/C2**CP41318C.

84. Yoo, D.-H.; Cuong, T. V.; Pham, V. H.; Chung, J. S.; Khoa, N. T.; Kim, E. J.; Hahn, S. H. Enhanced Photocatalytic Activity of Graphene Oxide Decorated on TiO$_2$ Films Under UV and Visible Irradiation. *Curr. Appl. Phys.* **2011**, *11* (3), **805–808. https:// doi.org/10.1016/j.cap.20**10.11.077.

85. Cruz, M.; Gomez, C.; Duran-Valle, C. J.; Pastrana-Martínez, L. M.; Faria, J. L.; Silva, A. M. T.; Faraldos, M.; Bahamonde, A. Bare TiO$_2$ and Graphene Oxide TiO$_2$ Photocatalysts on the Degradation of Selected Pesticides and Influence of the Water Matrix. *Appl. Surf. Sci.* **2017**, *416*, 10 **13–1021. https://doi.org/10.1016/j.apsusc.20**15.09.268.

86. Dong, S.; Sun, J.; Li, Y.; Yu, C.; Li, Y.; Sun, J. ZnSnO$_3$ Hollow Nanospheres/ Reduced Graphene Oxide Nanocomposites as High-Performance Photocatalysts for Degradation of Metronidazole. *Appl. Catal. B Environ.* **2014**, *144*, **386–393. https:// doi.org/10.1016/j.apcatb.20**13.07.043.

87. Cheng, X.; Deng, X.; Wang, P.; Liu, H. Coupling TiO$_2$ Nanotubes Photoelectrode with Pd Nano-Particles and Reduced Graphene Oxide for Enhanced Photocatalytic Decomposition of Diclofenac and Mechanism Insights. *Sep. Purif. Technol.* **2015**, *154*, **51–59. https://doi.org/10.1016/j.seppur.20**15.09.032.

88. Lin, L.; Wang, H.; Xu, P. Immobilized TiO_2-Reduced Graphene Oxide Nanocomposites on Optical Fibers as High Performance Photocatalysts for Degradation of Pharmaceuticals. *Chem. Eng. J.* **2017**, *310*, **389–398. https://doi.org/10.1016/j.cej.2016. 04.024.**

89. Peralta-Hernández, J. M.; Meas-Vong, Y.; Rodríguez, F. J.; Chapman, T. W.; Maldonado, M. I.; Godínez, L. A. In Situ Electrochemical and Photo-Electrochemical Generation of the Fenton Reagent: A Potentially Important New Water Treatment Technology. *Water Res.* **2006**, *40* (9), 17 **54–1762. https://doi.org/10.1016/j.watres.20 06.03.004.**

90. Peralta-Hernández, J. M.; Meas-Vong, Y.; Rodríguez, F. J.; Chapman, T. W.; Maldonado, M. I.; Godínez, L. A. Comparison of Hydrogen Peroxide-Based Processes for Treating Dye-Containing Wastewater: Decolorization and Destruction of Orange II Azo Dye in Dilute Solution. *Dyes Pigments.* **2008**, *76* (3), **656–662. https://doi.org/10.1016/j. dyepig.2**007.01.001.

91. Zhao, B.; Li, X.; Wang, P. Degradation of 2,4-Dichlorophenol with a Novel TiO_2/Ti-Fe-Graphite Felt Photoelectrocatalytic Oxidation Process. *J. Environ. Sci.* **2007**, *19* (8), 10 **20–1024. https://doi.org/10.1016/S1001-0742(07)60165-X.**

92. Philippidis, N.; Sotiropoulos, S.; Efstathiou, A.; Poulios, I. Photoelectrocatalytic Degradation of the Insecticide Imidacloprid Using TiO_2/Ti Electrodes. *J. Photochem. Photobiol. Chem.* **2009**, *204* (2), **129–136. https://doi.org/10.1016/j.jphotochem. 2**009.03.007.

93. Pelaez, M.; Nolan, N. T.; Pillai, S. C.; Seery, M. K.; Falaras, P.; Kontos, A. G.; Dunlop, P. S. M.; Hamilton, J. W. J.; Byrne, J. A.; O'Shea, K.; Entezari, M. H.; Dionysiou, D. D. A Review on the Visible Light Active Titanium Dioxide Photocatalysts for Environmental Applications. *Appl. Catal. B Environ.* **2012**, *125*, **331–349. https://doi.org/10.1016/j.apcatb.20**12. 05.036.

94. Sharma, D.; Steen, G. W.; Korterik, J. P.; Garcia-Iglesias, M.; Vazquez, P.; Torres, T.; Herek, J. L.; Huijser, J. M. Impact of the Anchoring Ligand on Electron Injection and Recombination Dynamics at the Interface of Novel Asymmetric Push-Pull Zinc Phthalocyanines and TiO_2. *J. Phys. Chem. C.* **2013**, *117* (48), 2539 **7–25404. https:// doi.org/10.1021/**jp410080a.

95. Li, S.; Xu, S.; He, L.; Xu, F.; Wang, Y.; Zhang, L. Photocatalytic Degradation of Polyethylene Plastic with Polypyrrole/TiO_2 Nanocomposite as Photocatalyst. *Polym.-Plast. Technol. Eng.* **2010**, *49* (4), **400–406. https://doi.org/10.1080/03602550**903532166.

96. Tian, L.; Chen, Q.; Jiang, W.; Wang, L.; Xie, H.; Kalogerakis, N.; Ma, Y.; Ji, R. A Carbon-14 Radiotracer-Based Study on the Phototransformation of Polystyrene Nanoplastics in Water Versus in Air. *Environ. Sci. Nano.* **2019**, *6* (9), 29 **07–2917. https://doi.org/10.1039/**C9EN00662A.

97. Lianou, A.; Frontistis, Z.; Chatzisymeon, E.; Antonopoulou, M.; Konstantinou, I.; Mantzavinos, D. Sonochemical Oxidation of Piroxicam Drug: Effect of Key Operating Parameters and Degradation Pathways: Sonodegradation of Piroxicam. *J. Chem. Technol. Biotechnol.* **2018**, *93* (1) , **28–34. https://doi.org/10.1002/**jctb.5346.

98. Wood, R. J.; Lee, J.; Bussemaker, M. J. A Parametric Review of Sonochemistry: Control and Augmentation of Sonochemical Activity in Aqueous Solutions. *Ultrason. Sonochem.* **2017**, *38*, **351–370. https://doi.org/10.1016/j.ultsonch.2**017.03.030.

99. Villaroel, E.; Silva-Agredo, J.; Petrier, C.; Taborda, G.; Torres-Palma, R. A. Ultrasonic Degradation of Acetaminophen in Water: Effect of Sonochemical Parameters and Water Matrix. *Ultrason. Sonochem.* **2014**, *21* (5), 17 **63–1769. https://doi.org/10.1016/j. ultsonch.20**14.04.002.

100. Elsayed, M. A. Ultrasonic Removal of Pyridine from Wastewater: Optimization of the Operating Conditions. *Appl. Water Sci.* **2015**, *5* (3), **221–227. https://doi.org/10.1007/ s13201-0**14-0182-x.

101. Tran, N.; Drogui, P.; Brar, S. K. Sonochemical Techniques to Degrade Pharmaceutical Organic Pollutants. *Environ. Chem. Lett.* **2015**, *13* (3), **251–268. https://doi.org/10.1007/s10311-0**15-0512-8.

102. González-García, J.; Sáez, V.; Tudela, I.; Díez-Garcia, M. I.; Deseada Esclapez, M.; Louisnard, O. Sonochemical Treatment of Water Polluted by Chlorinated Organocompounds. A Review. *Water.* **2010**, *2* (1) **, 28–74. https://doi.org/10.3390/**w2010028.

103. Katiyar, J.; Bargole, S.; George, S.; Bhoi, R.; Saharan, V. K. Advanced Technologies for Wastewater Treatment: New Trends. In *Handbook of Nanomaterials for Wastewater Treatment*; Elsevier, 2021; pp. **85–133. https://doi.org/10.1016/B978-0-12-821496-1.00011-8.**

104. Ince, N. H.; Tezcanli, G.; Belen, R. K.; Apikyan, İ. G. Ultrasound as a Catalyzer of Aqueous Reaction Systems: The State of the Art and Environmental Applications. *Appl. Catal. B Environ.* **2001**, *29* (3), **167–176. https://doi.org/10.1016/S0926-3373(00)**00224-1.

105. Kıdak, R.; Doğan, Ş. Medium-High Frequency Ultrasound and Ozone Based Advanced Oxidation for Amoxicillin Removal in Water. *Ultrason. Sonochem.* **2018**, *40*, **131–139. https://doi.org/10.1016/j.ultsonch.20**17.01.033.

106. Pinjari, D. V.; Pandit, A. B. Cavitation Milling of Natural Cellulose to Nanofibrils. *Ultrason. Sonochem.* **2010**, *17* (5), **845–852. https://doi.org/10.1016/j.ultsonch.20**10.03.005.

107. Serna-Galvis, E. A.; Botero-Coy, A. M.; Martínez-Pachón, D.; Moncayo-Lasso, A.; Ibáñez, M.; Hernández, F.; Torres-Palma, R. A. Degradation of Seventeen Contaminants of Emerging Concern in Municipal Wastewater Effluents by Sonochemical Advanced Oxidation Processes. *Water Res.* **2019**, *154*, **349–360. https://doi.org/10.1016/j.watres.20**19.01.045.

108. Adityosulindro, S.; Barthe, L.; González-Labrada, K.; Jáuregui Haza, U. J.; Delmas, H.; Julcour, C. Sonolysis and Sono-Fenton Oxidation for Removal of Ibuprofen in (Waste)Water. *Ultrason. Sonochem.* **2017**, *39*, **889–896. https://doi.org/10.1016/j.ultsonch.20**17.06.008.

109. Naddeo, V.; Landi, M.; Scannapieco, D.; Belgiorno, V. Sonochemical Degradation of Twenty-Three Emerging Contaminants in Urban Wastewater. *Desalination Water Treat.* **2013**, *51* (34–36), 66 **01–6608. https://doi.org/10.1080/19443994.20**13.769696.

110. Mahamuni, N. N.; Adewuyi, Y. G. Advanced Oxidation Processes (AOPs) Involving Ultrasound for Waste Water Treatment: A Review with Emphasis on Cost Estimation. *Ultrason. Sonochem.* **2010**, *17* (6), 9 **90–1003. https://doi.org/10.1016/j.ultsonch.20**09.09.005.

111. Güyer, G. T.; Ince, N. H. Degradation of Diclofenac in Water by Homogeneous and Heterogeneous Sonolysis. *Ultrason. Sonochem.* **2011**, *18* (1), **114–119. https://doi.org/10.1016/j.ultsonch.20**10.03.008.

112. Bethi, B.; Sonawane, S. H.; Bhanvase, B. A.; Gumfekar, S. P. Nanomaterials-Based Advanced Oxidation Processes for Wastewater Treatment: A Review. *Chem. Eng. Process.—Process Intensif.* **2016**, *109*, **178–189. https://doi.org/10.1016/j.cep.20**16.08.016.

113. Kanakaraju, D.; Glass, B. D.; Oelgemöller, M. Advanced Oxidation Process-Mediated Removal of Pharmaceuticals from Water: A Review. *J. Environ. Manage.* **2018**, *219*, **189–207. https://doi.org/10.1016/j.jenvman.20**18.04.103.

114. Joseph, C. G.; Li Puma, G.; Bono, A.; Krishnaiah, D. Sonophotocatalysis in Advanced Oxidation Process: A Short Review. *Ultrason. Sonochem.* **2009**, *16* (5), **583–589. https://doi.org/10.1016/j.ultsonch.20**09.02.002.

115. Mark, G.; Tauber, A.; Laupert, R.; Schuchmann, H.-P.; Schulz, D.; Mues, A.; von Sonntag, C. OH-Radical Formation by Ultrasound in Aqueous Solution—Part II:

Terephthalate and Fricke Dosimetry and the Influence of Various Conditions on the Sonolytic Yield. *Ultrason. Sonochem.* **1998**, *5* (2) , **41–52. https://doi.org/10.1016/ S1350-4177(98)**00012-1.

116. Isariebel, Q.-P.; Carine, J.-L.; Ulises-Javier, J.-H.; Anne-Marie, W.; Henri, D. Sonolysis of Levodopa and Paracetamol in Aqueous Solutions. *Ultrason. Sonochem.* **2009**, *16* (5), **610–616. https://doi.org/10.1016/j.ultsonch.20**08.11.008.

117. De Bel, E.; Janssen, C.; De Smet, S.; Van Langenhove, H.; Dewulf, J. Sonolysis of Ciprofloxacin in Aqueous Solution: Influence of Operational Parameters. *Ultrason. Sonochem.* **2011**, *18* (1), **184–189. https://doi.org/10.1016/j.ultsonch.20** 10.05.003.

118. Rahmani, H.; Gholami, M.; Mahvi, A. H.; Alimohammadi, M.; Azarian, G.; Esrafili, A.; Rahmani, K.; Farzadkia, M. Tinidazole Removal from Aqueous Solution by Sonolysis in the Presence of Hydrogen Peroxide. *Bull. Environ. Contam. Toxicol.* **2014**, *92* (3), **341–346. https://doi.org/10.1007/s00128-0**13-1193-2.

119. Ince, N. H. Ultrasound-Assisted Advanced Oxidation Processes for Water Decontamination. *Ultrason. Sonochem.* **2018**, *40*, **97–103. https://doi.org/10.1016/j. ultsonch.20**17.04.009.

120. Ghafoori, S.; Mowla, A.; Jahani, R.; Mehrvar, M.; Chan, P. K. Sonophotolytic Degradation of Synthetic Pharmaceutical Wastewater: Statistical Experimental Design and Modeling. *J. Environ. Manage.* **2015**, *150*, **128–137. https://doi.org/10.1016/j. jenvman.20**14.11.011.

121. Sutar, R. S.; Rathod, V. K. Ultrasound Assisted Enzyme Catalyzed Degradation of Cetirizine Dihydrochloride. *Ultrason. Sonochem.* **2015**, *24* , **80–86. https://doi. org/10.1016/j.ultsonch.20**14.10.016.

122. Neis, U. The Use of Power Ultrasound for Wastewater and Biomass Treatment. In *Power Ultrasonics*; Elsevier, 2015; pp. **973–996. https://doi.org/10.1016/ B978-1-78242-028-**6.00032-6.

123. Villegas-Guzman, P.; Silva-Agredo, J.; Giraldo-Aguirre, A. L.; Flórez-Acosta, O.; Petrier, C.; Torres-Palma, R. A. Enhancement and Inhibition Effects of Water Matrices during the Sonochemical Degradation of the Antibiotic Dicloxacillin. *Ultrason. Sonochem.* **2015**, *22*, **211–219. https://doi.org/10.1016/j.ultsonch.20**14.07.006.

124. Lastre-Acosta, A. M.; Cruz-González, G.; Nuevas-Paz, L.; Jáuregui-Haza, U. J.; Teixeira, A. C. S. C. Ultrasonic Degradation of Sulfadiazine in Aqueous Solutions. *Environ. Sci. Pollut. Res.* **2015**, *22* (2), **918–925. https://doi.org/10.1007/s11356-0**14-2766-2.

125. Serna-Galvis, E. A.; Montoya-Rodríguez, D.; Isaza-Pineda, L.; Ibáñez, M.; Hernández, F.; Moncayo-Lasso, A.; Torres-Palma, R. A. Sonochemical Degradation of Antibiotics from Representative Classes-Considerations on Structural Effects, Initial Transformation Products, Antimicrobial Activity and Matrix. *Ultrason. Sonochem.* **2019**, *50*, **157–165. https://doi.org/10.1016/j.ultsonch.20**18.09.012.

126. Nejumal, K. K.; Manoj, P. R.; Aravind, U. K.; Aravindakumar, C. T. Sonochemical Degradation of a Pharmaceutical Waste, Atenolol, in Aqueous Medium. *Environ. Sci. Pollut. Res.* **2014**, *21* (6), 42 **97–4308. https://doi.org/10.1007/s11356-0**13-2301-x.

127. Bhatt, J.; Rai, A. K.; Gupta, M.; Vyas, S.; Ameta, R.; Ameta, S. C.; Chavoshani, A.; Hashemi, M. Surfactants: An Emerging Face of Pollution. In *Micropollutants and Challenges*; Elsevier, 2020; pp. **145–178. https://doi.org/10.1016/ B978-0-12-818612-**1.00004-0.

128. Singla, R.; Grieser, F.; Ashokkumar, M. Kinetics and Mechanism for the Sonochemical Degradation of a Nonionic Surfactant. *J. Phys. Chem. A.* **2009**, *113* (12), 28 **65–2872. https://doi.org/10.1021/**jp808968e.

129. Singla, R.; Grieser, F.; Ashokkumar, M. The Mechanism of Sonochemical Degradation of a Cationic Surfactant in Aqueous Solution. *Ultrason. Sonochem.* **2011**, *18* (2), **484– 488. https://doi.org/10.1016/j.ultsonch.20**10.09.013.

130. Nakui, H.; Okitsu, K.; Maeda, Y.; Nishimura, R. Effect of Coal Ash on Sonochemical Degradation of Phenol in Water. *Ultrason. Sonochem.* **2007**, *14* (2), **191–196. https:// doi.org/10.1016/j.ultsonch.20**06.04.003.
131. Selli, E.; Bianchi, C.; Pirola, C.; Cappelletti, G.; Ragaini, V. Efficiency of 1,4-Dichlorobenzene Degradation in Water Under Photolysis, Photocatalysis on TiO$_2$ and Sonolysis. *J. Hazard. Mater.* **2008**, *153* (3), 11 **36–1141. https://doi.org/10.1016/j. jhazmat.20**07.09.071.
132. Jiang, Y.; Pétrier, C.; David Waite, T. Kinetics and Mechanisms of Ultrasonic Degradation of Volatile Chlorinated Aromatics in Aqueous Solutions. *Ultrason. Sonochem.* **2002**, *9* (6), **317–323. https://doi.org/10.1016/S1350-4177(02)**00085-8.
133. Tunc Dede, O.; Aksu, Z.; Rehorek, A. Sonochemical Degradation of C.I. Reactive Orange 107. *Environ. Eng. Sci.* **2019**, *36* (2), **158–171. https://doi.org/10.1089/ ees.**2018.0076.
134. Guzman-Duque, F.; Pétrier, C.; Pulgarin, C.; Peñuela, G.; Torres-Palma, R. A. Effects of Sonochemical Parameters and Inorganic Ions during the Sonochemical Degradation of Crystal Violet in Water. *Ultrason. Sonochem.* **2011**, *18* (1), **440–446. https://doi. org/10.1016/j.ultsonch.20**10.07.019.
135. Zúñiga-Benítez, H.; Soltan, J.; Peñuela, G. A. Application of Ultrasound for Degradation of Benzophenone-3 in Aqueous Solutions. *Int. J. Environ. Sci. Technol.* **2016**, *13* (1) , **77–86. https://doi.org/10.1007/s13762-0**15-0842-x.
136. Ismail, Z. Z.; Jasim, A. S. Ultrasonic Treatment of Wastewater Contaminated with Furfural. *IDA J. Desalination Water Reuse.* **2014**, *6* (3–4), **103–111. https://doi.org/1 0.1179/2051645214Y.0**000000028.
137. Chiha, M.; Hamdaoui, O.; Baup, S.; Gondrexon, N. Sonolytic Degradation of Endocrine Disrupting Chemical 4-Cumylphenol in Water. *Ultrason. Sonochem.* **2011**, *18* (5), **943–950. https://doi.org/10.1016/j.ultsonch.20**10.12.014.

5 Kinetic Modelling of the Photodegradation Process of Water-Soluble Polymers

*Nadeem A. Khan, Afzal Husain Khan,
Viola Vambol, Sergij Vambol,
Nabisab Mujawar Mubarak and Rama Rao Karri*

CONTENTS

5.1   Introduction 99
5.2   Reaction Mechanisms... 100
5.3   Kinetic Model Development.. 101
5.4   Solution Technique 103
5.5   Model Validation 104
5.6   Estimation of Kinetic Parameters... 104
5.7   Conclusion 105
References... 105

5.1 INTRODUCTION

Advanced oxidation processes (AOPs) are emerging alternative technologies for efficiently removing organic wastewater pollutants with high chemical stability and/or low biodegradability [1–3]. The environmental impacts of the aforementioned process are of great importance due to their high reaction rates and efficiencies, as well as their promotion of green chemistry by producing minimal hazardous by-products [4, 5]. Common to all AOPs is the formation of hydroxyl radicals ($^{\bullet}OH$), which are capable of acting as the main group of chemicals for decontaminating and/or biodegradation of highly toxic pollutants in many fields [6, 7].

Although a promising new technology, the mechanisms involved in the oxidation of organic compounds during AOPs are very complex. To date, there is insufficient information in open scientific sources on the kinetic modeling of the photo-oxidative water-soluble polymers' degradation in water during the UV/H_2O_2 process [8, 9]. The kinetics of such soluble polymers using AOT, considering the radicals and chemical compounds present in water, is of interest [10]. PEG was chosen for the model water-soluble polymer. Radical depolymerization of the photo-oxidative water-soluble synthetic polymers' degradation in an aqueous solution was modeled based on population balance equations (PBEs). The method of moments was used to solve the PBEs.

DOI: 10.1201/9781003247913-5

The transformation of integral-differential equations into ordinary differential equations (ODE) was performed using momentum operation to solve PBEs. This made it possible to obtain polymer photodegradation rate coefficients. Experimental data have been found from the open sources for the PEG photodegradation in an aqueous solution by UV/H_2O_2 [11]. These data were compared with the data obtained from the models for the average molecular weight and the number of chain breaks per molecule. A nonlinear least-squares objective function was applied to calculate the model parameters. This can become the basis for the creation of industrial photo-reactors.

5.2   REACTION MECHANISMS

The macromolecules' degradation is often described by three reversible elementary steps of a chemical reaction:

- radicals' initiation-breakage,
- depropagation (splitting, disintegration),
- hydrogen removal.

Since each "start-stop" step has a significant effect on the complete reaction velocity, using the long-chain approximation, these steps are neglected. Therefore, we will consider a random chain break as a degradation mechanism. Thus, UV irradiation generates oxidative radicals that initiate the reaction to form random and specific finite polymer radicals by abstracting hydrogen from the polymer molecule. Under the influence of UV photolysis with a wavelength of 254 nm, the decomposition of H_2O_2 with the formation of free radicals is observed (Table 5.1) [9].

The elimination of hydrogen from the polymer chain P_n with the formation of random radicals via $HO_2^{\bullet}$ and $^{\bullet}OH$ can be characterized by reactions (5.1) and (5.2):

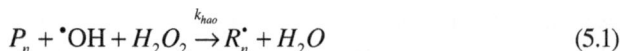

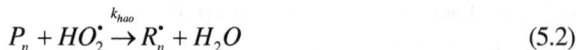

$$P_n + {}^{\bullet}OH + H_2O_2 \xrightarrow{k_{hao}} R_n^{\bullet} + H_2O \tag{5.1}$$

$$P_n + HO_2^{\bullet} \xrightarrow{k_{hao}} R_n^{\bullet} + H_2O \tag{5.2}$$

Where k_{hao} and k'_{hao} are the rate constants of hydrogen evolution to the formation of random radicals through oxidation of $^{\bullet}OH$ and $HO_2^{\bullet}$ radicals, accordingly. Radicals are stabilized by removing a hydrogen atom from a stable position.

The bilateral intermolecular elimination of hydrogen from the polymer chain with the formation of a random radical can be described through the reaction:

$$P_n \underset{kHa}{\overset{kha}{\underset{\longleftarrow}{\longrightarrow}}} R_n^{\bullet} \tag{5.3}$$

TABLE 5.1

Photodegradation Stages' Sequence in UV/H_2O_2 Process.

No.	Reaction	Rate Constant ($M^{-1}s^{-1}$)
R1	$H_2O_2 \xrightarrow{\varphi H_2O_2^{hv}} 2\,{}^{\bullet}OH$	0.5
R2	${}^{\bullet}OH + H_2O_2 \xrightarrow{k_2} HO_2^{\bullet} + H_2O$	2.7×10^7
R3	${}^{\bullet}OH + HO_2^- \xrightarrow{k_3} HO_2^{\bullet} + OH^-$	7.5×10^9
R4	$HO_2^{\bullet} + H_2O_2 \xrightarrow{k_4} {}^{\bullet}OH + H_2O + O_2$	3
R5	$O_2^{\bullet-} + H_2O_2 \xrightarrow{k_5} {}^{\bullet}OH + O_2 + OH^-$	0.13
R6	$O_2^{\bullet-} + H \xrightarrow{k_6} HO_2^{\bullet}$	$_{10}10$
R7	$HO_2^{\bullet} \xrightarrow{k_7} O_2^{\bullet-} + H^+$	1.58×10^5
R8	${}^{\bullet}OH + {}^{\bullet}OH \xrightarrow{k_8} H_2O_2$	5.5×10^9
R9	$HO_2^{\bullet} + HO_2^{\bullet} \xrightarrow{k_9} H_2O_2 + O_2$	8.3×10^5
R10	${}^{\bullet}OH + HO_2^{\bullet} \xrightarrow{k_{10}} O_2 + H_2O$	6.6×10^9
R11	$HO_2^{\bullet} + O_2^{\bullet-} \xrightarrow{k_{11}} HO_2^- + O_2$	9.7×10^7
R12	${}^{\bullet}OH + O_2^{\bullet-} \xrightarrow{k_{12}} O_2 + OH^-$	8×10^9
R13	$H_2O_2 \leftrightarrow H^+ + HO_2^-$	$pK_a = 11.6$

Where k_{ha} k_{Ha} are the rate constants of detaching hydrogen from the polymer with the formation of a random radical and reversible detaching hydrogen from a random radical, accordingly. The random radical propagation reaction is presented as follows:

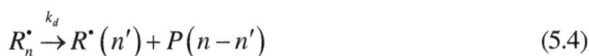

$$R_n^{\bullet} \xrightarrow{k_d} R^{\bullet}(n') + P(n - n') \tag{5.4}$$

where k_d is the degradation factor for rejecting random radical propagation.

5.3 KINETIC MODEL DEVELOPMENT

For the chemical species of the reacting polymer, n molecular radicals are presented as: P_n and $R_n^{\bullet}$, and their molecular weight distributions (MWDs) are presented as $p(n, t)$ and $r(n, t)$, respectively. In a continuous distribution model, it is not possible to distinguish between polymer reactants and rupture products. Therefore, the polymer mixture can be characterized by its molecular weight distribution $p(n, t)$ at any given time, while $r(n, t)$ can only describe the molecular weight range $(n, n+dn)$. It is common to distinguish between the MWD of a reactant and a product and describe this with separate differential equations for their behavior. For a binary fragmentation reaction, the rate of formation of binary rupture products is represented by Eq. (5.5), in which, to simplify calculations, the integral is used instead of the sum:

$$R(n) = \int_{0}^{\infty} k(n')\Omega(n,\ n')r(n',\ t)\ dn' \tag{5.5}$$

where $\Omega(n,\ n')$ is the stoichiometric coefficient. This reveals a reaction in which the molecule splits into two molecules of size n and $n'- n$, where $n' \geq n$ and $k(n')$. Therefore, it is the rate factor. The stoichiometric coefficient can be expressed by Eq. (5.6):

$$\Omega(n,\ n') = \frac{n^m (n'-n)^m \Gamma(2m+2)}{\Gamma(m+1)^2 (n')^{2m+1}} \tag{5.6}$$

Where Γ is the gamma function and m is a parameter indicating the shape of the discontinuity fragment distribution, respectively, with a random distribution $m = 0$ and $\Omega(n, n') = 1/n'$.

The preceding makes it possible to develop kinetic equations of basic chemical reactions in an aqueous solution. The oxidizing agent disappearing rate during straight photolysis can be described by Eq. (5.7). Since straight photolysis of PEG is negligible compared to straight photolysis, it has not been considered.

$$\frac{dC_{UV,\,H_2O_2}}{dt} = -\varphi_{H_2O_2} e_\lambda^a \tag{5.7}$$

where $\varphi_{H_2O_2}$ is the quantum yield of acid-base conjugates H_2O_2 and HO_2^- (mol Einstein^{-1}) and e_a^a is the local volumetric energy absorption rate (LVREA) (Einstein^{-1} s^{-1}), which is determined following the Beer-Lambert law based on the radiation balance as shown here:

$$e_\lambda^a = l_0 \left(\frac{\varepsilon_{H_2O_2}[H_2O_2] + \varepsilon_{HO2}^-[HO_2^-]}{\varepsilon_{H_2O_2}[H_2O_2] + \varepsilon_{HO2}^-[HO_2^-] + \varepsilon_{TOCl}[TOCl]} \right) \cdot \left(1 - e^{\left(-2.303b\left(\varepsilon_{H_2O_2}[H_2O_2] + \varepsilon_{HO2}^-[HO_2^-] + \varepsilon_{TOCl}[TOCl]\right)\right)} \right)$$

$$\tag{5.8}$$

where lo, the intensity of UV light absorbed by the aqueous solution, was determined by calculation as 1.63×10^{-5} Einstein L^{-1} s^{-1}, which is the incident photon radiation at a wavelength of 254 nm, calculated with a 96 W LP lamp placed above the photo-reactor. The fraction demonstrates the absorption percentage of UV radiation by hydrogen peroxide. Parameters $\varepsilon_{H_2O_2}$, $\varepsilon_{HO_2^-}$, and ε_{PEG} are the molar extinction coefficients of H_2O_2 (18.7 M^{-1} cm^{-1}), HO$_2^-$ (210 M^{-1} cm^{-1}), and PEG (118.8 M^{-1} cm^{-1}) at 254 nm, respectively. Parameter b is the effective path length of the photo-reactor (2 cm).

H_2O_2 and HO_2^- are interconnected through the equilibrium reaction (R13) in Table 5.1. The total H_2O_2 concentration is determined by:

$$[H_2O_2]_T = [H_2O_2] + [HO_2^-] \tag{5.9}$$

Therefore, the new equilibrium concentration for two separate species can be expressed by such equations:

$$[H_2O_2]_T = [H_2O_2] = \frac{[H_2O_2]_T}{1 + K_a[H^+]^{-1}} + [HO_2^-] \tag{5.10}$$

$$[HO_2^-] = \frac{[H_2O_2]_T}{1 + K_a^{-1}[H^+]} \tag{5.11}$$

where K_a is the rate constant of equilibrium.

These concentrations will be used in the development of the kinetic model. The remaining kinetic expressions can be derived by including decomposition and particle formation from free radical reactions.

The essential parts are reactions of hydrogen abstraction between ${}^\bullet OH$ and $HO_2^\bullet$ radicals and a polymer with any mass of molecules. In the model, pH is assumed to be constant since it has been vindicated that changes in pH within the process do not significantly affect model predictions. Therefore, the PBE for a polymer and polymer radicals can be represented as:

$$R_P(n,t) = -k(n)_{hao}[{}^\bullet OH]p(n,t) - k'(n)_{hao}[HO_2^\bullet]p(n,t) - k(n)_{ha}\,p(n,t) +$$
$$+ k(n)_{HA}\,r(n,t) + \int_x^\infty k_d(n')r(n',t)\Omega(n,n')dn' \tag{5.12}$$

$$R_r(n,t) = k(n)_{hao}[{}^\bullet OH]p(n,t) + k'(n)_{hao}[HO_2^\bullet]p(n,t) + k(n)_{ha}\,p(n,t) +$$
$$+ k(n)_{ha}\,p(n,t) - k(n)_{HA}\,r(n,t) + \int_x^\infty k_d(n')r(n',t)\Omega(n,n')dn' \tag{5.13}$$

If the reaction time is short, we can assume that all rate factors are constant since they do not depend on molecular weight.

5.4   SOLUTION TECHNIQUE

The integrodifferential equations were transformed into ODEs using the moment operation. Then, over the molecular weight x, the MWD moments were determined:

$$r^p(t) = \int_0^x r(n,t)x^p dn \tag{5.14}$$

The total molar concentration of the polymer, which depends on time, is zero moments ($p = 0$), while the mass concentration of the polymer is the first moment ($p = 1$). If moment operations are applied, then Eqs. (5.12) and (5.13) will have the form of an ordinary differential:

$$\frac{dp^{(p)}}{dt} = -k_{hao}[{}^\bullet OH]p^{(p)} - k'_{hao}[HO_2^\bullet]p^{(p)} - k_{ha}p^{(p)} + k_{HA}r^{(p)} - \frac{1}{n+1}k_d r^{(p)} \tag{5.15}$$

$$\frac{dr^{(p)}}{dt} = -k_{hao}[{}^\bullet OH]p^{(p)} + k'_{hao}[HO_2^\bullet]p^{(p)} + k_{ha}p^{(p)} - k_{HA}r^{(p)} - \frac{n}{n+1}k_d r^{(p)} \tag{5.16}$$

Using the quasi-stationary state approximation to random radicals (Eq. 5.16) for the moments zero and first, the following equations are obtained:

$$\frac{dp^{(0)}}{dt} = \left(k_A\left[\bullet OH\right]p^{(p)} - k_B'\left[HO_2^{\bullet}\right] + k_C\right)p^{(0)} \tag{5.17}$$

$$\frac{dp^{(1)}}{dt} = 0 \tag{5.18}$$

where $k_A = \dfrac{k_d\, k_{hao}}{k_{HA}}, k_B = \dfrac{k_d\, k_{hao}'}{k_{HA}}, k_C = \dfrac{k_d\, k_{ha}}{k_{HA}},$

The change rate of the first moment is 0 ($dp^{(1)}/dt = 0$), which reaffirms mass preservation. Therefore, the oxidative radicals change rate can be written in the form:

$$R_{\bullet OH} = 2R_{UV,H_2O_2} - k_2\left[\bullet OH\right]\left[H_2O_2\right] - k_3\left[\bullet OH\right]\left[HO_2^-\right] + k_4\left[HO_2^{\bullet}\right]\left[H_2O_2\right] + k_5\left[O_2^{\bullet-}\right]\left[H_2O_2\right] + $$
$$-k_8\left[\bullet OH\right]^2 - k_{10}\left[\bullet OH\right]\left[HO_2^{\bullet}\right] - k_{12}\left[\bullet OH\right]\left[O_2^{\bullet-}\right] - \left[\bullet OH\right]k_{hao}p^{(0)} \tag{5.19}$$

$$R_{HO_2^{\bullet}} = k_2\left[\bullet OH\right]\left[H_2O_2\right] + k_3\left[\bullet OH\right]\left[HO_2^-\right] - k_4\left[HO_2^{\bullet}\right]\left[H_2O_2\right] + k_6\left[O_2^{\bullet-}\right]\left[H^+\right] + $$
$$-k_7\left[HO_2^{\bullet}\right] - k_9\left[HO_2^{\bullet}\right]^2 - k_{10}\left[\bullet OH\right]\left[HO_2^{\bullet}\right] - k_{11}\left[HO_2^{\bullet}\right]\left[O_2^{\bullet-}\right] - \left[HO_2^{\bullet}\right]k_{hao}'p^{(0)} \tag{5.20}$$

5.5   MODEL VALIDATION

In order to describe the free radical degradation of PEG at different stages, a detailed mathematical model was developed. Verification of the model is confirmed by experimental data on the photodegradation of polyethylene glycol in the UV/H$_2$O$_2$ process published in the open literature. The predicted model for the number of chain breaks per molecule (S), which can be calculated from Eq. (5.21), is as follows:

$$S = M_n(0)/M_n(t) - 1 \tag{5.21}$$

where $M_n(0)$ and $M_n(t)$ are the number average molecular weight of the polyethylene glycol before and after the irradiation time (t) in minutes, accordingly.

5.6   ESTIMATION OF KINETIC PARAMETERS

Using an optimization algorithm to minimize the sum of squared errors, approximate kinetic parameters can be estimated in the absence of initial molecular weight results:

$$F(k) = \sum_i \left(y_{i,m} - y_i(k)\right)^2 \tag{5.22}$$

Where $y_{i,m}$ is the experimental data point from photodegradation experiments for i number of data points, and $y_i(k)$ is the model prediction with the kinetic parameter vector k.

So, the task is to get the k-minimizing F(k) values, which should be exposed to the equations of the kinetic model.

5.7 CONCLUSION

The mass balance of the main particles contained in an aqueous solution is the basis for developing a mathematical model of PEG photodegradation under the action of hydrogen peroxide and UV radiation. The mathematical model of polymer degradation includes reactions. The polymer degradation kinetic model was derived by applying polymer molecule population dynamics undergoing random chain scission.

For the model's numerical solution efficiency, moment operations were used. The convenience of this model is that it is a system of ODEs. The model was tested on experimental data from open scientific sources. During testing, the predictive results of the model correlated well with the available experimental data, both for the average molecular weight and the number of chain breaks per molecule.

The increase in the number of chain breaks per molecule and the exponential behavior of the zero-moment, depending on the time of radiation exposure, confirmed the assumption of a random chain break. Therefore, kinetic rate constants can be evaluated using any nonlinear optimization.

REFERENCES

1. Miklos, D.B., Remy, C., Jekel, M., Linden, K.G., Drewes, J.E., and Hübner, U., (2018). Evaluation of advanced oxidation processes for water and wastewater treatment–A critical review. *Water Research*. 139: 118–131.
2. Sillanpää, M., Ncibi, M.C., and Matilainen, A., (2018). Advanced oxidation processes for the removal of natural organic matter from drinking water sources: A comprehensive review. *Journal of Environmental Management*. 208: 56–76.
3. Pandis, P.K., Kalogirou, C., Kanellou, E., Vaitsis, C., Savvidou, M.G., Sourkouni, G., Zorpas, A.A., and Argirusis, C., (2022). Key points of advanced oxidation processes (AOPs) for wastewater, organic pollutants and pharmaceutical waste treatment: A mini review. *ChemEngineering*. 6(1): 8.
4. Bokare, A.D. and Choi, W., (2014). Review of iron-free Fenton-like systems for activating H_2O_2 in advanced oxidation processes. *Journal of Hazardous Materials*. 275: 121–135.
5. Katheresan, V., Kansedo, J., and Lau, S.Y., (2018). Efficiency of various recent wastewater dye removal methods: A review. *Journal of Environmental Chemical Engineering*. 6(4): 4676–4697.
6. Wang, J.L. and Xu, L.J., (2012). Advanced oxidation processes for wastewater treatment: Formation of hydroxyl radical and application. *Critical Reviews in Environmental Science and Technology*. 42(3): 251–325.
7. Cheng, M., Zeng, G., Huang, D., Lai, C., Xu, P., Zhang, C., and Liu, Y., (2016). Hydroxyl radicals based advanced oxidation processes (AOPs) for remediation of soils contaminated with organic compounds: A review. *Chemical Engineering Journal*. 284: 582–598.

8. Hamad, D., Mehrvar, M., and Dhib, R., (2019). Kinetic modeling of photodegradation of water-soluble polymers in batch photochemical reactor. *Kinetic Modeling for Environmental Systems.* 1: 38.

9. Ghafoori, S., Mehrvar, M., and Chan, P., (2012). Kinetic study of photodegradation of water soluble polymers. *Iranian Polymer Journal.* 21(12): 869–876.

10. Hamad, D., Mehrvar, M., and Dhib, R., (2018). Photochemical kinetic modeling of degradation of aqueous polyvinyl alcohol in a UV/H$_2$O$_2$ photoreactor. *Journal of Polymers and the Environment.* 26(8): 3283–3293.

11. Ghafoori, S., (2013). Modeling, simulation, and optimization of advanced oxidation processes for treatment of polymeric wastewater (Doctoral dissertation, Ryerson University). ISBN 9780499223395.

6 Application of UV/TiO$_2$ and UV/H$_2$O$_2$ Systems for Micropollutants' Treatment Process

Yuri Park and Allison L. Mackie

CONTENTS

6.1 Introduction 107
6.2 Principles of UV/H$_2$O$_2$ and UV/TiO$_2$ 108
    6.2.1 UV/H$_2$O$_2$ Advanced Oxidation Process (AOP) 109
    6.2.2 UV/TiO$_2$ Advanced Oxidation Process 111
    6.2.3 Other Photocatalytic Processes 113
6.3 Application of UV/H$_2$O$_2$ and UV/TiO$_2$ Process 114
    6.3.1 Degradation of Micropollutants by UV/H$_2$O$_2$ Process.................... 114
    6.3.2 Degradation of Micropollutants by UV/TiO$_2$ Process..................... 117
    6.3.3 Comparison of UV/H$_2$O$_2$ and UV/TiO$_2$.. 120
6.4 Challenges and Future Perspectives 121
6.5 Conclusion 121
Acknowledgments... 122
References... 122

6.1 INTRODUCTION

Ultraviolet (UV) radiation has been widely used for the treatment of water and wastewater around the world. Numerous studies show that UV treatment is useful for degrading many organic contaminants, including emerging pollutants such as pharmaceuticals, pesticides, and endocrine-disrupting compounds, frequently found at µg/L concentrations or lower in many waterways, sediments, and soil [1–4]. UV irradiation on its own is commonly used for disinfection and to remove organic contaminants in water, providing direct photolysis effects to disrupt chemical bonds. However, higher UV doses or fluences are required to treat micropollutants compared to disinfection processes [5]. The enhancement of UV treatment with hydrogen peroxide (H$_2$O$_2$) or titanium dioxide (TiO$_2$) has been shown to improve treatment outcomes for the removal of many micropollutants from water and wastewater streams. H$_2$O$_2$ addition can improve treatment outcomes upon absorption of UV light by generating strongly oxidizing hydroxyl radicals (HO•), while TiO$_2$ acts as a

DOI: 10.1201/9781003247913-6

photocatalyst to improve treatment efficacy of UV irradiation through the generation of HO• as well as direct oxidation effects as described later in the chapter.

UV/H_2O_2 and UV/TiO_2 treatment processes have been tested to remove many organic micropollutants, such as pharmaceuticals, pesticides, dyes, endocrine disrupting compounds (EDCs), and cyanobacterial toxins. AOPs using strongly oxidizing HO• radicals can result in complete oxidation/mineralization of micropollutants, though intermediate species are produced. These intermediate species can sometimes be more toxic than the original contaminant [6–11]. Both UV/H_2O_2 and UV/TiO_2 processes have been tested extensively at bench-scale, less so at pilot scale [12–18], and with some UV/H_2O_2 studies also taking place in full-scale installations [19–21].

This chapter describes the principles behind the use of UV/H_2O_2 and UV/TiO_2 for the treatment of water contaminated with micropollutants, provides an overview of currently available research into different applications and micropollutants, and discusses some of the challenges with the technology and future research directions.

6.2   PRINCIPLES OF UV/H_2O_2 AND UV/TiO_2

Irradiation of water samples using UV light alone can have low efficiency of degradation of micropollutants due to poor absorption of UV light by the specific micropollutant [4–5]. The addition of H_2O_2 can improve treatment by increasing the production of strongly oxidizing hydroxyl radicals (HO•) from the absorption of photons by H_2O_2 [4–5, 17, 22, 23]. Utilizing a substrate of TiO_2 particles or thin films also increases the production of HO• upon irradiation with UV light and provides some direct degradation effects [8, 24–25]. Hydroxyl radicals are powerful oxidizers that completely mineralize the target micropollutants to H_2O, CO_2, NH_4, NO_3, and/or SO_4 [6, 8, 26]. However, the generation of sometimes toxic intermediate by-products often occurs [6–11].

Background constituents in the water to be treated, such as dissolved organic carbon (DOC), nitrite, bicarbonate, phosphate, and metals, can interfere with treatment. Some background constituents can provide positive effects [8, 27–29], while most reduce treatment efficacy by attenuating UV light, scavenging the HO• radicals, or adsorbing onto the TiO_2 photocatalyst [8, 22, 23, 28, 30–33].

Mercury UV lamps have commonly been adopted to treat emerging micropollutants [23, 34–36]. These lamps have fixed emission patterns determined by the inherent properties of mercury, which are either monochromatic, typically at 254 nm, or polychromatic in the range of 200 to 300 nm, depending on the gas pressure (i.e. low pressure versus medium pressure). Vacuum UV (VUV) lamps, which can emit radiation at a wavelength of 185 nm in addition to 254 nm, have been shown to improve treatment outcomes [36–38]. The use of UV light-emitting diodes (UV-LEDs) with narrow wavelength distributions has been more recently investigated for use in AOPs as a lower cost, less energy-intensive alternative to conventional UV light sources [18, 39–43]. Additionally, UV-LEDs do not contain mercury, making them even more environmentally sustainable. Solar radiation has also been studied for use in AOPs [8, 12, 16, 24, 44–45], but applications are currently limited because the sun does not emit much of its radiation in the UV range.

6.2.1 UV/H₂O₂ ADVANCED OXIDATION PROCESS (AOP)

The absorption of photon energy (*hv*) from UV light by H_2O_2 results in the generation of HO• (Eq. 6.1), which is then available to oxidize and break apart the target micropollutant molecules. H_2O_2 can be continuously introduced as a liquid to the water or wastewater treatment process, typically in concentrations ranging from 1 to 10 mg/L [5, 17, 19–21, 23, 29, 34, 39, 46–51]. Higher concentrations between 20 and 680 mg/L [4, 14, 16, 22, 30, 34, 36, 52–60] and even 3.4 g/L [31] have also been tested. The degradation of many micropollutants increases with increasing doses of H_2O_2; however, H_2O_2 is a known HO• scavenger, and too high a dose can interfere with the treatment process [4, 30, 31, 53, 54, 57–59, 61, 62]. The studies testing the highest H_2O_2 concentrations found that lower concentrations were more optimal for treatment. Lutterbeck et al. [57] found that the lowest H_2O_2 concentration they tested (i.e., 330 mg/L) was optimal for the degradation of cyclophosphamide, and Hu et al. [30] found the highest concentration they studied (i.e., 680 mg/L) inhibited methyl tert-butyl ether (MTBE) degradation. Autin et al. [31] found an optimal H_2O_2 dose of 272 mg/L for the removal of metaldehyde. Additionally, some micropollutants are not affected by the addition of H_2O_2 due to the fact that they are more susceptible to direct UV photolysis than hydroxyl radical oxidation, for example, diclofenac [4–5, 17] and the main urinary metabolite of tetrahydrocannabinol, THC-COOH [23].

$$H_2O_2 + hv \rightarrow 2HO• \tag{6.1}$$

The UV/H₂O₂ treatment process generates intermediate species or transformation products, which can be more or less toxic than the parent molecule [11, 48, 51, 56, 58, 63, 64]. Wang and colleagues [10] have published an excellent review of the common degradation pathways resulting from HO• oxidation. They classify the HO• oxidation process into four categories, dependent on the chemical structure of the target micropollutant: hydroxylation, dealkylation, decarboxylation, and deamination. If the reactions are allowed to go to completion, i.e., with an excess of HO•, complete mineralization of the micropollutants occurs.

Background constituents in the water stream such as dissolved organic matter (DOM), bicarbonate, phosphate, and nitrate can interfere with the treatment process by scavenging HO• radicals, requiring increased UV intensity and/or increased H_2O_2 dose to degrade the target micropollutants [23, 31, 32, 37, 60, 62, 65–66]. However, some studies have shown improved micropollutant degradation when treated in the presence of other ions or natural waters compared to pure water [29, 43, 60, 65, 67]. Zhang and Li [67] found that an increased concentration of some metal ions, especially Fe, Ag, and Cu, in waste improved sludge removal of endocrine-disrupting compounds (EDCs). Humic acid (HA) was also shown to improve treatment. However, it was noted that too high of a HA concentration would be detrimental due to competition for HO•. Bennett et al. [29] found improved removal of EDCs in aquaculture wastewater compared to pure water, potentially due to the higher pH of the wastewater and/or the presence of DOM and nitrate. Nitrate has been shown to have a weak positive effect on clofibric acid degradation [65] and to improve the removal of sulfamonomethoxine (SMM) [60]. Nitrate can generate hydroxyl radicals upon irradiation with UV light. Cai et al. [43] showed that trace copper ions could improve the degradation of carbamazepine.

Environmental factors such as temperature and pH are important elements that should be considered to enhance the UV/H_2O_2 process. Solution pH can affect the treatment efficacy of the UV/H_2O_2 AOP, with more acidic pH being shown to increase treatment efficacy partially due to the higher stability of H_2O_2 at lower pH [30, 54, 65, 67, 68]. The pH effect also depends on the molecular structure of the target micropollutant and any background constituents; improved oxidation is typically seen at pH closer to the pKa for the contaminant. Li et al. [60] found that alkaline conditions promoted SMM degradation due to the transformation to its anionic form at higher pH (i.e. 9 to 11). Yuan et al. [62] found inconsistent effects of pH on the treatment of several acid pharmaceuticals, with degradation of some being favored at neutral pH and some at acidic pH. Few studies have examined the effect of temperature on the UV/H_2O_2 process. Sanz et al. [61] showed that increasing temperature to 60 °C improved linear alkylbenzene sulphonate degradation. Li et al. [65] found improved removal of clofibric acid with increased temperature to 30 °C. Galindo and Kalt [68] found no effect of temperature on monoazo dyes' degradation between 22 and 45 °C. Yuan et al. [62] found that increasing temperatures from 10 to 30 °C positively impacted the degradation of acid pharmaceuticals. The temperature was noted to play a considerably more important role than the solution pH.

The production of HO• from H_2O_2 using UV irradiation requires a significant input of energy to overcome the low absorption of UV by H_2O_2 and scavenging and attenuation of UV by components of the water [5, 31, 32, 50]. UV fluences of up to 3,000 mJ/cm^2 are typically studied for use in UV-AOP treatment processes [14, 17, 22, 23, 29, 31, 32, 34, 50, 56], with many studies reporting lamp power and irradiation time instead of fluence. Increasing UV dose or fluence generally increases the degradation of micropollutants, with optimal UV fluences typically ranging between 500 and 1,000 mJ/cm^2. UV-C light concentrated at a peak wavelength of 254 nm, using a low pressure (LP) lamp, or in the range of 200 to 300 nm, using a medium pressure (MP) lamp, is most frequently used in the UV/H_2O_2 process. UV irradiation at 254 nm is known to target biological and other organic contaminants which strongly absorb UV light at this wavelength. Some researchers have also studied the efficacy of other UV wavelengths for micropollutant degradation in the UV/H_2O_2 process [30, 39–41, 51, 64]. Hu et al. [30] found that UV/H_2O_2 degradation of MTBE was improved at 254 nm compared to 365 nm. UV-B radiation at a wavelength of 312 nm was found to be effective for the degradation of microcystin-LR (MCLR) [51]. Both ciprofloxacin [39] and chloramphenicol [41] were found to be optimally degraded at a wavelength of 280 using UV-LEDs. UV-LEDs can be more easily tailored to emit narrow band wavelengths. Vacuum UV (VUV) can improve the degradation of micropollutants in the UV/H_2O_2 process due to its ability to emit a second band of radiation in the VUV range at 185 nm [36, 37, 69]. More recently, the UV activation of other chemical oxidizers such as chlorine (Cl_2) has been investigated and shown to improve treatment outcomes compared to UV/H_2O_2 [13, 21, 45, 70–72]. This improvement is mainly attributed to the increased reactivity of Cl_2 to higher UV wavelengths compared to H_2O_2.

6.2.2 UV/TiO₂ ADVANCED OXIDATION PROCESS

The UV treatment process can also be improved by adding TiO_2, which absorbs energy from UV light and acts as a photocatalyst. Upon absorption of UV light, TiO_2 can generate HO• radicals and can also provide direct oxidation and reduction effects [8, 24]. In the UV/TiO₂ treatment process, micropollutants are first adsorbed onto the TiO_2 surface, then degraded by the photocatalytic reactions, and finally desorbed from the TiO_2 surface [8, 25]. UV irradiation of TiO_2 with photon energy (hv) greater than that of the band gap for TiO_2 ($E_g = 3.2$ eV) results in the promotion of electrons (e⁻) to the conduction band. It leaves electron gaps or holes (h⁺) in the valence band of the TiO_2 molecules (Eq. 6.2). The positive holes allow for the generation of HO• from adsorbed water (Eq. 6.3) and hydroxyl ions (OH⁻), while direct oxidation or reduction reactions can also occur when the micropollutants react with the positive holes or bound electrons [8, 24–25]. Hydroxyl radicals are additionally produced via the reduction of O_2 (Eqs. 6.4 through 6.8), which also prevents recombination of the electron/hole pairs, meaning sufficient oxygenation is required for efficient treatment [8, 25]. Figure 6.1 details these processes [8]. More detailed explanations of these reactions can be found in the literature [6, 8, 24, 25, 73].

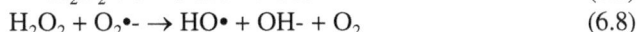

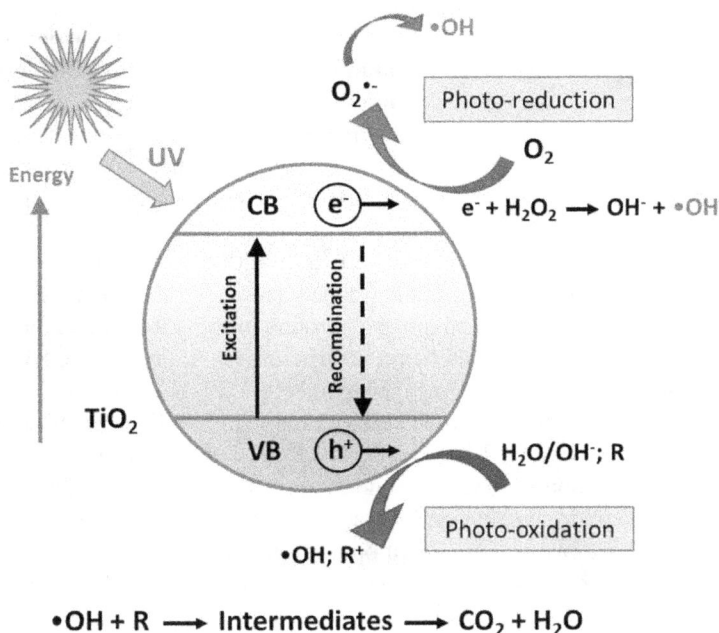

$$TiO_2 + hv \rightarrow TiO_2 + e\text{-} + h^+ \quad (6.2)$$
$$H_2O + h^+ \rightarrow H^+ + HO\bullet \quad (6.3)$$
$$O_2 + e\text{-} \rightarrow O_2\bullet\text{-} \quad (6.4)$$
$$O_2\bullet\text{-} + H^+ \rightarrow HO_2 \quad (6.5)$$
$$HO_2 + HO_2 \rightarrow H_2O_2 + O_2 \quad (6.6)$$
$$H_2O_2 + e\text{-} \rightarrow HO\bullet + OH^- \quad (6.7)$$
$$H_2O_2 + O_2\bullet\text{-} \rightarrow HO\bullet + OH\text{-} + O_2 \quad (6.8)$$

FIGURE 6.1 Diagram of TiO_2 photocatalytic process. Reproduced with permission [8].

Similar to UV/H$_2$O$_2$ treatment processes, UV/TiO$_2$ treatment processes not only result in complete mineralization of micropollutants, if allowed to go to completion, but can also produce intermediate by-products of increased or decreased toxicity compared to the parent compound [7–10, 28, 57, 74–80]. For example Elghniji et al. [74] found decreased toxicity of intermediates of 4-chlorophenol degradation, while Ma et al. [80] found highly toxic intermediates upon degradation of pentachlorophenol using UV/TiO$_2$ AOP.

Many factors impact the efficacy of UV/TiO$_2$ treatment processes, including the configuration, concentration, and composition of the TiO$_2$; solution pH; concentration of micropollutant; presence of competing ions in the water stream; and wavelength and intensity of light [8, 24, 25, 44]. Many different configurations of treatment processes have been studied to introduce the TiO$_2$ photocatalyst to the water stream, but, in general, TiO$_2$ can be introduced to the treatment process as a suspension of micro- or nano-sized particles or immobilized onto a surface as a thin film [8, 18, 24, 25]. The efficacy of TiO$_2$-coated glass or other beads and zeolite particles has also been studied [8, 12, 18, 81, 82]. Although using a suspension of TiO$_2$ particles or TiO$_2$-coated beads increases the available surface area for reactions to take place, thereby making the process more efficient, it presents challenges due to the required separation of the suspended particles from the treated water and also reduces the penetration of UV light into the solution [8, 24]. Increasing the photocatalyst concentration tends to increase treatment efficacy until a point where the TiO$_2$ begins to reduce UV light penetration into the solution.

The efficacy of different crystalline forms of TiO$_2$ has been investigated, with the anatase form found to be more reactive than the rutile form [8, 25, 83]. Research has also been done on engineering different structural configurations of TiO$_2$ particles to make the treatment process more efficient and/or to make the TiO$_2$ more easily recoverable [83, 84]. This can be done using different TiO$_2$ particle synthesis techniques and/or calcination temperatures. In addition, doping of the titanium with metallic (e.g. Co, Zn, Ni, Fe) or non-metallic elements (e.g. Ge, N, P, S) has also been investigated for improving the treatment process. Improvements through doping of the TiO$_2$ occur mainly through a reduction of the band gap in the material and/or a reduction in the recombination of the electron/hole (e$^-$/h$^+$) pairs [24, 25, 74, 79, 82, 85].

The solution pH influences treatment efficacy, with the optimal pH for the degradation of a given micropollutant being dependent on its chemical characteristics and characteristics of the TiO$_2$ and any background ions present due to the pH effect on molecular surface charge and chemical structure [8, 9, 26, 30, 76, 86–91]. TiO$_2$ has a point of zero charge (pzc) of approximately 6.25, having a positive surface charge below this pH, and a negative surface charge at alkaline pH [8]. Jafari et al. [91] found that more acidic pH improved MCLR degradation, which was attributed to the stronger attraction of MCLR to the more positively charged TiO$_2$ surface at lower pH. The degradation rate constants for several micropollutants were found to be reduced both above and below an optimal pH of 5.0 [76]. Removal of phenol by UV/TiO$_2$ process has been found to have a maximum at pH 9, with poor performance at acidic pH attributed to a lower concentration of HO• and at higher pH due to both phenol and TiO$_2$ having strong negative charges, as well as recombination of e$^-$/h$^+$ pairs [89]. Degradation of 5-fluorouracil was found to be reduced both above and below the optimal pH of 5.8, attributed again to repulsive forces at alkaline pH and reduced photocatalytic activity at acidic pH [9]. Therefore, optimal treatment pH should be determined for each water stream and micropollutant to be treated. Temperature can also have an effect on the treatment efficacy of the UV/TiO$_2$ process, with increasing temperature generally improving treatment outcomes [86, 87, 89, 92–94]. Improved degradation of azo dyes was found up to 40 °C [94], 45 °C [87], and 60 °C [93], with lowered efficacy at 70 °C [93]. Increasing temperature from 25 to 55 °C also improved UV/TiO$_2$ degradation of chlorfenapyr [92].

Micropollutant concentration can also have a varying effect on treatment efficacy. In general, increasing micropollutants concentration results in decreasing treatment efficacy. However, for some micropollutants at low initial concentrations, increasing their concentration has increased treatment efficacy to a point after which any further increases result in reduced efficacy [8]. The concentration of competing ions in solution can also have a varying effect on treatment efficacy, but, in most cases, increasing the concentration of competing ions in the water reduces treatment efficacy. Interfering ions can reduce treatment efficacy by adsorbing onto the TiO$_2$ surface and competing for active sites on the TiO$_2$ with the micropollutant, reacting with the HO• radicals, and reducing UV penetration by increasing turbidity [8, 28, 30–33]. However, low concentrations of ions such as Fe^{3+} and Cu^{2+} have been shown to improve the degradation of some micropollutants by preventing the recombination of the e$^-$/h$^+$ pairs by scavenging the electrons [8, 27]. Additionally, the presence of bicarbonate was shown to improve the degradation of methotrexate due to the generation of carbonate radicals [28].

Increasing the intensity or power of the UV light improves treatment outcomes to an upper limit [8]. The wavelength of UV light used also has an impact on the reactions. The UV-A wavelength range with a peak at 365 nm is typically used for AOPs involving TiO$_2$ [18, 26, 85, 95, 96], which is higher than that typically used for direct photolysis or UV/H$_2$O$_2$ AOP treatment of organic pollutants in water (i.e. 254 nm). However, some UV/TiO$_2$ studies have been performed at a peak wavelength of 254 nm [9, 30, 32, 91], and other wavelengths of UV light have also been investigated for use in the process [8]. UV-LED wavelengths between 250 and 365 nm were tested for degradation of four pharmaceuticals and MCLR cyanobacterial toxin with shorter wavelengths improving treatment outcomes [97]. Chen et al. [40] tested UV-LED at a wavelength of 280 nm to remove 1*H*-benzotriazole. The use of solar energy for photocatalytic activation of TiO$_2$ has also been investigated [8, 12, 24, 44], with improvements being seen by using TiO$_2$ doped with other elements to reduce the band gap energy, as discussed earlier.

6.2.3 OTHER PHOTOCATALYTIC PROCESSES

The combined UV/TiO$_2$/H$_2$O$_2$ process has been studied for the degradation of various contaminants [26, 82, 89, 94–96, 98]. The addition of H$_2$O$_2$ provides another electron acceptor and increases HO• generation. Other electron acceptors (e.g. K$_2$S$_2$O$_8$, KBrO$_3$) have also been studied with UV alone or in combination with TiO$_2$ [8, 99–101]. Semiconductors such as ZnO, CeO$_2$, and Fe$_2$O$_3$, among many others, have been studied in place of or in combination with TiO_2 [8, 25, 44, 77, 102, 103]. Metal sulphides such as SnS_2 and CdS have also been investigated as photocatalysts with higher reactivity in the visible light spectrum compared to semiconductors due to their narrower band gaps [104].

6.3   APPLICATION OF UV/H$_2$O$_2$ AND UV/TIO$_2$ PROCESS

There is much need for more research toward the presence and treatment of emerging contaminants such as pharmaceuticals and personal care products (PPCPs) in the aquatic environment. Pharmaceuticals are synthetic or natural chemicals found in prescription medicines, over-the-counter therapeutic drugs, and veterinary drugs in households and healthcare facilities. The occurrence of pharmaceuticals in the environment at trace levels (i.e. in the range of nanograms to low micrograms per litre) is largely confirmed in municipal wastewater influent and effluent as well as receiving environments. The main issue with these emergent contaminants is that they are not effectively removed through conventional wastewater treatment plant (WWTP) processes; therefore, searching for efficient and cost-effective treatment strategies from wastewater systems is imperative.

Although it is widely used to treat water contaminants, UV irradiation is mostly applicable to water types containing photosensitizing compounds and low levels of chemical oxygen demand (COD) (e.g. river and drinking water). The addition of hydrogen peroxide (H_2O_2) or titanium dioxide (TiO_2) to the UV-based photodegradation process can achieve higher removal efficiency of organic molecules. The following is a summary of UV/H_2O_2 and UV/TiO_2 applications for treating current emerging contaminants, particularly focusing on pharmaceuticals and personal care products, including antibiotics and endocrine disrupting compounds (EDCs).

6.3.1   DEGRADATION OF MICROPOLLUTANTS BY UV/H$_2$O$_2$ PROCESS

The UV/H_2O_2 process is an AOP that can produce powerful reactive oxygen species (ROS), including hydroxyl radicals. The use of H_2O_2 in combination with UV light is an effective method for treating water that emerging contaminants have contaminated. Many researchers have examined the degradation behavior of pharmaceuticals by the UV/H_2O_2 process under different conditions [4, 47, 62, 66, 105–108]. Benitez et al. [106] examined the degradation of four specific pharmaceutical compounds (metoprolol, naproxen, amoxicillin, phenacetin) in deionized (DI) water via monochromatic UV irradiation (λ = 254 nm). The higher oxidation rate of target compounds (e.g. first-order rate constants (k) for amoxicillin = 68.3×10^3 to 92.2×10^3 min^{-1}) was observed by increasing the concentration of H_2O_2 from 1×10^{-4} to 5×10^{-5} M, showing the positive effect of the combined UV/H_2O_2 system in relation to the photodegradation via UV radiation alone (k = 37.0×10^3 min^{-1}). The influence of the water type on photo-oxidation for amoxicillin, naproxen, and phenacetin was also evaluated. The fastest degradation took place in mineral water (68.3×10^3, 62.2×10^3, 24.9×10^3 min^{-1}), followed by groundwater (56.6×10^3, 46.7×10^3, 16.0×10^3 min^{-1}), and, finally, in reservoir water (47.7×10^3, 42.0×10^3, 16.0×10^3 min^{-1}) at an initial H_2O_2 concentration of 1×10^{-4} M. This sequence of reactivities can be explained by the natural organic matter (NOM) content in the three water systems; the pharmaceutical removal rate increased with decreasing content of organic and inorganic compounds which can scavenge hydroxyl radicals during the treatment process. Yuan et al. [62] also investigated the degradation of acid pharmaceuticals in water environments and the effect of the presence of humic acid (HA). HA acted as a photosensitizer and hydroxyl radical scavenger and showed a significant inhibitory effect on the degradation of six target compounds. The presence of anions commonly found in natural waters (e.g. chloride, bicarbonate, and nitrate) also affected the degradation of these pharmaceuticals, with the removal efficiency tending to be lower in natural water samples than in DI water samples. A similar observation was reported by Rivas et al. [66] for the degradation of five pharmaceuticals in DI water and secondary effluent with low pressure (LP) UV lamp. Also, their study showed that regardless of the operation conditions, the initial pharmaceutical concentration affects the reactivity of contaminants by UV/H$_2$O$_2$. When the initial pharmaceutical concentration was at 15 mg/L, the reactivity of contaminants followed the order: caffeine < acetaminophen ≈ ketorolac < doxycycline ≪ antipyrine, while the observed reactivity differed following the order: doxycycline < caffeine < ketorolac < acetaminophen ≪ antipyrine when 5 mg/L of initial pharmaceutical concentration was spiked.

Several studies reported on the efficiency of the UV/H$_2$O$_2$ process toward pharmaceuticals, using either monochromatic LP or polychromatic MP radiation sources. Pereira et al. [109, 110] examined the degradation of six pharmaceuticals for both LP and MP lamps in surface and DI waters. Their studies proved that MP lamps are more efficient in maximizing the bench-scale degradation of the compounds by both UV irradiation as well as UV/H$_2$O$_2$ from surface water, except for iohexol. Wols et al. [34] further examined the effect of UV and UV/H$_2$O$_2$ treatment on a large group of pharmaceuticals (more than 40 substances), including popular pharmaceuticals that are often researched in other studies (e.g. carbamazepine, naproxen, paracetamol, sulfamethoxazole) as well as pharmaceuticals which are not commonly reported (e.g. metformin, paroxetine, pindolol, sotalol, venlafaxine, etc.), using both LP and MP lamps in different water matrices. Again, they found improved degradation rate constants of the combined photolysis and oxidation process for MP lamps compared to the LP lamps, and that UV fluence (from MP lamp) of 500 to 1,000 mJ/cm^2 could degrade most of the compounds by 90%. For natural water matrices, the degradation was markedly lower due to the presence of hydroxyl radical scavenging (i.e., bicarbonate and NOM). Hence, the removal efficiency of pharmaceuticals and personal care products (PPCPs) using the UV/H$_2$O$_2$ process is very much dependent on UV process conditions, H$_2$O$_2$ concentrations, water matrix, and specific target organic compounds.

The degradation of benzotriazoles (BTs) and benzothiazoles (BTHs), which conventional WWTPs poorly eliminate, was examined via direct UV photolysis and UV/H$_2$O$_2$ over the pH range of 4 to 12 [108]. UV irradiation alone was not an efficient method to remove BTs and BTHs from impacted waters, and all BTs and BTHs, except for benzothiazole, exhibited pH-dependent direct photo-transformation rate constants and quantum yields in accordance with their acid–base speciation (7.1 < pKa < 8.0). UV/H$_2$O$_2$ process was proven to be more efficient for removing BTs and

BTHs in both wastewater and river water. UV/H_2O_2 was applied for the degradation of $1H$-benzotriazole (1H-BTA), the most commonly examined compound among BTs, as it is a potential EDC [40]. 1H-BTA degradation followed pseudo-first-order degradation kinetics with k of 1.63×10^{-3} s^{-1} ($t_{0.5}$ at 7.09 min^{-1}) in low power UV/H_2O_2 system ($\lambda = 280$ nm). Even with incomplete mineralization, the use of hydroxyl radical oxidation was shown to have the potential for the degradation and detoxification of $1H$-BTA [40]. Other endocrine disruptors such as 17β-estradiol (E2) and its metabolites in wastewaters were effectively removed by the UV-based AOP system not only at the bench scale but also at the pilot scale [14, 29]. Adding H_2O_2 oxidant at 30 and 50 mg/L with UV photolysis at 423 and 520 mJ/cm^2 removed 80% of initial estrogenic compounds and estrogenic activity [14].

Recently, UV/H_2O_2 processes have increasingly been applied to remove antibiotics from wastewater. The occurrence of antibiotics in wastewater promotes the proliferation of antibiotic resistant bacteria (ARB). It enriches the abundance of antibiotic resistance genes (ARGs), impacting terrestrial and aquatic organisms and even human health through drinking water [101, 111, 112]. The antibiotics commonly reported include sulfamethoxazole [53, 107, 114], cefixime [115], amoxicillin [55, 101], SMM [60], phenicol antibiotics (chloramphenicol, florfenicol, and thiamphenicol) [116, 117], and tetracycline antibiotics (oxytetracycline, doxycycline, ciprofloxacin) [46]. Many studies have proved the great potential of UV-based AOP in the treatment of antibiotic pollutants. Jung et al. [55] reported the 99% degradation of β-lactam antibiotic amoxicillin (AMX) by UV/H_2O_2 photolytic process within 20 minutes with the addition of 10 mM H_2O_2. However, less than 22% of TOC removal was achieved, indicating the formation of by-products in the solution. Zhang et al. [101] attempted to remove AMX in DI water and wastewater. The degradation rate constant (k, min^{-1}) of AMX by UV/H_2O_2 in DI water, drinking water, and secondary effluent was observed to be 0.73 min^{-1}, 0.65 min^{-1}, 0.35 min^{-1}, respectively, at an H_2O_2 concentration of 0.5 mM and pH 7. This indicates that AMX degradation efficiency decreases significantly in the secondary effluent because of the complex water matrix (TOC = 10.6 mg/L, turbidity = 5.3 NTU), leading to shielding of the UV light, inhibiting the UV photon absorption, and resulting in the decrease of HO• generation. Among the inorganic anions, the presence of Cl^{-1} showed the most adverse effect on AMX degradation with UV/H_2O_2 treatment.

Given the consideration of the maximum enhancement of the degradation rate constants, different experimental conditions (light source path length and hydrogen peroxide concentration) should be optimized. In particular, an optimal initial concentration of H_2O_2 is important in order to produce HO• radicals that can enhance the degradation of pharmaceuticals while minimizing H_2O_2 scavenging of light that would otherwise be available to degrade the compounds. A kinetic model has been applied to predict the removal of pharmaceuticals by the UV/H_2O_2 process associated with both LP and MP lamps [118, 119]. Since the efficiency of the UV/H_2O_2 process is proven, more realistic studies of UV-based AOP applications are being conducted at the pilot scale [13–17] and during full-scale operations [19–21], where it is noted that H_2O_2 addition could be a major cost [14]. Based on the results presented in the literature, the UV/H_2O_2 process has shown a strong capacity to remove many organic contaminants which are poorly removed during the conventional wastewater treatment processes, and, by combining the appropriate concentration of H$_2$O$_2$ and UV fluence, it could be possible to design a cost-effective treatment for the removal of micropollutants in WWTPs.

6.3.2 DEGRADATION OF MICROPOLLUTANTS BY UV/TiO$_2$ PROCESS

Heterogeneous photocatalysis is a promising and highly efficient technique that can be used for the degradation of recalcitrant pharmaceuticals and other contaminants present in various water sources as well as for the complete mineralization of the pollutants without the involvement of carcinogenic compounds. The success of TiO$_2$ as a photocatalyst depends on its physicochemical properties as a semiconductor. The two common forms of crystalline TiO$_2$, anatase and rutile, have similar band gap energies. Still, the photocatalytic activity of TiO$_2$ with a high anatase content (e.g. Degussa P25, which typically consists of a 75:25 ratio of anatase: rutile) and pure anatase photocatalysts such as Hobikat UV100 has been shown to be higher during the degradation of several common pharmaceuticals from water [95, 120–123]. Several pharmaceuticals and MCLR, a cyanobacterial toxin, were also degraded by a similar photocatalytic reaction using LEDs as a source of UV radiation [97]. Shorter wavelength UV in order of UVC > UVB > UVA was more effective for the decomposition of the target pharmaceutical compounds acetaminophen, diclofenac, ibuprofen, and sulfamethoxazole. However, the utilization of solar irradiation is still limited for activating the supported TiO$_2$ due to its broad band gap, as mentioned previously [8, 12, 24, 44, 125]. TiO$_2$ has shown remarkable efficiency in removing the pain-relieving drug acetaminophen, also known as paracetamol [125–128]. Moreover, at pH 9, nontoxic by-products formed in wastewater and drinking water samples, suggesting the high removal efficiency of acetaminophen by UV/TiO$_2$ [127]. TiO$_2$ has also shown the capacity to reduce the antibiotic chlorhexidine digluconate when used as a suspension or when supported by calcium alginate beads [129]. UV-based AOP systems (i.e., UV, UV/H$_2$O$_2$, UV/TiO$_2$) were applied to secondary effluent spiked with anti-cancer drugs etoposide, paclitaxel, cyclophosphamide and ifosfamide at 500 µg/L initial concentration. After 10 minutes of irradiation, >98% removal (k = 0.46 min^{-1}) of etoposide and paclitaxel was achieved, while the two other compounds investigated were not removed after 180 minutes with TiO$_2$ catalyst or H$_2$O$_2$ [130]. The inefficiency of cyclophosphamide (CP) and ifosfamide degradation by UV/TiO$_2$ (P25) or UV/H$_2$O$_2$ is due to their chemical properties [5]. However, other works reported successful degradations of highly stable and resilient CP by TiO$_2$ (P25) photocatalysis at similar pollutants levels (~100 µg/L at pH 5.8 within 2 hour reaction time) [9]. Also, an increase of catalyst concentrations from 100 to 500 mg/L of TiO$_2$ (P25), leading to an increased number of catalyst active sites, resulted in an enhancement in the efficiency of CP mineralization (58.4 to 89.6%) [57]. However, there is no further improvement in mineralization by increasing the dosage of TiO$_2$ to 1,000 mg/L; an excess concentration of catalyst can hinder or reflect the penetrating light, leading to the loss of e$^-$/h$^+$ pairs in the TiO$_2$, which mediates photocatalysis. Similar observations were reported in other studies involving pharmaceuticals [95, 120, 132]. The amount of TiO$_2$ used may also significantly affect the degradation rate. The apparent first-order rate constant for amoxicillin degradation (initial concentration = 10 mg/L) increased from 0.0172 min^{-1} to 0.0237 min^{-1} when the suspended TiO_2 concentration was increased from 0.2 g/L to 0.8 g/L [132].

The pH of the solution can affect TiO_2-based photocatalysis. The surface charge of the photocatalyst can be presented negatively or as positively depending on the ambient pH. TiO_2 has a pzc at an approximate pH value of 6.3 [8, 133], and its chemical structure at operational pH influences the adsorption behavior of pharmaceuticals onto the TiO_2 surface. For example, the degradation of three sulfa pharmaceuticals, sulfachlorpyridazine, sulfapyridine, and sulfisoxazole, as zwitterionic compounds was explored, and it was found that the initial pH value has different influences on the photocatalytic degradation rate constants on the basis of the form of the pharmaceuticals in the solution [134]. Hu et al. [135] also explored the use of TiO_2-based photocatalysis for the photodegradation of ciprofloxacin and found that pH conditions significantly influence the target group for oxidation. Alvarez-Corena et al. [76] tested a pH range of 3.0 to 9.0 and found an optimal degradation rate for several micropollutants at a pH value of 5.0. Hence, controlling pH is an important factor, but in the real wastewater environment, pH is variable, and it is challenging to control a specific pH.

TiO_2 has also been reported to be efficient in treating a stream containing a combination of four different pesticides: diuron, alachlor, isoproturon, and atrazine. While the catalyst was shown to be effective in treating both pure and natural water, the presence of other pollutants in natural water resulted in a noticeable decrease in the efficiency of TiO_2. This can be attributed to the inhibition of forming radicals, which quickens the catalytic reaction [136]. Thus, the influence of matrix conditions on the degradation rate should be considered. This matrix effect was discussed earlier in the UV/H_2O_2 system; the presence of NOM acts not only as an inner filter which has been shown to absorb roughly 3.5% of the incident radiation but also acts as a hydroxyl radical scavenger, reducing the production of hydroxyl radicals [137]. Adding electron acceptors to the photodegradation mixture significantly enhances the photodegradation rates of substrates. Oxygen saturation has also increased degradation rates in the UV/TiO_2 treatment process. Paul et al. [138] observed a marked improvement in the degradation of ciprofloxacin and other fluoroquinolones when bromate ion (BrO_3^-) was added as an electron acceptor. Electron acceptors promote micropollutant degradation by preventing a recombination of the e$^-$/h$^+$ pairs, and the absence of BrO_3^- or molecular oxygen prevented any significant photo-oxidation in that study. Ozone or hydrogen peroxide addition to the TiO_2 system promotes photodegradation and mineralization rates.

Additionally, intermediates or transformation by-products can be formed via UV/TiO_2 and added ozone could degrade these by-products, improving the effluent quality from the UV/TiO_2 treatment process [76, 139]. Furthermore, the photodegradation pathways and extent of mineralization for TiO_2-photocatalysed micropollutants have been widely reported [57, 101, 140]. A proposed degradation pathway for CP is shown in Figure 6.2 [56]. Transformation by-products formed after the UV/TiO_2 process can be more toxic than the parent compound itself, and their environmental toxicity evaluation becomes more important.

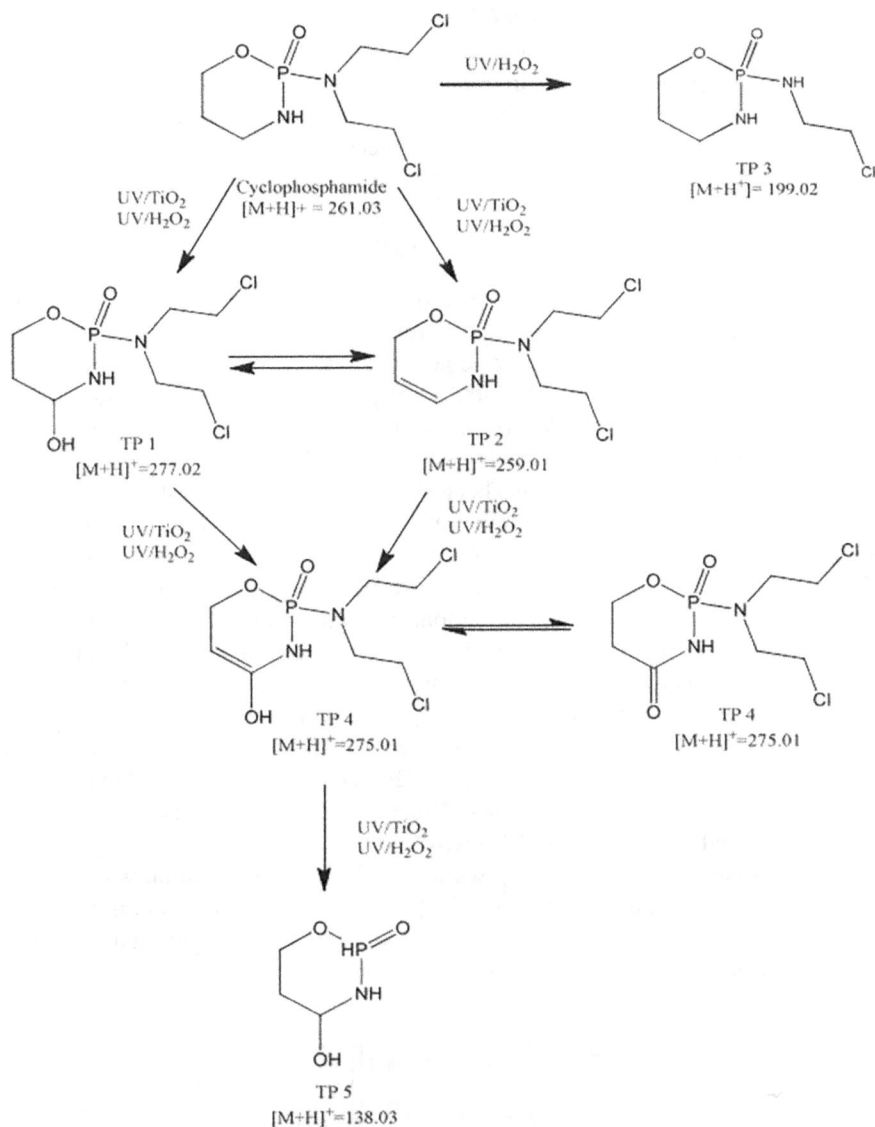

FIGURE 6.2 Proposed degradation pathway of cyclophosphamide (CP) photodegradation by UV-based AOP processes. Reproduced from Reference [56].

As observed with the data collected in these previous studies, it is often impossible to predict the removal of a particular pharmaceutical concentration without computer-aided modelling to determine optimal conditions for the photocatalytic reaction, which a few studies have addressed [141–143]. Such modelling is necessary to determine the optimal quantity of TiO₂ required for efficient photocatalysis in a particular WWTP, given that the expected concentration of certain pharmaceuticals in the wastewater is known.

These findings suggest that UV/TiO$_2$ processes can successfully remove many emerging contaminants, which have become a threat to humans, aquatic species, and ecosystems. An appropriate treatment timeframe should be applied, and the suggested operational parameters are especially important to consider when treating wastewater suspected of being contaminated with these frequently occurring micropollutants.

6.3.3   COMPARISON OF UV/H$_2$O$_2$ AND UV/TiO$_2$

The UV/H$_2$O$_2$ process can provide improved degradation effects over UV/TiO$_2$ photocatalysis. Janssens et al. [130] found a dramatically improved degradation rate of the molecular probe para-chlorobenzoic acid (pCBA) using UV/H$_2$O$_2$ compared to UV/TiO$_2$, which was partially attributed to aggregation of the TiO$_2$ catalyst. UV/H$_2$O$_2$ also outperformed UV/TiO$_2$ for the degradation of ifosfamide, while close to 100% removal of etoposide and paclitaxel was found for both AOPs (Figure 6.3) [130]. The UV/H$_2$O$_2$ process has also been shown to degrade CP faster than the UV/TiO$_2$ process; however, mineralization was more complete using UV/TiO$_2$ after a 256-minute reaction time (72.2% DOC removal for H$_2$O$_2$ versus 89.6% for TiO$_2$) [57]. Pablos et al. [96] also found that the UV-C/H$_2$O$_2$ process was more efficient for disinfection and micropollutant degradation. In contrast, the use of the UV-A/TiO$_2$ process was able to remove the tested pharmaceuticals more completely and could be advantageous where solar radiation is abundant. The degradation rate of 1H-BTA was found to be slightly but not significantly faster with UV/TiO$_2$ versus UV/H$_2$O$_2$ treatment (i.e. $1.87 \pm 0.20 \times 10^{-3}$ s^{-1} versus $1.63 \pm 0.23 \times 10^{-3}$ s^{-1}). Still, mineralization was found to be higher with UV/H$_2$O$_2$ (72% TOC removal versus 60% for UV/TiO$_2$ with 30-minute reaction time) [40]. Ninety-eight per cent and 80% removal of MTBE was found using optimized UV/H$_2$O$_2$ and UV/TiO$_2$ processes, respectively, after a 60-minute reaction time [30], while metaldehyde degradation was found to be identical between the two processes [31]. Background constituents in the water stream were found to affect the UV/TiO$_2$ process more extensively than UV/H$_2$O$_2$ due to their adsorption onto the TiO$_2$ surface [31, 32].

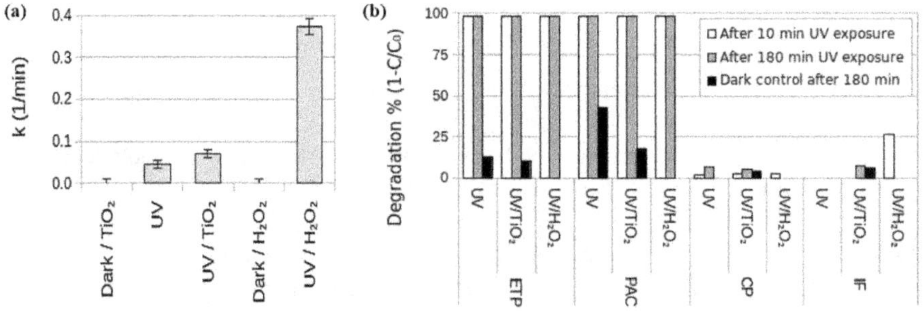

FIGURE 6.3   Comparison of UV/TiO$_2$ (100 mg/L) and UV/H$_2$O$_2$ (50 mg/L) for the degradation of anti-cancer drugs (e.g. etoposide (ETP), paclitaxel (PAC), cyclophosphamide (CP), and ifosfamide (IF)) spiked at 500 µg/L in secondary wastewater effluent. Reproduced from Reference [130].

6.4   CHALLENGES AND FUTURE PERSPECTIVES

Considering all aspects of UV-based AOPs, these technologies are proven beneficial in environmental remediation and in treatment of micropollutants found in wastewater. In many cases, UV-AOP can be successfully applied to remove pollutants present in wastewater by chemically degrading the parent compounds into their complete mineralization products, producing H$_2$O and CO$_2$ without generating any toxic by-products. However, much of the reported literature has been applied at bench or pilot scales, and their applicability at the industrial scale is still limited. In order to increase the feasibility of these AOPs to be implemented at a large scale, several aspects, in addition to operating parameters such as UV dose and pH, should be considered. With such aim, further research should focus on the efficiency of AOPs and the costs of the process, the toxicity of effluents and by-products, photocatalytic technology, and reactor design.

Among the different alternatives to generate hydroxyl radicals in AOPs, heterogeneous photocatalysis appears as a promising solution in wastewater treatment. A photocatalyst and light are the only requirements of this process, and TiO$_2$ is the most widely used metal oxide in AOPs because of its inexpensiveness, commercial availability, and nontoxic and stable nature. However, lowering the operational costs of UV-based AOPs is a critically important factor to overcome as it is commonly accepted that photocatalysis (mainly using TiO$_2$) is a relatively expensive AOP despite its high removal efficiency of pollutants. Moreover, UV/TiO$_2$ achieves higher mineralization efficiency as compared with the UV/H$_2$O$_2$ process, and, to increase the applicability of these processes in the full-scale application of AOP in treatment plants, the proper design of the reactor should be taken into consideration in the future, with the goal of yielding higher degradation efficiency. Improving mixing and mass transfer ability in photocatalytic processes will improve the overall yield. The filtration step should be additionally examined to avoid the presence of TiO$_2$ nanoparticles in the treated effluent.

6.5   CONCLUSION

This chapter investigates the basic principles of UV-based AOPs, specifically UV/H$_2$O$_2$ and UV/TiO$_2$ processes, and their applications for the removal of micropollutants present in waters. In order to maximize their efficiency for the removal of micropollutants present in wastewaters, several operating conditions (i.e. selection of UV lamps, exposure time, UV wavelength) and environmental factors (i.e. pH, temperature, water matrix, radical scavengers, co-presence of pollutants, and pollutant concentrations) should be considered. After several decades of research on AOPs, these highly efficient technologies are proven to be applicable at the industrial scale. However, implementing AOPs at the full scale is still challenging, mainly due to their high operational costs. In this context, using a broader part of the light spectrum instead of UV will reduce operating costs, and this will follow the current carbon-neutral policy being implemented worldwide. Also, searching for alternatives to pure TiO$_2$ photocatalysts will make UV/photocatalytic processes more affordable. Thus, the operating cost per unit mass of pollutants can be improved, and the industrial implementation of these technologies will become much more viable and attractive for water and wastewater treatment industries.

ACKNOWLEDGMENTS

Financial support from National Research Foundation of Korea (NRF) grant funded by the Korea government (MSIT) (No. 2022R1A2C1008696) is kindly acknowledged.

REFERENCES

1. M. Pera-Titus, V. García-Molina, M.A. Baños, J. Giménez, and S. Esplugas, *Degradation of chlorophenols by means of advanced oxidation processes: A general review*, Applied Catalysis B: Environmental 47, no. 4 (2004), pp. 219–256.
2. K. Ikehata, M.G. El-Din, and S.A. Snyder, *Ozonation and advanced oxidation treatment of emerging organic pollutants in water and wastewater*, Ozone: Science and Engineering 30, no. 1 (2008), pp. 21–26.
3. M. Klavarioti, D. Mantzavinos, and D. Kassinos, *Removal of residual pharmaceuticals from aqueous systems by advanced oxidation processes*, Environment International 35, no. 2 (2009), pp. 402–417.
4. Z. Shu, J.R. Bolton, M. Belosevic, and M.G. El Din, *Photodegradation of emerging micropollutants using the medium-pressure UV/H_2O_2 advanced oxidation process*, Water Research 47, no. 8 (2013), pp. 2881–2889.
5. I. Kim, N. Yamashita, and H. Tanaka, *Photodegradation of pharmaceuticals and personal care products during UV and UV/H_2O_2 treatments*, Chemosphere 77, no. 4 (2009), pp. 518–525.
6. A. Houas, H. Lachheb, M. Ksibi, E. Elaloui, C. Guillard, and J.-M. Herrmann, *Photocatalytic degradation pathway of methylene blue in water*, Applied Catalysis B: Environmental 31, no. 2 (2001), pp. 145–157.
7. I.K. Konstantinou, and T.A. Albanis, *TiO_2-assisted photocatalytic degradation of azo dyes in aqueous solution: Kinetic and mechanistic investigations: A review*, Applied Catalysis B: Environmental 49, no. 1 (2004), pp. 1–14.
8. S. Ahmed, M.G. Rasul, R. Brown, and M.A. Hashib, *Influence of parameters on the heterogeneous photocatalytic degradation of pesticides and phenolic contaminants in wastewater: A short review*, Journal of Environmental Management 92, no. 3 (2011), pp. 311–330.
9. H.H.-H. Lin, and A.Y.-C. Lin, *Photocatalytic oxidation of 5-fluorouracil and cyclophosphamide via UV/TiO_2 in an aqueous environment*, Water Research 48 (2014), pp. 559–568.
10. W.-L. Wang, Q.-Y. Wu, N. Huang, Z.-B. Xu, M.-Y. Lee, and H.-Y. Hu, *Potential risks from UV/H_2O_2 oxidation and UV photocatalysis: A review of toxic, assimilable, and sensory-unpleasant transformation products*, Water Research 141 (2018), pp. 109–125.
11. T. Perondi, W. Michelon, P. Reis Junior, P.M. Knoblauch, M. Chiareloto, R. de Fátima Peralta Muniz Moreira, R.A. Peralta, E. Düsman, and T.S. Pokrywiecki, *Advanced oxidative processes in the degradation of 17β-estradiol present on surface waters: Kinetics, by-products and ecotoxicity*, Environmental Science and Pollution Research 27, no. 17 (2020), pp. 21032–21039.
12. N. Miranda-García, M. Ignacio Maldonado, J.M. Coronado, and S. Malato, *Degradation study of 15 emerging contaminants at low concentration by immobilized TiO_2 in a pilot plant*, Catalysis Today 151, no. 1–2 (2010), pp. 107–113.

13. D. Wang, J.R. Bolton, S.A. Andrews, and R. Hofmann, *UV/chlorine control of drinking water taste and odour at pilot and full-scale*, Chemosphere 136 (2015), pp. 239–244.

14. B. Cédat, C. de Brauer, H. Métivier, N. Dumont, and R. Tutundjan, *Are UV photolysis and UV/H$_2$O$_2$ process efficient to treat estrogens in waters? Chemical and biological assessment at pilot scale*, Water Research 100 (2016), pp. 357–366.

15. X. Chu, Y. Xiao, J. Hu, E. Quek, R. Xie, T. Pang, and Y. Xing, *Pilot-scale UV/H$_2$O$_2$ study for emerging organic contaminants decomposition*, Reviews on Environmental Health 31, no. 1 (2016), pp. 71–74.

16. S. Miralles-Cuevas, D. Darowna, A. Wanag, S. Mozia, S. Malato, and I. Oller, *Comparison of UV/H$_2$O$_2$, UV/S$_2$O$_8$$^{2-}$, solar/Fe(II)/H$_2O_2$ and solar/Fe(II)/S$_2$O$_8$$^{2-}$ at pilot plant scale for the elimination of micro-contaminants in natural water: An economic assessment*, Chemical Engineering Journal 310 (2017), pp. 514–524.

17. D.B. Miklos, R. Hartl, P. Michel, K.G. Linden, J.E. Drewes, and U. Hübner, *UV/H$_2$O$_2$ process stability and pilot-scale validation for trace organic chemical removal from wastewater treatment plant effluents*, Water Research 136 (2018), pp. 169–179.

18. I. Menezes, J. Capelo-Neto, C.J. Pestana, A. Clemente, J. Hui, J.T.S. Irvine, H.Q. Nimal Gunaratne, et al. *Comparison of UV-A photolytic and UV/TiO$_2$ photocatalytic effects on Microcystis aeruginosa PCC7813 and four microcystin analogues: A pilot scale study*, Journal of Environmental Management 298 (2021), p. 113519.

19. J.C. Kruithof, P.C. Kamp, and B.J. Martijn, *UV/H$_2$O$_2$ treatment: A practical solution for organic contaminant control and primary disinfection*, Ozone: Science and Engineering 29, no. 4 (2007), pp. 273–280.

20. J.C. Kruithof, and B.J. Martijn, *UV/H$_2$O$_2$ treatment: An essential process in a multi barrier approach against trace chemical contaminants*, Water Science and Technology: Water Supply 13, no. 1 (2013), pp. 130–138.

21. C. Wang, N. Moore, K. Bircher, S. Andrews, and R. Hofmann, *Full-scale comparison of UV/H$_2$O$_2$ and UV/Cl$_2$ advanced oxidation: The degradation of micropollutant surrogates and the formation of disinfection by-products*, Water Research 161 (2019), pp. 448–458.

22. M. Nihemaiti, D.B. Miklos, U. Hübner, K.G. Linden, J.E. Drewes, and J.-P. Croué, *removal of trace organic chemicals in wastewater effluent by UV/H$_2$O$_2$ and UV/PDS*, Water Research 145 (2018), pp. 487–497.

23. Y. Park, A.L. Mackie, S.A. MacIsaac, and G.A. Gagnon, *Photo-oxidation of 11-nor-9-carboxy-Δ 9-tetrahydrocannabinol using medium-pressure UV and UV/H$_2$O$_2$–a kinetic study*, Environmental Science: Water Research & Technology 4, no. 9 (2018), pp. 1262–1271.

24. A.A. Adesina, *Industrial exploitation of photocatalysis: Progress, perspectives and prospects*, Catalysis Surveys from Asia 8, no. 4 (2004), pp. 265–273.

25. M.R. Al-Mamun, S. Kader, M.S. Islam, and M.Z.H. Khan, *Photocatalytic activity improvement and application of UV-TiO$_2$ photocatalysis in textile wastewater treatment: A review*, Journal of Environmental Chemical Engineering 7, no. 5 (2019), p. 103248.

26. E.S. Elmolla, and M. Chaudhuri, *Photocatalytic degradation of amoxicillin, ampicillin and cloxacillin antibiotics in aqueous solution using UV/TiO$_2$ and UV/H$_2$O$_2$/TiO$_2$ photocatalysis*, Desalination 252, no. 1–3 (2010), pp. 46–52.

27. J. Chen, Z. Hu, D. Wang, C. Gao, and R. Ji, *Photocatalytic mineralization of dimethoate in aqueous solutions using TiO$_2$: Parameters and by-products analysis*, Desalination 258, no. 1–3 (2010), pp. 28–33.

28. W.W.-P. Lai, M.-H. Hsu, and A. Y.-C. Lin, *The role of bicarbonate anions in methotrexate degradation via UV/TiO$_2$: Mechanisms, reactivity and increased toxicity*, Water Research 112 (2017), pp. 157–166.

29. J.L. Bennett, A.L. Mackie, Y. Park, and G.A. Gagnon, *Advanced oxidation processes for treatment of 17β-Estradiol and its metabolites in aquaculture wastewater*, Aquacultural Engineering 83 (2018), pp. 40–46.

30. Q. Hu, C. Zhang, Z. Wang, Y. Chen, K. Mao, X. Zhang, Y. Xiong, and M. Zhu, *Photodegradation of methyl tert-butyl ether (MTBE) by UV/H₂O₂ and UV/TiO₂*, Journal of Hazardous Materials 154, no. 1–3 (2008), pp. 795–803.

31. O. Autin, J. Hart, P. Jarvis, J. MacAdam, S.A. Parsons, and B. Jefferson, *Comparison of UV/H₂O₂ and UV/TiO₂ for the degradation of metaldehyde: Kinetics and the impact of background organics*, Water Research 46, no. 17 (2012), pp. 5655–5662.

32. O. Autin, J. Hart, P. Jarvis, J. MacAdam, S.A. Parsons, and B. Jefferson, *The impact of background organic matter and alkalinity on the degradation of the pesticide metaldehyde by two advanced oxidation processes: UV/H₂O₂ and UV/TiO₂*, Water Research 47, no. 6 (2013), pp. 2041–2049.

33. T. Tang, G. Lu, W. Wang, R. Wang, K. Huang, Z. Qiu, X. Tao, and Z. Dang, *Photocatalytic removal of organic phosphate esters by TiO₂: Effect of inorganic ions and humic acid*, Chemosphere 206 (2018), pp. 26–32.

34. B.A. Wols, C.H.M. Hofman-Caris, D.J.H. Harmsen, and E.F. Beerendonk, *Degradation of 40 selected pharmaceuticals by UV/H₂O₂*, Water Research 47, no. 15 (2013), pp. 5876–5888.

35. D. Bertagna Silva, A. Cruz-Alcalde, C. Sans, J. Gimenez, and S. Esplugas, *Performance and kinetic modelling of photolytic and photocatalytic ozonation for enhanced micropollutants removal in municipal wastewaters*, Applied Catalysis B: Environmental 249 (2019), pp. 211–217.

36. M. Li, W. Li, D. Wen, J.R. Bolton, E.R. Blatchley III, and Z. Qiang, *Micropollutant degradation by the UV/H₂O₂ process: Kinetic comparison among various radiation sources*, Environmental Science & Technology 53, no. 9 (2019), pp. 5241–5248.

37. W. Li, S. Lu, Z. Qiu, and K. Lin, *UV and VUV photolysis vs. UV/H₂O₂ and VUV/H₂O₂ treatment for removal of clofibric acid from aqueous solution*, Environmental Technology 32, no. 10 (2011), pp. 1063–1071.

38. C. Duca, G. Imoberdorf, and M. Mohseni, *Effects of inorganics on the degradation of micropollutants with vacuum UV (VUV) advanced oxidation*, Journal of Environmental Science and Health, Part A 52, no. 6 (2017), pp. 524–532.

39. H. Ou, J. Ye, S. Ma, C. Wei, N. Gao, and J. He, *Degradation of ciprofloxacin by UV and UV/H₂O₂ via multiple-wavelength ultraviolet light-emitting diodes: Effectiveness, intermediates and antibacterial activity*, Chemical Engineering Journal 289 (2016), pp. 391–401.

40. Y. Chen, J. Ye, C. Li, P. Zhou, J. Liu, and H. Ou, *Degradation of 1 H-benzotriazole by UV/H₂O₂ and UV/TiO₂: Kinetics, mechanisms, products and toxicology*, Environmental Science: Water Research & Technology 4, no. 9 (2018), pp. 1282–1294.

41. M. Wu, Y. Tang, Q. Liu, Z. Tan, M. Wang, B. Xu, S. Xia, S. Mao, and N. Gao, *Highly efficient chloramphenicol degradation by UV and UV/H₂O₂ processes based on LED light source*, Water Environment Research 92, no. 12 (2020), pp. 2049–2059.

42. D. Bertagna Silva, G. Buttiglieri, and S. Babić, *State-of-the-art and current challenges for TiO₂/UV-LED photocatalytic degradation of emerging organic micropollutants*, Environmental Science and Pollution Research 28, no. 1 (2021), pp. 103–120.

43. A. Cai, J. Deng, T. Zhu, C. Ye, J. Li, S. Zhou, Q. Li, and X. Li, *Enhanced oxidation of carbamazepine by UV-LED/persulfate and UV-LED/H₂O₂ processes in the presence of trace copper ions*, Chemical Engineering Journal 404 (2021), p. 127119.

44. S. Sakthivel, B. Neppolian, M.V. Shankar, B. Arabindoo, M. Palanichamy, and V. Murugesan, *Solar photocatalytic degradation of azo dye: Comparison of photocatalytic efficiency of ZnO and TiO₂*, Solar Energy Materials and Solar Cells 77, no. 1 (2003), pp. 65–82.

45. Z. Shu, C. Li, M. Belosevic, J.R. Bolton, and M.G. El-Din, *Application of a solar UV/ chlorine advanced oxidation process to oil sands process-affected water remediation*, Environmental Science & Technology 48, no. 16 (2014), pp. 9692–9701.

46. F. Yuan, C. Hu, X. Hu, D. Wei, Y. Chen, and J. Qu, *Photodegradation and toxicity changes of antibiotics in UV and UV/H$_2$O$_2$ process*, Journal of Hazardous Materials 185, no. 2–3 (2011), pp. 1256–1263.

47. O.S. Keen, S. Baik, K.G. Linden, D.S. Aga, and N.G. Love, *Enhanced biodegradation of carbamazepine after UV/H$_2$O$_2$ advanced oxidation*, Environmental Science & Technology 46, no. 11 (2012), pp. 6222–6227.

48. W. Zong, F. Sun, and X. Sun, *Oxidation by-products formation of microcystin-LR exposed to UV/H$_2$O$_2$: Toward the generative mechanism and biological toxicity*, Water Research 47, no. 9 (2013), pp. 3211–3219.

49. D. Gerrity, Y. Lee, S. Gamage, M. Lee, A.N. Pisarenko, R.A. Trenholm, U. Von Gunten, and S.A. Snyder, *Emerging investigators series: Prediction of trace organic contaminant abatement with UV/H$_2$O$_2$: Development and validation of semi-empirical models for municipal wastewater effluents*, Environmental Science: Water Research & Technology 2, no. 3 (2016), pp. 460–473.

50. Y. Lee, D. Gerrity, M. Lee, S. Gamage, A. Pisarenko, R.A. Trenholm, S. Canonica, S.A. Snyder, and U. Von Gunten, *Organic contaminant abatement in reclaimed water by UV/H$_2$O$_2$ and a combined process consisting of O$_3$/H$_2$O$_2$ followed by UV/H$_2$O$_2$: Prediction of abatement efficiency, energy consumption, and by-product formation*, Environmental Science & Technology 50, no. 7 (2016), pp. 3809–3819.

51. B.-R. Moon, T.-K. Kim, M.-K. Kim, J. Choi, and K.-D. Zoh, *Degradation mechanisms of Microcystin-LR during UV-B photolysis and UV/H$_2$O$_2$ processes: By-products and pathways*, Chemosphere 185 (2017), pp. 1039–1047.

52. A. Afzal, T. Oppenländer, J.R. Bolton, and M.G. El-Din, *Anatoxin-a degradation by advanced oxidation processes: Vacuum-UV at 172 nm, photolysis using medium pressure UV and UV/H$_2$O$_2$*, Water Research 44, no. 1 (2010), pp. 278–286.

53. Y. Lester, D. Avisar, and H. Mamane, *Photodegradation of the antibiotic sulphamethoxazole in water with UV/H$_2$O$_2$ advanced oxidation process*, Environmental Technology 31, no. 2 (2010), pp. 175–183.

54. X. He, M. Pelaez, J.A. Westrick, K.E. O'Shea, A. Hiskia, T. Triantis, T. Kaloudis, M.I. Stefan, A. Armah, and D.D. Dionysiou, *Efficient removal of microcystin-LR by UV-C/H$_2$O$_2$ in synthetic and natural water samples*, Water Research 46, no. 5 (2012), pp. 1501–1510.

55. Y.J. Jung, W.G. Kim, Y. Yoon, J.-W. Kang, Y.M. Hong, and H.W. Kim, *Removal of amoxicillin by UV and UV/H$_2$O$_2$ processes*, Science of the Total Environment 420 (2012), pp. 160–167.

56. X. He, G. Zhang, A.A. de la Cruz, K.E. O'Shea, and D.D. Dionysiou, *Degradation mechanism of cyanobacterial toxin cylindrospermopsin by hydroxyl radicals in homogeneous UV/H$_2$O$_2$ process*, Environmental Science & Technology 48, no. 8 (2014), pp. 4495–4504.

57. C.A. Lutterbeck, Ê.L. Machado, and K. Kümmerer, *Photodegradation of the antineoplastic cyclophosphamide: A comparative study of the efficiencies of UV/H$_2$O$_2$, UV/ Fe^{2+}/H$_2$O$_2$ and UV/TiO$_2$ processes*, Chemosphere 120 (2015), pp. 538–546.

58. S. Verma, and M. Sillanpää, *Degradation of anatoxin-a by UV-C LED and UV-C LED/H$_2$O$_2$ advanced oxidation processes*, Chemical Engineering Journal 274 (2015), pp. 274–281.

59. C. Afonso-Olivares, C. Fernández-Rodríguez, R.J. Ojeda-González, Z. Sosa-Ferrera, J.J. Santana-Rodríguez, and J.M. Doña Rodríguez, *Estimation of kinetic parameters and UV doses necessary to remove twenty-three pharmaceuticals from pre-treated urban wastewater by UV/H$_2$O$_2$*, Journal of Photochemistry and Photobiology A: Chemistry 329 (2016), pp. 130–138.

60. Y. Li, L. Yang, X. Chen, Y. Han, and G. Cao, *Transformation kinetics and pathways of sulfamonomethoxine by UV/H₂O₂ in swine wastewater*, Chemosphere 265 (2021), pp. 129125.

61. J. Sanz, J.I. Lombraña, and A. de Luis, *Temperature-assisted UV/H₂O₂ oxidation of concentrated linear alkylbenzene sulphonate (LAS) solutions*, Chemical Engineering Journal 215 (2013), pp. 533–541.

62. H. Yuan, X. Zhou, and Y.-L. Zhang, *degradation of acid pharmaceuticals in the UV/H₂O₂ process: Effects of humic acid and inorganic salts*, CLEAN–Soil, Air, Water 41, no. 1 (2013), pp. 43–50.

63. Y. Liu, J. Ren, X. Wang, and Z. Fan, *Mechanism and reaction pathways for micro cystin-LR degradation through UV/H₂O₂ treatment*, PLoS One 11, no. 6 (2016), p. e0156236.

64. J. Liu, J. Ye, H. Ou, and J. Lin, *Effectiveness and intermediates of microcystin-LR degradation by UV/H₂O₂ via 265 nm ultraviolet light-emitting diodes*, Environmental Science and Pollution Research 24, no. 5 (2017), pp. 4676–4684.

65. W. Li, S. Lu, Z. Qiu, and K. Lin, *Clofibric acid degradation in UV254/H₂O₂ process: Effect of temperature*, Journal of Hazardous Materials 176, no. 1–3 (2010), pp. 1051–1057.

66. J. Rivas, O. Gimeno, T. Borralho, and J. Sagasti, *UV-C and UV-C/peroxide elimination of selected pharmaceuticals in secondary effluents*, Desalination 279, no. 1–3 (2011), pp. 115–120.

67. A. Zhang, and Y. Li, *Removal of phenolic endocrine disrupting compounds from waste activated sludge using UV, H₂O₂, and UV/H₂O₂ oxidation processes: Effects of reaction conditions and sludge matrix*, Science of the Total Environment 493 (2014), pp. 307–323.

68. C. Galindo, and A. Kalt, *UV–H₂O₂ oxidation of monoazo dyes in aqueous media: A kinetic study*, Dyes and Pigments 40, no. 1 (1999), pp. 27–35.

69. M.V. Ngouyap Mouamfon, W. Li, S. Lu, N. Chen, Z. Qiu, and K. Lin, *Photodegradation of sulfamethoxazole applying UV-and VUV-based processes*, Water, Air, & Soil Pollution 218, no. 1 (2011), pp. 265–274.

70. D. Wang, J.R. Bolton, and R. Hofmann, *Medium pressure UV combined with chlorine advanced oxidation for trichloroethylene destruction in a model water*, Water Research 46, no. 15 (2012), pp. 4677–4686.

71. K. Guo, Z. Wu, S. Yan, B. Yao, W. Song, Z. Hua, X. Zhang, X. Kong, X. Li, and J. Fang, *Comparison of the UV/chlorine and UV/H₂O₂ processes in the degradation of PPCPs in simulated drinking water and wastewater: Kinetics, radical mechanism and energy requirements*, Water Research 147 (2018), pp. 184–194.

72. T. Chen, C. Wang, S. Andrews, and R. Hofmann, *Effects of UV Light Path Length and Wavelength on UV/Chlorine versus UV/H₂O₂ Efficacy*, ACS ES&T Water 1, no. 5 (2021), pp. 1145–1152.

73. R. Thiruvenkatachari, S. Vigneswaran, and I.S. Moon, *A review on UV/TiO₂ photocatalytic oxidation process (Journal Review)*, Korean Journal of Chemical Engineering 25, no. 1 (2008), pp. 64–72.

74. K. Elghniji, O. Hentati, N. Mlaik, A. Mahfoudh, and M. Ksibi, *Photocatalytic degradation of 4-chlorophenol under P-modified TiO₂/UV system: Kinetics, intermediates, phytotoxicity and acute toxicity*, Journal of Environmental Sciences 24, no. 3 (2012), pp. 479–487.

75. A. Tong, R. Braund, D. Warren, and B. Peake, *TiO₂-assisted photodegradation of pharmaceuticals—a review*, Open Chemistry 10, no. 4 (2012), pp. 989–1027.

76. J.R. Alvarez-Corena, J.A. Bergendahl, and F.L. Hart, *Advanced oxidation of five contaminants in water by UV/TiO₂: Reaction kinetics and by-products identification*, Journal of Environmental Management 181 (2016), pp. 544–551.

77. H. Gong, and W. Chu, *Determination and toxicity evaluation of the generated products in sulfamethoxazole degradation by UV/CoFe₂O₄/TiO₂*, Journal of Hazardous Materials 314 (2016), pp. 197–203.

78. N. Jallouli, K. Elghniji, O. Hentati, A.R. Ribeiro, A.M.T. Silva, and M. Ksibi, *UV and solar photodegradation of naproxen: TiO₂ catalyst effect, reaction kinetics, products identification and toxicity assessment*, Journal of Hazardous Materials 304 (2016), pp. 329–336.

79. X. Jin, X. Zhou, P. Sun, S. Lin, W. Cao, Z. Li, and W. Liu, *Photocatalytic degradation of norfloxacin using N-doped TiO₂: Optimization, mechanism, identification of intermediates and toxicity evaluation*, Chemosphere 237 (2019), p. 124433.

80. H.-Y. Ma, L. Zhao, L.-H. Guo, H. Zhang, F.-J. Chen, and W.-C. Yu, *Roles of reactive oxygen species (ROS) in the photocatalytic degradation of pentachlorophenol and its main toxic intermediates by TiO₂/UV*, Journal of Hazardous Materials 369 (2019), pp. 719–726.

81. C.J. Pestana, J. Portela Noronha, J. Hui, C. Edwards, H.Q. Nimal Gunaratne, J.T.S. Irvine, P.K.J. Robertson, J. Capelo-Neto, and L.A. Lawton, *Photocatalytic removal of the cyanobacterium Microcystis aeruginosa PCC7813 and four microcystins by TiO₂ coated porous glass beads with UV-LED irradiation*, Science of the Total Environment 745 (2020), pp. 141–154.

82. K. Badvi, and V. Javanbakht, *Enhanced photocatalytic degradation of dye contaminants with TiO₂ immobilized on ZSM-5 zeolite modified with nickel nanoparticles*, Journal of Cleaner Production 280 (2021), pp. 124518.

83. J. Carbajo, A. Bahamonde, and M. Faraldos, *Photocatalyst performance in wastewater treatment applications: Towards the role of TiO₂ properties*, Molecular Catalysis 434 (2017), pp. 167–174.

84. S.L. Gora, R. Liang, Y.N. Zhou, and S.A. Andrews, *Photocatalysis with easily recoverable linear engineered TiO₂ nanomaterials to prevent the formation of disinfection by-products in drinking water*, Journal of Environmental Chemical Engineering 6, no. 1 (2018), pp. 197–207.

85. S.Y. Chun, W.J. Chung, S.S. Kim, J.T. Kim, and S.W. Chang, *Optimization of the TiO₂/Ge composition by the response surface method of photocatalytic degradation under ultraviolet-A irradiation and the toxicity reduction of amoxicillin*, Journal of Industrial and Engineering Chemistry 27 (2015), pp. 291–296.

86. J. Saien, and S. Khezrianjoo, *Degradation of the fungicide carbendazim in aqueous solutions with UV/TiO₂ process: Optimization, kinetics and toxicity studies*, Journal of Hazardous Materials 157, no. 2–3 (2008), pp. 269–276.

87. J. Saien, and A.R. Soleymani, *Degradation and mineralization of Direct Blue 71 in a circulating upflow reactor by UV/TiO₂ process and employing a new method in kinetic study*, Journal of Hazardous Materials 144, no. 1–2 (2007), pp. 506–512.

88. R. Wang, D. Ren, S. Xia, Y. Zhang, and J. Zhao, *Photocatalytic degradation of Bisphenol A (BPA) using immobilized TiO₂ and UV illumination in a horizontal circulating bed photocatalytic reactor (HCBPR)*, Journal of Hazardous Materials 169, no. 1–3 (2009), pp. 926–932.

89. R.G. Saratale, H.S. Noh, J.Y. Song, and D.S. Kim, *influence of parameters on the photocatalytic degradation of phenolic contaminants in wastewater using TiO₂/UV system*, Journal of Environmental Science and Health, Part A 49, no. 13 (2014), pp. 1542–1552.

90. S. Mortazavian, A. Saber, and D.E. James, *Optimization of photocatalytic degradation of Acid Blue 113 and acid Red 88 textile dyes in a UV-C/TiO₂ suspension system: Application of response surface methodology (RSM)*, Catalysts 9, no. 4 (2019), p. 360.

91. N. Jafari, A. Ebrahimi, K. Ebrahimpour, and A. Abdolahnejad, *Optimization and Modeling of Microcystin-LR Degradation by TiO₂ Photocatalyst Using Response Surface Methodology*, Journal of Environmental Health and Sustainable Development 5, no. 3 (2020), pp. 1063–1076.

92. Y. Cao, J. Chen, L. Huang, Y. Wang, Y. Hou, and Y. Lu, *Photocatalytic degradation of chlorfenapyr in aqueous suspension of TiO₂*, Journal of Molecular Catalysis A: Chemical 233, no. 1–2 (2005), pp. 61–66.

93. S. Mozia, M. Tomaszewska, and A.W. Morawski, *Photocatalytic degradation of azo-dye Acid Red 18*, Desalination 185, no. 1–3 (2005), pp. 449–456.

94. V.K. Gupta, R. Jain, A. Mittal, T.A. Saleh, A. Nayak, S. Agarwal, and S. Sikarwar, *Photocatalytic degradation of toxic dye amaranth on TiO₂/UV in aqueous suspensions*, Materials Science and Engineering: C 32, no. 1 (2012), pp. 12–17.

95. A. Achilleos, E. Hapeshi, N.P. Xekoukoulotakis, D. Mantzavinos, and D. Fatta-Kassinos, *Factors affecting diclofenac decomposition in water by UV-A/TiO₂ photocatalysis*, Chemical Engineering Journal 161, no. 1–2 (2010), pp. 53–59.

96. C. Pablos, J. Marugan, R. van Grieken, and E. Serrano, *Emerging micropollutant oxidation during disinfection processes using UV-C, UV-C/H₂O₂, UV-A/TiO₂ and UV-A/TiO₂/H₂O₂*, Water Research 47, no. 3 (2013), pp. 1237–1245.

97. M.R. Eskandarian, H. Choi, M. Fazli, and M.H. Rasoulifard, *Effect of UV-LED wavelengths on direct photolytic and TiO₂ photocatalytic degradation of emerging contaminants in water*, Chemical Engineering Journal 300 (2016), pp. 414–422.

98. E.T. Wahyuni, R. Roto, M. Sabrina, V. Anggraini, N.F. Leswana, and A.C. Vionita, *Photodegradation of detergent anionic surfactant in wastewater using UV/TiO₂/H₂O₂ and UV/Fe²⁺/H₂O₂ processes*, American Journal of Applied Chemistry 4, no. 5 (2016), pp. 174–180.

99. J. Chen, and P. Zhang, *Photodegradation of perfluorooctanoic acid in water under irradiation of 254 nm and 185 nm light by use of persulfate*, Water Science and Technology 54, no. 11–12 (2006), pp. 317–325.

100. R. Hazime, Q.H. Nguyen, C. Ferronato, A. Salvador, F. Jaber, and J-M. Chovelon, *Comparative study of imazalil degradation in three systems: UV/TiO₂, UV/K₂S₂O₈ and UV/TiO₂/K₂S₂O₈*, Applied Catalysis B: Environmental 144 (2014), pp. 286–291.

101. Y. Zhang, Y. Xiao, Y. Zhong, and T-T. Lim, *Comparison of amoxicillin photodegradation in the UV/H₂O₂ and UV/persulfate systems: Reaction kinetics, degradation pathways, and antibacterial activity*, Chemical Engineering Journal 372 (2019), pp. 420–428.

102. N. Boussatha, M. Gilliot, H. Ghoualem, and J. Martin, *Formation of nanogranular ZnO ultrathin films and estimation of their performance for photocatalytic degradation of amoxicillin antibiotic*, Materials Research Bulletin 99 (2018), pp. 485–490.

103. S.M. Saleh, *ZnO nanospheres based simple hydrothermal route for photocatalytic degradation of azo dye*, Spectrochimica Acta Part A: Molecular and Biomolecular Spectroscopy 211 (2019), pp. 141–147.

104. X. Li, J. Zhu, and H. Li, *Comparative study on the mechanism in photocatalytic degradation of different-type organic dyes on SnS₂ and CdS*, Applied Catalysis B: Environmental 123 (2012), pp. 174–181.

105. M.M. Huber, S. Canonica, G-Y. Park, and U. Von Gunten, *Oxidation of pharmaceuticals during ozonation and advanced oxidation processes*, Environmental Science & Technology 37, no. 5 (2003), pp. 1016–1024.

106. F.J. Benitez, F.J. Real, J.L. Acero, and G. Roldan, *Removal of selected pharmaceuticals in waters by photochemical processes*, Journal of Chemical Technology & Biotechnology: International Research in Process, Environmental & Clean Technology 84, no. 8 (2009), pp. 1186–1195.

107. K. Lekkerkerker-Teunissen, M.J. Benotti, S.A. Snyder, and H.C. Van Dijk, *Transformation of atrazine, carbamazepine, diclofenac and sulfamethoxazole by low and medium pressure UV and UV/H₂O₂ treatment*, Separation and Purification Technology 96 (2012), pp. 33–43.

108. S. Bahnmüller, C.H. Loi, K.L. Linge, U. Von Gunten, and S. Canonica, *Degradation rates of benzotriazoles and benzothiazoles under UV-C irradiation and the advanced oxidation process UV/H$_2$O$_2$*, Water Research 74 (2015), pp. 143–154.

109. V.J. Pereira, H.S. Weinberg, K.G. Linden, and P.C. Singer, *UV degradation kinetics and modeling of pharmaceutical compounds in laboratory grade and surface water via direct and indirect photolysis at 254 nm*, Environmental Science & Technology 41, no. 5 (2007), pp. 1682–1688.

110. V.J. Pereira, K.G. Linden, and H.S. Weinberg, *Evaluation of UV irradiation for photolytic and oxidative degradation of pharmaceutical compounds in water*, Water Research 41, no. 19 (2007), pp. 4413–4423.

111. L. Rizzo, C. Manaia, C. Merlin, T. Schwartz, C. Dagot, M.C. Ploy, I. Michael, and D. Fatta-Kassinos, *Urban wastewater treatment plants as hotspots for antibiotic resistant bacteria and genes spread into the environment: A review*, Science of the Total Environment 447 (2013), pp. 345–360.

112. S. Li, S. Zhang, C. Ye, W. Lin, M. Zhang, L. Chen, J. Li, and X. Yu, *Biofilm processes in treating mariculture wastewater may be a reservoir of antibiotic resistance genes*, Marine Pollution Bulletin 118, no. 1–2 (2017), pp. 289–296.

113. P. Li, Y. Wu, Y. He, B. Zhang, Y. Huang, Q. Yuan, and Y. Chen, *Occurrence and fate of antibiotic residues and antibiotic resistance genes in a reservoir with ecological purification facilities for drinking water sources*, Science of The Total Environment 707 (2020), pp. 135276.

114. S.K. Alharbi, and W.E. Price, *Degradation and fate of pharmaceutically active contaminants by advanced oxidation processes*, Current Pollution Reports 3, no. 4 (2017), pp. 268–280.

115. I. Belghadr, G.S. Khorramabadi, H. Godini, and M. Almasian, *The removal of the cefixime antibiotic from aqueous solution using an advanced oxidation process (UV/H$_2$O$_2$)*, Desalination and Water Treatment 55, no. 4 (2015), pp. 1068–1075.

116. L. Rizzo, G. Lofrano, C. Gago, T. Bredneva, P. Iannece, M. Pazos, N. Krasnogorskaya, and M. Carotenuto, *Antibiotic contaminated water treated by photo driven advanced oxidation processes: Ultraviolet/H$_2$O$_2$ vs ultraviolet/peracetic acid*, Journal of Cleaner Production 205 (2018), pp. 67–75.

117. K. Yin, L. Deng, J. Luo, J. Crittenden, C. Liu, Y. Wei, and L. Wang, *Destruction of phenicol antibiotics using the UV/H$_2$O$_2$ process: Kinetics, by-products, toxicity evaluation and trichloromethane formation potential*, Chemical Engineering Journal 351 (2018), pp. 867–877.

118. B.A. Wols, D.J.H. Harmsen, E.F. Beerendonk, and C.H.M. Hofman-Caris, *Predicting pharmaceutical degradation by UV (LP)/H$_2$O$_2$ processes: A kinetic model*, Chemical Engineering Journal 255 (2014), pp. 334–343.

119. B.A. Wols, D.J.H. Harmsen, E.F. Beerendonk, and C.H.M. Hofman-Caris, *Predicting pharmaceutical degradation by UV (MP)/H$_2$O$_2$ processes: A kinetic model*, Chemical Engineering Journal 263 (2015), pp. 336–345.

120. E. Hapeshi, A. Achilleos, M.I. Vasquez, C. Michael, N.P. Xekoukoulotakis, D. Mantzavinos, and D. Kassinos, *Drugs degrading photocatalytically: Kinetics and mechanisms of ofloxacin and atenolol removal on titania suspensions*, Water Research 44, no. 6 (2010), pp. 1737–1746.

121. C. Dai, X. Zhou, Y. Zhang, Y. Duan, Z. Qiang, and T.C. Zhang, *Comparative study of the degradation of carbamazepine in water by advanced oxidation processes*, Environmental Technology 33, no. 10 (2012), pp. 1101–1109.

122. S. Carbonaro, M.N. Sugihara, and T.J. Strathmann, *Continuous-flow photocatalytic treatment of pharmaceutical micropollutants: Activity, inhibition, and deactivation of TiO$_2$ photocatalysts in wastewater effluent*, Applied Catalysis B: Environmental 129 (2013), pp. 1–12.

123. T. Aissani, I. Yahiaoui, F. Boudrahem, S. Ait Chikh, F. Aissani-Benissad, and A. Amrane, *The combination of photocatalysis process (UV/TiO₂ (P25) and UV/ZnO) with activated sludge culture for the degradation of sulfamethazine*, Separation Science and Technology 53, no. 9 (2018), pp. 1423–1433.
124. J. Tian, Y. Leng, Z. Zhao, Y. Xia, Y. Sang, P. Hao, J. Zhan, M. Li, and H. Liu, *Carbon quantum dots/hydrogenated TiO₂ nanobelt heterostructures and their broad spectrum photocatalytic properties under UV, visible, and near-infrared irradiation*, Nano Energy 11 (2015), pp. 419–427.
125. X. Zhang, F. Wu, X.W. Wu, P. Chen, and N. Deng, *Photodegradation of acetaminophen in TiO₂ suspended solution*, Journal of Hazardous Materials 157, no. 2–3 (2008), pp. 300–307.
126. L. Yang, E.Y. Liya, and M.B. Ray, *Degradation of paracetamol in aqueous solutions by TiO₂ photocatalysis*, Water Research 42, no. 13 (2008), pp. 3480–3488.
127. E. Moctezuma, E. Leyva, C.A. Aguilar, R.A. Luna, and C. Montalvo, *Photocatalytic degradation of paracetamol: Intermediates and total reaction mechanism*, Journal of Hazardous Materials 243 (2012), pp. 130–138.
128. J. Chun-Te Lin, M.D.G. de Luna, G.L. Aranzamendez, and M-C. Lu, *Degradations of acetaminophen via a K₂S₂O₈-doped TiO₂ photocatalyst under visible light irradiation*, Chemosphere 155 (2016), pp. 388–394.
129. S. Sarkar, S. Chakraborty, and C. Bhattacharjee, *Photocatalytic degradation of pharmaceutical wastes by alginate supported TiO₂ nanoparticles in packed bed photo reactor (PBPR)*, Ecotoxicology and Environmental Safety 121 (2015), pp. 263–270.
130. R. Janssens, M.B. Cristovao, M.R. Bronze, J.G. Crespo, V.J. Pereira, and P. Luis, *Coupling of nanofiltration and UV, UV/TiO₂ and UV/H₂O₂ processes for the removal of anti-cancer drugs from real secondary wastewater effluent*, Journal of Environmental Chemical Engineering 7, no. 5 (2019), pp. 103351.
131. A.L. Giraldo, G.A. Penuela, R.A. Torres-Palma, N.J. Pino, R.A. Palominos, and H.D. Mansilla, *Degradation of the antibiotic oxolinic acid by photocatalysis with TiO₂ in suspension*, Water Research 44, no. 18 (2010), pp. 5158–5167.
132. L. Rizzo, S. Meric, M. Guida, D. Kassinos, and V. Belgiorno, *Heterogenous photocatalytic degradation kinetics and detoxification of an urban wastewater treatment plant effluent contaminated with pharmaceuticals*, Water Research 43, no. 16 (2009), pp. 4070–4078.
133. R.A. French, A.R. Jacobson, B. Kim, S.L. Isley, R.L. Penn, and P.C. Baveye, *Influence of ionic strength, pH, and cation valence on aggregation kinetics of titanium dioxide nanoparticles*, Environmental Science & Technology 43, no. 5 (2009), pp. 1354–1359.
134. H. Yang, G. Li, T. An, Y. Gao, and J. Fu, *Photocatalytic degradation kinetics and mechanism of environmental pharmaceuticals in aqueous suspension of TiO₂: A case of sulfa drugs*, Catalysis Today 153, no. 3–4 (2010), pp. 200–207.
135. X. Hu, X. Hu, Q. Peng, L. Zhou, X. Tan, L. Jiang, C. Tang et al., *Mechanisms underlying the photocatalytic degradation pathway of ciprofloxacin with heterogeneous TiO₂*, Chemical Engineering Journal 380 (2020), p. 122366.
136. M. Cruz, C. Gomez, C.J. Duran-Valle, L.M. Pastrana-Martinez, J.L. Faria, A.M.T. Silva, M. Faraldos, and A. Bahamonde, *Bare TiO₂ and graphene oxide TiO₂ photocatalysts on the degradation of selected pesticides and influence of the water matrix*, Applied Surface Science 416 (2017), pp. 1013–1021.
137. D. Awfa, M. Ateia, M. Fujii, and C. Yoshimura, *Photocatalytic degradation of organic micropollutants: Inhibition mechanisms by different fractions of natural organic matter*, Water Research 174 (2020), p. 115643.
138. T. Paul, P.L. Miller, and T.J. Strathmann, *Visible-light-mediated TiO₂ photocatalysis of fluoroquinolone antibacterial agents*, Environmental Science & Technology 41, no. 13 (2007), pp. 4720–4727.

139. L. Kovalova, H. Siegrist, U. Von Gunten, J. Eugster, M. Hagenbuch, A. Wittmer, R. Moser, and C.S. McArdell, *Elimination of micropollutants during post-treatment of hospital wastewater with powdered activated carbon, ozone, and UV*, Environmental Science & Technology 47, no. 14 (2013), pp. 7899–7908.

140. C.A. Lutterbeck, M.L. Wilde, E. Baginska, C. Leder, Ê.L. Machado, and K. Kümmerer, *Degradation of 5-FU by means of advanced (photo) oxidation processes: UV/H$_2$O$_2$, UV/Fe^{2+}/H$_2$O$_2$ and UV/TiO$_2$—comparison of transformation products, ready biodegradability and toxicity*, Science of the Total Environment 527 (2015), pp. 232–245.

141. R. Hazime, Q.H. Nguyen, C. Ferronato, T.K.X. Huynh, F. Jaber, and J-M. Chovelon, *Optimization of imazalil removal in the system UV/TiO$_2$/K$_2$S$_2$O$_8$ using a response surface methodology (RSM)*, Applied Catalysis B: Environmental 132 (2013), pp. 519–526.

142. M. Ateia, M.G. Alalm, D. Awfa, M.S. Johnson, and C. Yoshimura, *Modeling the degradation and disinfection of water pollutants by photocatalysts and composites: A critical review*, Science of The Total Environment 698 (2020), p. 134197.

143. Z. Jiang, J. Hu, X. Zhang, Y. Zhao, X. Fan, S. Zhong, H. Zhang, and X. Yu, *A generalized predictive model for TiO$_2$–Catalyzed photodegradation rate constants of water contaminants through artificial neural network*, Environmental Research 187 (2020), p. 109697.

7 Emerging Materials in Advanced Oxidation Processes for Micropollutant Treatment Process

Anup Singh, Jashandeep Singh, Ajay Vasishth,
Ashok Kumar and Shyam Sundar Pattnaik

CONTENTS

7.1 Introduction 133
7.2 Emerging Materials in Microwave-stimulated AOP 134
7.3 Emerging Materials in the Electrochemical Advanced Oxidation Process.... 136
7.4 Emerging Materials in Chemical Advance Oxidation Process for
Micropollutants Treatment 139
7.5 Emerging Materials in Photochemical Advanced Oxidation Processes 142
7.6 Emerging Materials in Sonochemical Advance Oxidation Processes 144
7.7 Other Materials... 147
7.8 Conclusion 149
References... 149

7.1 INTRODUCTION

Advanced oxidation processes (AOPs) have intensively been investigated due to the limitations associated with removing micropollutants by conventional wastewater treatment methods. The primary mechanism of AOP involves the making of hydroxyl radicals. Afterwards, these radicals combine with compounds with an abundance of electrons and help them degrade. Hydroxyl radical (•OH), which has a substantial redox potential (2.8 V) and works as an oxidant to disintegrate refractory molecules, is generated in situ during AOP processes [1]. Wastewater from the distillery, agriculture sectors, industries (like textile, pulp and paper), oilfields, hospitals and slaughterhouses, pathogen and pharmaceutical residue treatment plants, etc., can be treated through AOP implementation. Heavy metals (chromium and arsenic) may also be removed from water by following the same process [2]. AOPs are redox techniques, which include several processes

DOI: 10.1201/9781003247913-7

133

like microwave, ozonation with H_2O_2, UV light, Fenton and similar reactions, semiconductor-stimulated photocatalysis like TiO_2, sonolysis, electrosono-chemical oxidation, and other combinations of them, which form a more efficient mechanism for the degradation of emerging contaminants. AOPs are based on extremely reactive oxygen species (ROS) generation in addition to the main oxidizing agent, such as hydroxyl radicals. Other ROS (e.g., hydroperoxyl radicals, superoxide radical anions, hydrogen peroxide, sulphate radicals) can be used as pre- and post-treatment agents to remove emerging organic and inorganic contaminants [3]. The AOPs encompass all catalytic and non-catalytic processes and differ in the manner by which these oxidizing radicals are produced. The latest advances in analytic approaches have resulted in the detection and persistence level of those micropollutants which have been assumed to be present in the environment. Therefore, the deterioration of the emerging micropollutants in the environment requires novel materials and their combination with hybrid AOPs.

7.2 EMERGING MATERIALS IN MICROWAVE-STIMULATED AOP

Microwave induction has gained popularity among AOPs because of its superior degrading efficiencies, faster reaction times, and uniform and controlled heating. Moreover, progress and improvements are increasing in the sector of microwave reactors [4]. Pollutant degradation increases dramatically through microwave irradiation as microwave (MW) energy may be absorbed using a catalyst, which further develops hotspots to prompt more radicals [5]. The microwave's action excites micropollutant molecules to their upper rotational and vibrational energy levels by which their activity enhances. However, the research community targets developing advanced MW absorbing materials, wherein the interaction of irradiation, including hot spots or polarization, generates heating and nonheating impacts, leading to the formation of pairs of electron and hole or reactive oxygen products. H_2O_2 is mostly activated on the catalyst surface in microwave-aided AOPs. As a consequence, the surface area is crucial to determining catalytic activity. Several nanoparticles have subsequently been developed to enhance catalytic performance.

For example, MWs improved the Fenton-like (FL) system made from CuO/Cu_2O/Cu nanoparticles, and the production of hydroxyl radicals was enhanced for the selective and rapid pesticide (methomyl) treatment in wastewater. These systems were also utilized to boost catalyst activities [6]. Microwave catalytic oxidation degradation (MCOD) for nitrophenols (NPs) wastewater was created using green catalysts such as graphitic carbon nitride (g-C_3N_4)-$Bi_2O_2CO_3$ and SiC-$B_2iO_2CO_3$. This technique is environmentally friendly since the primary substance used is nontoxic. $B_2iO_2CO_3$, g-C_3N_4, and SiC do not contain any heavy metal pollution. g-C_3N_4-$Bi_2O_2CO_3$ and SiC-$B_2iO_2CO_3$ exhibited outstanding catalytic activity for para-NP degradation and mineralization with optimum para-NP degradation ratios of 95.20% and 98.96%, respectively. It appears to be a viable method for efficiently removing additional refractory organic micropollutants from industrial wastewater [7].

Magnetite (Fe_3O_4) is deployed as the most common catalyst in advanced oxidative processes. Rhodamine B (RhB) deterioration was chosen as an ideal reaction to examine MW effects with nano Fe_3O_4. When MW was irradiated under appropriate operational circumstances for AOP applications, Fe_3O_4 nanoparticles were unveiled as the coherent and effective heterogeneous catalyst and displayed outstanding efficiency in the short reusability [8]. The degradation of 4 nitrophenol (4-NP) was accounted for by utilizing a microwave oxidation reaction process based on a new combination between two catalysts: commercially available activated carbon (AC) and Mn_2O_3 nanoparticles. The microwave irradiation AOP, with the modified (AC/Mn_2O_3), is an effective catalyst for the degradation of 4-NP wastewater with a removal rate of up to 99.56 %, equal to 93.5 % total organic removal (TOC) [9].

Wastewater from sectors incorporating cooking, medicines, oil refining, and petrochemicals contains phenol and phenolic compounds. Before disposing of, these compounds should be treated because they are primary pollutants with substantial toxicity even at very low concentrations. Therefore, a nanomaterials combination CuOx/granular activated carbon (GAC) catalyst was prepared and used as a heterogeneous Fenton-like catalyst for phenol degradation in the MW-induced H_2O_2 catalytic oxidation process, achieving a degradation rate of 99.96%, corresponding to 88.6 % chemical oxygen demand (COD) reduction under optimal circumstances with MW power 400 W and CuOx/GAC dosage of 3 g/L, H_2O_2 dosage of 2 mL/L, reaction time 4 min, and pH 4. Figure 7.1 shows the COD removal rate in comparison to different integrated materials.

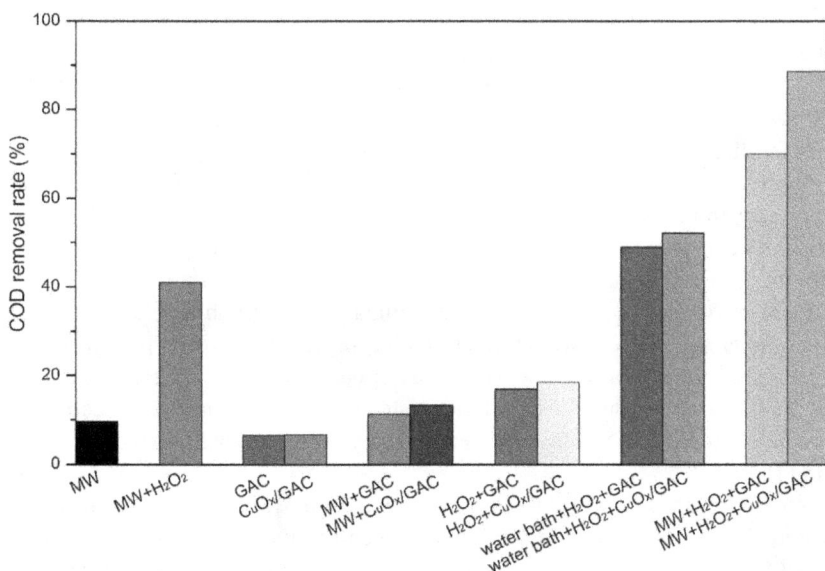

FIGURE 7.1 Comparison of COD elimination by different treatment techniques of phenol wastewater [10].

7.3   EMERGING MATERIALS IN THE ELECTROCHEMICAL ADVANCED OXIDATION PROCESS

Electro-Fenton is the latest headway in the field of electrochemical advanced oxidation processes (AOPs), which has seen much development somewhat recently. The electrochemical advanced oxidation process (EAOP) is an innovative technique for removing micropollutants from wastewater, and it is clean, efficient, and cost-effective [11]. In the electro-Fenton process, heterogeneous catalysis has been widely used.

For the treatment of biologically pretreated coal gasification wastewater, a new catalytic particle electrode (CPE) was designed using a three-dimensional electro-Fenton (3D EF). Pollution from coal gasification wastewater is a potential ecological issue that is exacerbated by the fact that the majority of the pollutants, such as phenolic compounds, nitrogenous heterocyclic compounds (NHCs), polynuclear aromatic hydrocarbons (PAHs), long-chain hydrocarbons, and ammonia, are toxic and difficult to degrade using conventional treatments. Sludge-derived activated carbon-supported iron oxide (SAC-Fe) significantly enhanced treated wastewater's biodegradability and served as catalyst and particle electrodes in a three-dimensional electrochemical reactor. Figure 7.2(a) shows that the TOC removal efficiencies of SAC-Fe in CPEs developed when the underlying pH was raised from 2 to 3 but declined when it was raised from 3 to 7. The reduction in removal capacity, on the other hand, was minimal [12].

In EAOPs for wastewater treatment, pure forms of graphene, for example, monolayer (G_{mono}), graphene multilayer (G_{multi}), and graphene foam (G_{foam}), were used, which provided excellent results. The Fe^{2+} regeneration rate and H_2O_2 production yield were adequate for all types of graphene to support the synthesis of •OH radicals via the Fenton reaction, resulting in the deterioration and mineralization of phenol, which was employed as a model pollutant. With degradation rates of 0.0081 min^{-1} and mineralization rates of 0.0818 h^{-1}, G_{foam} outperformed the competition in terms of H_2O_2 electrocatalytic generation efficiency as well as degradation and mineralization efficacy and proved to be a competitor of three-dimensional carbon-based cathode components like carbon felt (C_{felt}), as C_{felt} is the most widely utilized material in the EF research [13].

The electro-Fenton process was examined using graphite electrodes coated with Fe_2O_3 nanoparticles, with aspects including response time, pH, current density, H_2O_2/Fe^{2+} molar ratios, and volumetric percent of H_2O_2 to alcoholic effluent. Reducing COD from industrial alcoholic wastewater with the improved electro-Fenton approach was up to 66.15 %, as opposed to 51.8 % with the traditional electro-Fenton method [14].

Based on their redox reactivity, ferromagnetic iron oxide Fe_3O_4 nanoparticles have exhibited significant activity in oxidation processes among heterogeneous catalysts. The Fe_3O_4 nanoparticles are abundant and have a significant solubility rate relative to certain other iron oxides. It is the reason for its remarkable features. Furthermore, Fenton reaction performance is improved because of the existence and accessibility of Fe^{2+} in their octahedral structure. Synthesized material was employed purposely as a heterogeneous electro-Fenton catalyst as it can decolorize color index (CI) acid

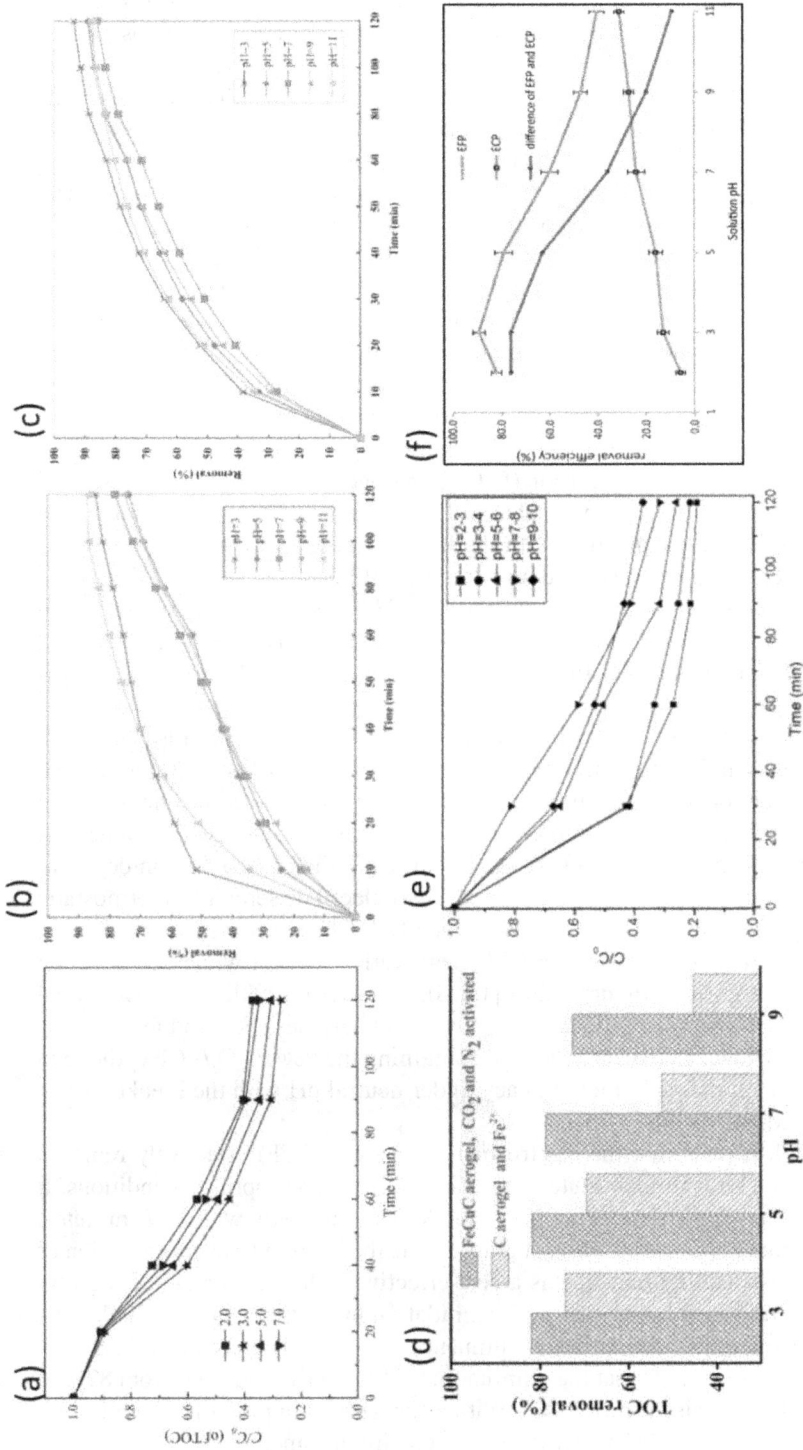

FIGURE 7.2 (a) By changing the pH from 2 to 7, the impact of pH impacting mineralization in the 3D EF using SAC-Fe for CPEs [12], (b) The effect of pH on the percentage of acid blue 92 dye degradation, (c) acid red 14 dye removal rate as a function of pH [15], (d) the impact of pH upon MB removal efficiency in heterogeneous and homogeneous EF systems [17], (e) electro-Fenton process using Fe@Fe$_2$O$_3$/ACF under varying pH [18], and (f) MG degradation in EFP and ECP as a function of solution pH [20].

blue 92 and CI acid red 14, and the results were astounding. Fe_3O_4 as a heterogeneous Fenton reagent has the advantage of being capable of being used across a wide pH range. For both acidic dyes (Figure 7.2(b) and (c)), there is a significant dispute between pH 11 and 3 to get the best decolorization point. When the pH is basic, the stable location of Fe in Fe_3O_4 blocks the production of Fe $(OH)_3$ sludge, which would otherwise occur. Fenton catalyst activity is continuously depleted when this sludge is generated at a pH over 5, decreasing removal efficiency. Even at lower concentrations, the dye removal efficacy was better and quicker than Na_2SO_4 [15].

Solid pyrite was exploited as a heterogeneous catalyst for the formation of hydroxyl radicals in the electro-Fenton process. In this study, it was termed pyrite electro-Fenton. It analyses 4-amino-3-hydroxy-2-p-tolylazo-naphthalene-1-sulfonic acid (AHPS) for oxidative degradation and mineralization. The findings revealed that over 90% mineralization was accomplished for TOC removal efficiency, low cost, and reusability, and no external acidification was required to reach pH 3, which is a substantial improvement over the traditional EF approach [16].

An innovative iron–copper–carbon (FeCuC) aerogel was activated with CO_2 and N_2 and used inside a heterogeneous electro-Fenton technology with superior mineralization effectiveness compared with the homogeneous EF method. Using CO_2- and N_2-activated FeCuC aerogel and a wide range of pH levels, high TOC removal of organic contaminants was obtained. Over a pH range of 3 to 9, the activated FeCuC aerogel had a considerable TOC removal rate (80%), as seen in Figure 7.2(d). In contrast, the homogeneous EF system's degradation rate declined as the pH climbed [17].

If the pH is greater than 4, then the Fenton system's oxidative limit is decreased. Accordingly, it's vital to have the Fenton reaction work effectively at neutral pH. Instead of conventional Fe_2O_3, nanostructured $Fe@Fe_2O_3$ was used for the first time as an oxygen diffusion cathode with activated carbon fiber inside (ACF) in a heterogeneous electro-Fenton oxidation system. To carry out the electro-Fenton degradation of rhodamine B (RhB), researchers used a two-electrode setup in a thermostatic cell split into specific sections. When $Fe@Fe_2O_3/ACF$ is used in the E-Fenton method, considerable changes in the RhB degradation are not witnessed with the variable pH, as shown in Figure 7.2(e) [18]. In comparison with the conventional E-Fenton systems with conventional zero valent iron particles (Fe^0) and ferrous ions (Fe^{2+}), the heterogeneous E-Fenton system containing the $Fe@Fe_2O_3/ACF$ cathode in water exhibited considerably high efficacy under neutral pH with the breakdown of dye pollutant RhB [18, 19].

Recent work reveals that the electro-Fenton process (EFP) effectively removes malachite green (MG) from a diluted aqueous solution under optimal conditions. It was observed that approximately 1000 mg/L MG decomposes within 30 minutes. The most prominent process for MG degradation is the hydroxyl radical oxidation of the compound, indicating that EFP is a cost-effective technique for extracting MG effluent containing such impurities. MG degradation in the ECP and EFP with variable pH was explored to determine the influence of pH at different pH values. It is observed from Figure 7.3(f) that the elimination of MG in EFP went up from 82% to 89.2% when pH was raised from 2 to 3. With a further rise in pH to 11, the elimination of MG dropped to 40.3% under the specified circumstances [20].

7.4 EMERGING MATERIALS IN CHEMICAL ADVANCE OXIDATION PROCESS FOR MICROPOLLUTANTS TREATMENT

Heterogeneous solid catalysts for Fenton-like reactions have been studied extensively for a considerable pH range in the chemical, AOP. This heterogeneous Fenton-like approach prevents iron ions from leaching into the water, resulting in the generation of iron hydroxide. A plethora of nanomaterials was explored with remarkable success due to the huge surface area and porosity leading to enhanced production of •OH radicals.

Fe_3O_4/TiO_2/reduced graphene oxide (RGO) composite was identified as a high-performance Fenton-like catalyst capable of catalyzing the degradation of H_2O_2 in the pH range of 5–9 for the decolorization of MB dye. The major contributor to the appropriate pH change approaching neutral should be TiO_2. When rinsing with a water/ethanol solution, Fe_3O_4/TiO_2/RGO may be effectively regenerated [21].

Salicylic acid (SA), a popular pharmaceutical raw ingredient, has been found in various environmental water sources, posing a serious threat to the atmosphere and public health. Two-dimensional (2D) CeO_2 nanomaterials are gaining popularity due to their exceptional ability to disperse/stabilize beneficial properties and more coordinatively unsaturated cerium (Ce) molecules for producing extra active sites/oxygen vacancies that could boost the effectiveness of Ce-based catalysts [22, 23]. Undersized smaller valence ions induce the mechanism of vacancy compensation as a dopant which significantly improves the quantity of Ce^{3+} on the surface. However, increasing the Fe^{3+} content reduces the amount of Ce^{3+}, which results in the disappearance of the sheet-like morphological characteristics owing to either interstitial doping or an additional amount. The hindrance to the reusability of the 2Fe-CeO_2 catalyst is the significant adsorption of partially decomposed light molecular weight organic acids through the active areas. Nonetheless, the catalytic performance could be recovered with a simple calcination approach [24].

One of the primary contaminants in wastewater effluent, benzoic acid, contributes up to 75% of the COD. Using the physical activation method, 10% Fe-TiO_2 catalysts were placed onto a walnut shell, and activated carbon (AC) was extracted to make Fe-TiO_2/AC (FTAC). Fenton-like oxidation had an efficiency of 40%, whereas catalytic moist air oxidation had an efficiency of 70%, and photocatalytic oxidation had an efficiency of 90%, according to parametric studies. In the existence of the N-FTAC catalyst, the photo-Fenton-like oxidation seemed to have a maximum degradation efficiency of 95%. Carbon, titanium, iron, zinc, and chlorine were found in the N-FTAC catalyst, according to Energy Dispersive X-Ray analysis [25].

Bisphenol A (BPA) is a widely utilized industrial chemical in polycarbonate plastics. Significant amounts of BPA can be released into the natural surroundings and surface waters during manufacturing, processing, and application [26]. There has been considerable interest in innovative, effective, and economically sustainable BPA treatment solutions, as BPA is a persistent organic compound with limited biodegradability. Inorganic contaminants can be degraded using multiferroic visible-light photocatalysts such as $BiFeO_3$ (BFO), which has a low band gap and high chemical stability. Doping the A-site of BFO with magnetic nanoparticles (MNPs) reduces leakage current and enhances the multiferroic photocatalytic activity [27, 28]. When exposed to visible light for 15 minutes, 10% Mn-doped BFO MNPs completely degraded BPA without any kind of metal ions leaching. This is significantly quicker than many other combinations like BFO and H_2O_2 [29].

2,4-Dichlorophenoxyacetic acid is used unrestrictedly due to its availability and ease of usage to manage weeds in the agriculture industry. 2,4-Dichlorophenoxyacetic is extensively mobile and long-lasting in aqueous systems despite being chemically stable and nonbiodegradable [30]. High catalytic performance is obtained through the oxidative disintegration of 2,4-dichlorophenol, zinc-substituted nickel ferrite and 2,4-dichlorophenoxy acetic acid. Oxidant concentration level, pollutant quantity, catalyst dosage, temperature, pollutant to oxidant molar proportion, duration, etc., directly impact the pace of the oxidation process. Researchers found that the catalyst has good structural and compositional stability, with little leaching, indicating a heterogeneous method for the catalytic peroxide oxidation process. Temperature, dopant quantity, and dose were all highly linked with catalyst degradation. Using nickel–zinc ferrite composite materials against the detoxification of chlorinated organic compounds could be an effective approach [31].

The usage of 3D ferrum manganese oxide nanosheets (Fe–Mn–O NSs)/wood carbon (WC) composite in a wastewater micropollutant treatment system was a breakthrough. The wood carbon had lengthy and uneven routes that served as a 3D substrate for the nanosheets and became the main reason for making Fe–Mn–O hybrid nanosheets from it. The defect richness of these thin hybrid nanosheets revealed extra catalytically active areas. The nanosheet structure and characteristics could be successfully adjusted by nano compositing alien-metal ions, increasing Fenton activity and stability for catalytic oxidation. Consequently, this framework considerably improves catalytic performance for gases and wastewater treatment by facilitating the diffusion of contaminants. Figure 7.3(a) illustrates that FeOOH/WC had better Fenton catalytic activity than iron-based catalysts, eliminating 93.0 % tetracycline (TC) in 40 minutes. The system's catalytic performance was further enhanced by light illumination [32].

The high iron percentage in iron mining waste is hypothesized to act as a catalyst in Fenton or Fenton-like reactions. Iron mineral tailings from dams were used as impetuses for wastewater treatment in a constant stream mode to eliminate natural colors utilizing methylene blue as a source of the perspective atom. When Fe^{3+} is present, the breakdown rate of the substrate is accelerated [33, 34]. Reduced calcined semi-sphere pellets (RCSSPs) have undergone a thermochemical treatment which leads to a change in the iron oxide phase from hematite to magnetite; the phase change is analyzed with the help of Mossbauer spectroscopy. The iron ore tailings utilized in this study comprised significant amounts of SiO_2 (46.8%), Fe (32.0%), and Al_2O_3 (3.89%). In a nonstop stream framework utilizing semi-spherical pellets after thermal treatment, i.e., CSSP, and pellets following a thermochemical treatment, i.e., reduced calcined semi-sphere pellets, Fenton-like experiments were carried out for CSSP and RCSSP that exhibited 40% and 49% dye adsorption after 30 minutes, respectively. When an oxidizing agent (H_2O_2) was added to both tests, dye degradation was accelerated, and about 95 and 99% of dye degradation took place across CSSP and RCSSP substrates following 180 minutes of reaction, as illustrated in Figure 7.3(b). There is substantial catalytic activity (approximately 75% dye removal) in the previously

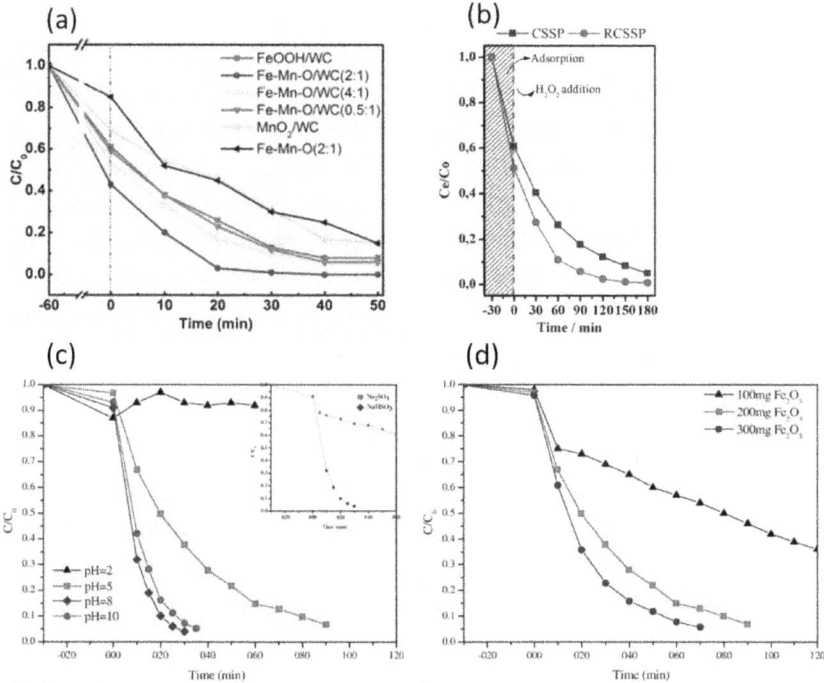

FIGURE 7.3 (a) Assessment of catalysts for TC decomposition using H_2O_2 at room temperature and stating pH = 4 [32], (b) methylene blue dye degraded across the CSSP and RCSSP materials, in the Fenton-like experiments [35], (c) under dark circumstances and at 25 °C, the degradation of orange II by $NaHSO_3$ and Fe_2O_3 at varied initial pH [36], and (d) Orange II deterioration by $NaHSO_3/Fe_2O_3$ with varying $NaHSO_3$ amounts and a starting pH of 5.0 [36].

calcined iron ore tailing dam without the need for extensive modification; however, surface activation can improve this activity. Thus, this methodology proposed an ecologically practical choice for treating natural defiled wastewater in consistent response frameworks utilizing catalytic agents produced during iron metal tailings from tailings dams [35].

Orange II dye is widely used in the textile sector and is nonbiodegradable; thus, for the degradation of this nonbiodegradable orange II dye (compound of azo-dye), a unique Fenton-like catalytic system based on Fe_2O_3 and $NaHSO_3$ was developed. The experiment was performed for a variety of pH values ranging from 2 to 10. Figure 7.3(c) revealed that the degradation efficiency is high at an initial pH of 8. The ratio of $NaHSO_3$ was also varied, which demonstrated that raising the concentration of $NaHSO_3$ from 25 mM to 100 mM under the same circumstances while maintaining Fe_2O_3 constant at 200 mg enhanced the degrading efficiency of orange II from 75% to 90%, as observed in Figure 7.3(d). Since $NaHSO_3$ is a source of sulfate radicals, its growing study stimulates the development of more sulfate-oriented revolutionary materials, consequently enhancing the decomposition of nonbiodegradable dyes. Therefore, this Fenton-like framework may be utilized effectively [36].

Because of its outstanding characteristics, graphene is gaining popularity as a carbon catalyst. Although graphene has high catalytic efficiency, it produces a lot of complicated intermediates during the phenol reaction. A continuous fixed-bed reactor, on the other hand, might overcome these issues [37]. Despite this, adding graphene powder to a process might significantly increase the catalyst's bed pressure drop, reducing reaction efficiency [38]. In order to resolve this problem, graphene was coated on the surface of paper-like sintered stainless-steel fibers (PSSF) to develop such a novel microfiber composite solely with a 3D network structure. Chemical vapor deposition (CVD) deposits the graphene layer directly on the PSSF. The catalytic behavior of the PSSF-monolayer graphene film (Gr) composite catalyst is then examined. Afterwards, the influence of feed flow rate, operating temperature, and long-term stability are investigated. The PSSF offered support and substantial porosity, which may boost mass and transfer of heat, thereby improving the efficiency of catalytic performance in a continuous fixed-bed framework.

Under optimum circumstances, the PSSF-Gr catalyst degraded all of the phenol in 72 hours, and the TOC conversion was excellent [39]. Researchers are now investigating ways to reduce pollutants in water, using the catalytic wet peroxide oxidation (CWPO) method combined with the ease-of-use graphene-based nanopowder catalyst free of metals. The catalytic behavior, from the nanopowders to graphene nanoplatelets (GNPs) and GO, is examined, and the most effective 3D printing of pristine porous graphene monoliths graphene-based nanostructure is chosen and then utilized in CWPO procedures. The findings demonstrate that the 3D graphene-based structures are catalytically active. However, they deactivate gradually owing to organic matter deposition. Thermal treatments can regenerate graphene-based monoliths, allowing them to be used on the industrial scale via reaction-thermal regeneration cycles [40].

Catechol and reactive red 195 (RR195) were debased in water, embracing another $CuFeO_2/Al_2O_3$ catalyst set up for the elimination of anionic toxins in CWPO AOP. At pH 3, this novel nanocomposite eliminated catechol and RR195 with 96.46% and 99.75% adequacy separately. Further, catechol and reactive red 195 (RR195) were degraded in water by adopting a new $CuFeO_2/Al_2O_3$ catalyst to eliminate and degrade anionic contaminants in CWPO. At pH 3, this novel nanocomposite removed catechol and RR195 with 96.46% and 99.75% effectiveness, respectively. Numerous reaction media, such as binary pollutant systems and wastewaters, showed promising results with $CuFeO_2/Al_2O_3$. Even after five cycles, the catalyst remained stable and leached just a tiny amount of material. It showed promise as a nanocomposite to remove and degrade anionic contaminants in CWPO [41].

7.5 EMERGING MATERIALS IN PHOTOCHEMICAL ADVANCED OXIDATION PROCESSES

For generating transitory species, the photochemical AOPs employ light energy. They are said to be smooth, reasonably cheap, easy, and usually more efficient than chemical AOPs. For micropollutant degradation, many semiconductors are devised along with ultraviolet (UV) (as semiconductor photocatalysis). There are materials with high-explicit surface regions, high adsorption capacity, and attractive magnetic properties which can act as good sorbents for micropollutant reduction reactions [42].

A magnetic nanocomposite $FeNi_3$@SiO_2@ZnO was developed to remove tamoxifen (TMX) from wastewater. TMX shows high bacterial biodegradation tolerance and has small molecules, making it difficult to eliminate by traditional wastewater treatment methods [43, 44]. But the complete TMX-degradation became possible when simulated sunlight rays were applied under streamlined states of pH 7 with reaction response length 60 min, beginning TMX quantity 10 mg/L, and $FeNi_3$@SiO_2@ZnO dose 0.01 g/L. Moreover, only mild chemicals were the by-products of the TMX after the photocatalytic treatment process [45]. Another work uncovered that the photocatalytic cycle, which utilizes $FeNi_3$@SiO_2@ZnO MNC as a catalyst and is illuminated with UV light, can be proposed as a successful technique for 100% expulsion of TMX from wastewater in an efficient, safe, and harmless way [46].

It is notable that numerous pharmaceutical mixtures destructively affect human lives, creatures, and sea-going life and contaminate the climate, particularly water bodies. These pharmaceutical mixtures may break down or blend in different substances in the climate to make harmful, cancer-causing or mutagenic side effects [47, 48]. Nanocomposite nickel ferrite/chitosan/bismuth (III) is utilized in combination with solar illuminations to remove medicinal chemicals such as metronidazole (MTZ). Bismuth (III) oxy-iodide (BiOI) is chemically stable and has a confined bandgap range (1.7–1.9 eV). It proved itself to be an effective photocatalyst in visible light in comparison to ZnO and TiO_2 despite being quite recent material. Since magnetic nanoparticles are integrated with catalytic materials, hence effective catalysts with the best treatment and post-treatment recovery characteristics may be developed. As per the outcomes, this catalyst has appropriate capacities, surface morphology, edges, and molecule size. A full MTZ debasement was accomplished at the operating parameters pH 7, $FeNi_3$/chitosan/BiOI 0.04 g/L, 20 mg/L MTZ, and the 200 minutes reaction response period. The impetus can again be utilized multiple times with a decrease in the evacuation pace of only 7%, demonstrating itself as a dependable photocatalyst [49].

A new magnetic photocatalytic nano-system made of Ag-$CuFe_2O_4$@WO_3 was tested for its photodegradation proficiency for dynamic drug toxins gemfibrozil (GEM) and TMX as well as the effects of ultraviolet (UV) light exposure. The effect of parameters like pH, initial drug concentration, photocatalyst dosage, H_2O_2 dosage, photocatalyst stability and reuse, and antibacterial properties has been studied. The photocatalyst has a huge potential for eliminating drug contaminations and microbes from water. The catalyst seems to have a promising future for the filtration of multi-micropollutant water streams [50].

Carbon nanomaterials (CNTs) are viewed as a promising candidate due to their superb electrical conductance and large surface area, among other progressed nanomaterials [51]. For filtration applications, they can easily be molded into porous 3D networks. The CNT can act as excellent in-situ catalysts for H_2O_2 generation through oxygen-reducing reactions (ORRs) [52]. Carbon nanotubes and composites of carbon nanotubes are effective materials for photocatalysis.

The TiO_2/CNT and Fe/CNT nanocomposites were utilized recently in photocatalytic AOP and emerged as excellent catalysts for sulfamethoxazole (SMX) degradation. The joined effect of advanced oxidation cycles and catalytic agents was invaluable for sulfamethoxazole breakdown. The results uncovered that the CNT and Fe/CNT impetuses are the supreme among all. Photocatalytic ozonation within sight of CNT eliminated all SMX quickly and showed huge mineralization, wiping out 93% of TOC following 3 hours of response. Photocatalytic ozonation with H_2O_2 within sight of Fe/CNT completely decomposed SMX following 30 minutes of response. They showed critical enhancements in TOC and poisonousness expulsion, accomplishing total mineralization and almost complete harmfulness evacuation following 3 hours of response time [53].

Single-wall carbon nanotube (SWNT) flexible membranes (or bucky paper, BP) were produced and evaluated for application in AOPs as effective substrates for chemical vapor deposition with thin tungsten trioxide layers (WO_3), which produce WO_3/BP composite membranes. In comparison to an analogous homogeneous WO_3-based reactor, the photocatalytic efficiency of the WO_3/BP composite membrane was evaluated on model pollutants (methylene blue, indigo carmine, and diclofenac sodium). Results allowed heterogeneous photo-catalytical processes with effortless catalyst recovery and high reusability [54]. In addition, because of the facilitated electron transfer between carbon nanostructures and catalyst nanoparticles and the reduced recombination between electrons and holes, WO_3/BP provided higher photo-catalytical efficiency [55].

7.6   EMERGING MATERIALS IN SONOCHEMICAL ADVANCE OXIDATION PROCESSES

Sonochemical advance oxidation processes (SAOPs) are an effective degrading technique of refractory pollutants that rely on free radicals (•OH and •H), which are produced by the pyrolysis of water under high intensity and pressure [56]. In sono-hybrid systems, the degradation rate of different micropollutants is greater than in traditional procedures. Ultrasound (US) oxidation and US combinations with different additives provide an important branch of AOPs. The pollutant-degrading procedures developed in the US entail using ultrasonic waves in the water to create reactive radicals and reactive zone by cavitation, resulting in the creation of microbubbles, which in turn produce high pressure and temperature [57].

The potential uses of developing persulfate-based AOPs in micropollutants' degradation or wastewater treatment have captivated the research community's interest for the last decade. In paper printing and material industries, micropollutants of azo colors are often encountered. Azo dyes are biologically resistant to oxidize. The decontamination by persulphate (PS) actuated with Fe^0 aggregate particles through PS and Fe^0 framework for azo color, CI direct red 23 (DR23), was obtained. The efficiency of discoloration has been further enhanced by employing other compositions such as (PS/Fe^0/US) or heat (PS/Fe^0/55 °C). The combined US/heat framework was utilized as a progressive approach in the PS oxidation system. The Fe^0 particles have proven to be an incentive for decolorization in the PS oxidation process. The rate for color depletion in PS/Fe^0/US was twice that without the US in the framework, and at

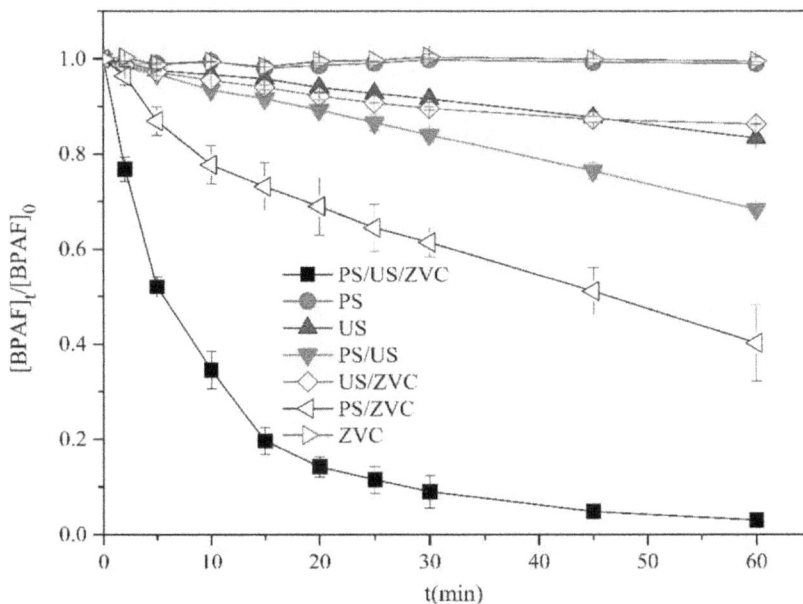

FIGURE 7.4 PS/ZVC/US shows better degradation of BPAF in comparison to other systems for initial pH 4 [59].

higher temperatures, DR23 depletion was accelerated even more [58]. According to a separate investigation of ultrasound activation of the PS/zero-valent copper (ZVC) framework for the debasement of bisphenol AF (BPAF), the sono-advanced PS/ZVC may be utilized as a compelling technique for the removal of BPAF via an oxidation process. It was assessed that BPAF clearance in a PS/ZVC/US system with expulsion adequacy up to 97.0% is impressively more effective at pH 4 than in the PS/ZVC process or other systems, as demonstrated in Figure 7.4 [59].

The $Co_3O_4/NiCo_2O_4$ derivatives of zeolitic imidazolate frame-67 (ZIF-67) were manufactured and used in the form of catalyst for organic synthetic compound bisphenol A (BPA) removal by peroxydisulfate (PDS) activation. The findings indicated the higher effectiveness of $Co_3O_4/NiCo_2O_4$ double shell nanocage (DSNCs) for BPA removal, which surpassed the efficiencies of nanocages (NC) Co_3O_4 and $NiCo_2O_4$. The PDS coupled with the inner and the outer shells of Co_3O_4 generates complex surfaces, and $NiCo_2O_4$ induces electron abstraction to cause nonradical oxidation of BPA. Furthermore, the double hollow structure contributed to the immediate concentration of reactants within the $Co_3O_4/NiCo_2O_4$ DSNC empty areas, thereby boosting BPA breakdown. This showed the excellent application possibility of improved oxidation technology based on persulfate-generated nanomaterials from a metal-organic framework (MOF) with a hollow structure [60].

Periodate (PI) is an emerging oxidant for micropollutant degradation treatment that has demonstrated its potential for reducing organic contaminants such as triethanolamine, 4-chlorophenol, and dyes, as well as lowering COD in industrial effluent. A variety of reactive species has been demonstrated to occur during the photoactivation of periodates, for example, radical OH, periodylic radical ($IO_3\bullet$), radical iodide ($IO_4\bullet$), peroxide hydrogen (H_2O_2), atomic oxygen O (3 P), and ozone [61, 62]. In addition, the breakdown of toxic chemical compounds in UV/TiO_2 systems has been demonstrated to improve at wavelengths approaching that of sunlight [63]. In a study on the degeneration of brilliant blue R (BBR), the combined impact of high frequency ultrasound and periodates has been explored. The results showed an efficient and quick procedure for treating bio-pollutants such as dyes under the union of high frequency ultrasound with periodate. Adding periodate significantly increased BBR's sono-chemical deterioration, and the reaction rate with 10 mM periodate was 2.4 times quicker than with ultrasound [64].

Another study has assessed periodate-sonochemical treatment efficacy for perfluorooctanoic acid (PFOA) breakdown. The undesirable prevalence of perfluorooctanoic acid PFOA in all-around surroundings caused rising concerns. The PFOA is very stable to chemical and thermal changes and is regarded in the natural environment as being nondegradable [65–67]. Periodate is a well-known oxidant that breaks down into highly reactive intermedia via photolysis or sonolysis with a potential of +1.60 V. The study revealed that POAF contamination in wastewater is treated efficiently and rapidly in the PI + US system. After 120 minutes of ultrasound, 96.5% of PFOA was removed with 95.7% defluorination using 45 mM PI, which was faster than US only. Further degradation and defluorination rates found in diverse compositions such as ultrasound periodate, ultrasound, ultrasound/tert-butyl alcohol (t-BuOH) and periodate/ultrasound/t-BuOH are shown in Figure 7.5 [68].

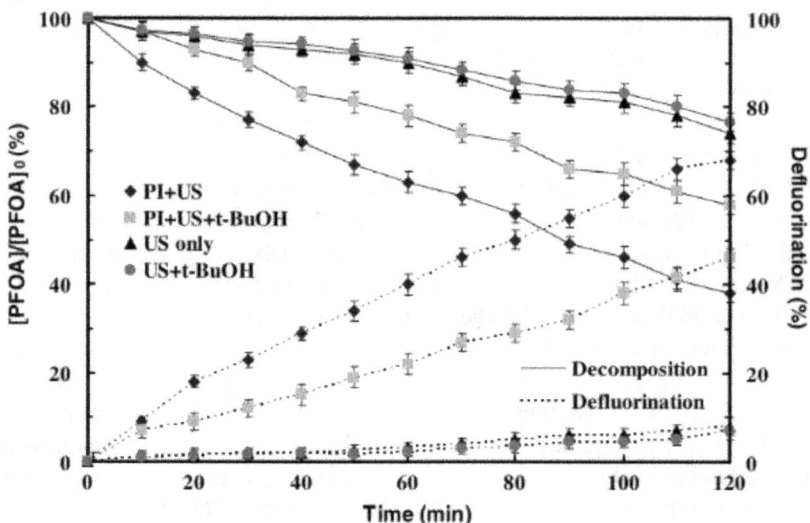

FIGURE 7.5   PFOA decomposition and defluorination efficiency with and without t-BuOH [68].

The new approach used the ultrasound-aided ZVC nanoparticles by extracting biodegradable hibiscus rosa-sinensis abstract as a reductive and stabilization material. The research community has recently focused on ZVC because of its strong mediator nature, controllable size and shape, and simplicity of operation [69, 70]. Electron donors are activated by zero-valent metals and are involved in the deterioration process by generating reactive OH and O_2 radicals [71]. However, aggregation of the reactive molecules on the nanoparticle surface reduces the radical efficiency of degradation/mineralization [72, 73]. Therefore, ultrasound has been utilized to enhance the surface response by cleaning the top catalyst area and generating new active sites. According to the catalyst analysis, incorporating nanoparticles ZVC and UV light during sound analytics increases the rates of reactions due to the production of OH radicals. As a result, approximately 91.3% and 93.2% of the pharmaceutical degradation efficacy were observed, and different operational factors, including a catalyst dose, starting solution pH-dynamics, and oxidant concentration, were involved [74].

7.7   OTHER MATERIALS

The transition metal carbide, nitride, or boride with the general formula $M_{n+1}X_nT_x$, where T belongs to the surface functionality terminated group; X is nitrogen, carbon, or boron; and M belongs to the transition metal, is called an MXene. MXenes are ready-to-use revolutionary adsorbents to remove cationic dyes from wastewater because of their vast area, layered structure, and negatively charged pore volume [75, 76]. Alkali-treated- and polymer functionalized-MXenes have also been reported for dye removal [77]. MXenes may be efficiently integrated into complex chemicals, such as urea, dimethyl sulfoxide, isopropylamine, hydroxide tetrabutylammonium, choline hydroxide, n-butylamine, and cationic dyes, as well as other two-dimensional materials because of their molecular structure [78]. Hence, MXenes and their composites are suggested as effective adsorbent materials for treating wastewater micropollutants. In recent studies, many organic dyes have been mentioned to be rapidly degraded by Ti_3C_2 MXene and Fe_3O_4-based magnetic nanocomposites. Methylene blue (MB), congo red (CR), methyl orange (MO), and rhodamine B (RB) have recorded degradation efficiencies of 96.00%, 31.78%, 81.00%, and 92.03% within 1 minute, respectively [79]. Multiple important aspects, such as catalyst concentration, pH, temperature and H_2O_2 concentration, may influence degradation efficiency. As a result of these determinant factors, the degradation efficiency for the heterogeneous Fenton reaction by the same catalyst for MB has enhanced in 1 minute as the pH value increases. When Ti_3C_2 MXene nanocomposite was tested for the first time, the first cycle's degradation efficiency for MB was 100% in 6 minutes and 94% after five cycles, which shows its high reusability potential [80]. Another research has evaluated the adsorbent effectiveness of delaminated Ti_3C_2-MXenes for six organic dyes in the water medium with different pH levels and ionic strengths, such as MB and methyl violet (MV), CR, methyl red (MR), MO, and orange G (OG). The electrostatic dye–MXene interactions strongly impacted dye adsorption on MXenes. The feasibility of using these Ti_3C_2-MNPs in AOPs for organic dye degradation opens the door to further work on MXenes-based versatile composites, which exhibit

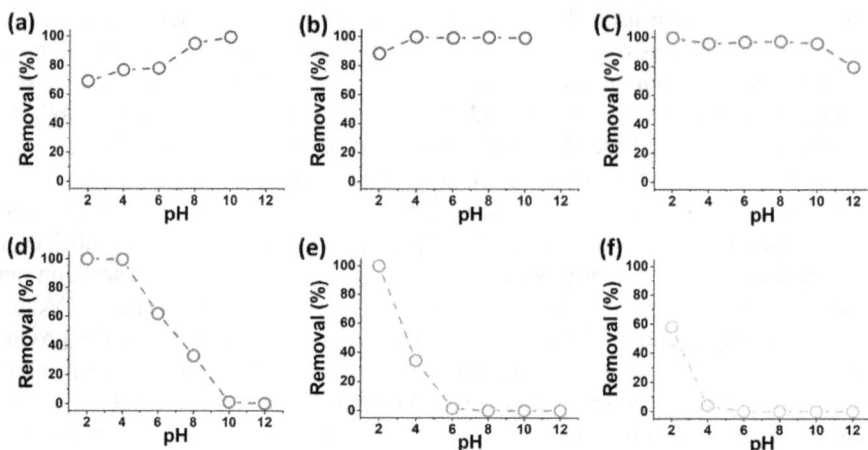

FIGURE 7.6 pH effects on the efficiency of dye adsorption using MXene in the pH range 2–12 for 1 hour for the six organic dyes: (a) MB, (b) MV, (c) CR, (d) MR, (e) MO, and (f) OG [81].

exceptionally significant AOP performance for micropollutant removal. Figure 7.6 shows pH effects on dye removal performance of MXene in the pH range of 2–12. Discoloration and precipitation occur for MB when they are treated at pH 12 [81].

A facile calcination technology for improved photocatalytic decomposing efficiency was used to successfully graze rice-like crust ZnO onto $Ti_3C_2T_x$ MXene. The optimized $ZnO/Ti_3C_2T_x$ hybrid structure's photocatalytic efficiency was three times greater at room temperature than those of pure ZnO under solar lighting. Its excellent recyclability has also been shown for rhodamine B (RhB) and methyl orange (MO), which is three times greater than pristine ZnO under solar light. The efficiencies of MO and RhB are estimated to be 99.7% (in 50 minutes) and 99.8% (in 70 minutes), respectively, for the best hybrid structure. The experimental results are therefore expected to expand the application areas of MXene for photocatalytic micropollutant treatment [82].

A new single-step chemical co-precipitation technique with ammonium bifluoride has been used to manufacture MXene. The prepared magnetic titanium carbide ($Ti_3C_2T_x$) was analyzed for the photocatalytic deterioration of diclofenac (DCF) by UV/chlorine. To evaluate its efficiency, it was then compared with other treatment strategies, such as chlorine, UV, H_2O_2, UV, and H_2O_2/UV. The effects of operating parameters like magnetic dosage of $Ti_3C_2T_x$, pH, and initial chlorine concentration for DCF degradation were assessed, and 100% degradation was observed with the magnetic material and UV/chlorine process within 30 min. The obtained $Ti_3C_2T_x$ photocatalyst is a prominent material for DCF removal in UV/chlorine AOP [83].

APOs such as UV photo-catalytical degradation, H_2O_2-assisted photocatalytic degradation, H_2O_2 catalytic oxidation, and adsorption process were evaluated for the degradation of direct red 80 and MB (anionic and cationic azo dyes, respectively) by new nanocomposites bentonite/TiO_2/Ag. After comparing the color degradation efficiency of several approaches, it was found to be the highest for hydrogen-peroxide-stimulated photocatalytic degradation followed by UV photocatalytic process. The H_2O_2 catalytic oxidation process attained the least degradation. The advanced oxidation method for 0.03 g of the nanocomposite with 0.25% silver component resulted in the greatest dye removal capabilities of 77 and 100% for direct red 80 and MB [84].

The attempts were made to assess kinetic and mechanistic details of peracetic acid (PAA) reactions and the potential for micropollutant degradation of this novel AOP with Fe (II). This study showed that PAA reactivity to Fe (II) is considerably greater (>650 times) than H_2O_2. Therefore, PAA may readily surpass H_2O_2 in the Fe (II) oxidation process. Radicals focused on carbon (CH_3C (O)O•, CH_3C(O)•, and •CH_3) were probably significant in the first stage for micropollutant degradation, and OH was crucial in the second stage of the reaction. The work shows that Fe(II)/PAA is a viable technology for AOPs and significantly increases insight into the organic PAA and Fe(II) response [85].

7.8   CONCLUSION

The rapid pace of industrialization has led to the emergence of hazardous organic and inorganic micropollutants into the environment. Among the remediation methods, AOPs have been gaining progressive attention as one of the most promising technical methods for effective micropollutant treatment. Recently, many materials for AOPs have been under extensive study for the degradation of micropollutants present in the environment, yet to promote the use of AOP in practical systems, issues such as biological safety evaluations, scaling-up studies, and market feasibility studies need to be considered.

Many emerging materials are used in AOP, out of which nanocomposites of carbon and MXene have been assessed as the most promising advanced materials because of their extraordinary properties like long-term stability, remarkable catalytic performance, and better degradation efficiency. However, further research is needed before these advanced materials can be used in practical applications. The integration of emerging materials with hybrid AOPs is also required to be explored to obtain an ultimate AOP with easy equipment setup, ease of operation, high oxidation ability, quicker response rate, and complete pollutant elimination of organic and inorganic pollutants in a unified framework. In addition, integrating advanced materials with hybrid processes can prevent excessive expenditures in terms of time, equipment installations, and reagents.

REFERENCES

 1. Jia-Qian Jiang, Z. Zhou and V.K. Sharma, "Occurrence, transportation, monitoring and treatment of emerging micropollutants in wastewater—A review from global". *Microchemical Journal*, 110, 292–300 (2013). Doi: 10.1016/j.microc.2013.04.014.
 2. B.A. Wols and C.H.M. Hofman-Caris, "Review of photochemical reaction constants of organic micropollutants required for UV advanced oxidation processes in water", *Water Research*, 46, 2815–2827 (2012). Doi.org/10.1016/j.watres.2012.03.036.

3. Petros Kokkinos, Danae Venieri and Dionissios Mantzavinos, "Advanced oxidation processes for water and wastewater viral disinfection. A systematic review", *Food and Environmental Virology*, 13, 283–302 (2021). Doi: 0.1007/s12560-021-09481-1.

4. Priyanshu Verma and Sujoy Kumar Samanta, "Microwave-enhanced advanced oxidation processes for the degradation of dyes in water", *Environmental Chemistry Letters*, 16, 969–1007 (2018). Doi: 10.1007/s10311-018-0739-2.

5. Yin Wang, Yun Wang, Lan Yu, Jiayuan Wang, Baobao Du and Xiaodong Zhang, "Enhanced catalytic activity of templated-double perovskite with 3D network structure for salicylic acid degradation under microwave irradiation: Insight into the catalytic mechanism", *Chemical Engineering Journal*, 386, 115–128 (2019). Doi.org/10.1016/j.cej.2019.02.174.

6. Tony Maha and Mansour Shehab, "Microwave-assisted catalytic oxidation of methomyl pesticide by $Cu/Cu_2O/CuO$ hybrid nanoparticles as a Fenton-like source", *International Journal of Environmental Science and Technology*, 17, 161–174 (2020). Doi. 10.1007/s13762–019–02436-x.

7. Qiu Yin and Zhou Jicheng, "Highly effective and green microwave catalytic oxidation degradation of nitrophenols over $Bi_2O_2CO_3$ based composites without extra chemical additives", *Chemosphere*, 214, 319–229 (2019). Doi.10.1016/j.chemosphere.2018.09.125.

8. Yasmin Vieira, Siara Silvestri, Jandira Leichtweis, Sérgio Luiz Jahn, Érico Marlon de Moraes Flore, Guilherme Luiz Dotto and Edson Luiz Foletto, "New insights into the mechanism of heterogeneous activation of nano–magnetite by microwave irradiation for use as Fenton catalyst", *Journal of Environmental Chemical Engineering*, 8, 103787 (2020). Doi.org/10.1016/j.jece.2020.103787.

9. Cheng Yin, Jinjun Cai, Lingfei Gao, Jingya Yin and Jicheng Zhou, "Highly efficient degradation of 4-nitrophenol over the catalyst of Mn_2O_3/AC by microwave catalytic oxidation degradation method", *Journal of Hazardous Materials*, 305, 515–20 (2016). Doi.org/10.1016/j.jhazmat.2015.11.028.

10. Zailiang Liu, Hailing Meng, Hui Zhang, Jiashun Cao, Ke Zhou and Jianjun Lian, "Highly efficient degradation of phenol wastewater by microwave induced H_2O_2-CuO_x/GAC catalytic oxidation process", *Separation and Purification Technology*, 193, 45–57 (2018). Doi.org/10.1016/j.seppur.2017.11.010.

11. Elin Marlina and P. Purwanto, "Electro-Fenton for industrial wastewater treatment: A review", *E3S Web of Conferences*, 125, 03003 (2019). Doi.10.1051/e3sconf/201912503003.

12. Baolin Hou, Hongjun Han, Shengyong Jia, Haifeng Zhuang, Peng Xu and Kun Li, "Three-dimensional heterogeneous electro-Fenton oxidation of biologically pretreated coal gasification wastewater using sludge derived carbon as catalytic particle electrodes and catalyst", *Journal of the Taiwan Institute of Chemical Engineers*, 60, 1–9 (2015). Doi.10.1016/j.jtice.2015.10.032.

13. Emmanuel Mousset, Zuxin Wang, Joshua Hammaker and Olivier Lefebvre, "Physico-chemical properties of pristine graphene and its performance as electrode material for electro-Fenton treatment of wastewater", *Electrochimica Acta*, 214, 1–42 (2016). Doi.10.1016/j.electacta.2016.08.002.

14. Reza Davarnejad and Jamal Azizi, "Alcoholic wastewater treatment using electro-Fenton technique modified by Fe_2O_3 nanoparticles", *Journal of Environmental Chemical Engineering*, 4, 2342–2349 (2016). Doi.10.1016/j.jece.2016.04.009.

15. Zahra Es'haghzade, Elmira Pajootan, Hajir Bahrami and Mokhtar Arami, "Facile synthesis of Fe_3O_4 nanoparticles via aqueous based electro chemical route for heterogeneous electro-Fenton removal of azo dyes", *Journal of the Taiwan Institute of Chemical Engineers*, 71, 1–15 (2016). Doi.10.1016/j.jtice.2016.11.015.

16. Lazhar Labiadh, Mehmet A. Oturan, Marco Panizza, Nawfel Ben Hamadi and Salah Ammar, "Complete removal of AHPS synthetic dye from water using new electro-fenton oxidation catalyzed by natural pyrite as heterogeneous catalyst", 297, 34–41 (2015). Doi.10.1016/j.jhazmat.2015.04.062.

17. Hongying Zhao, Lin Qian, Xiaohong Guan, Deli Wu and Guohua Zhao, "Continuous bulk FeCuC aerogel with ultradispersed metal nanoparticles: An efficient 3D heterogeneous electro-fenton cathode over a wide range of pH 3–9", *Environmental Science & Technology*, 50, 1–33 (2016). Doi.10.1021/acs.est.6b00265.

18. Jinpo Li, Zhihui Ai and Lizhi Zhang, "Design of a neutral electro-Fenton system with Fe@Fe$_2$O$_3$/ACF composite cathode for wastewater treatment", *Journal of Hazardous Materials*, 164, 18–25 (2009). Doi: 10.1016/j.jhazmat.2008.07.109.

19. George P. Anipsitakis and Dionysios D. Dionysiou, "Degradation of organic contaminants in water with sulfate radicals generated by the conjunction of peroxymonosulfate with cobalt", *Environmental Science & Technology*, 37, 4790–4797 (2003). Doi: 10.1021/es026379.

20. Maryam Teymori, Hassan Khorsandi, Ali Ahmad Aghapour, Seyed Javad Jafari and Ramin Maleki, "Electro-Fenton method for the removal of malachite green: Effect of operational parameters", *Applied Water Science*, 10, 1–14 (2020).

21. Chao Sun, Sheng-Tao Yang, Zhenjie Gao, Shengnan Yang, Ailimire Yilihamu, Qiang Ma, Ru-Song Zhao and Fumin Xue, "Fe$_3$O$_4$/TiO$_2$/reduced graphene oxide composites as highly efficient Fenton-like catalyst for the decoloration of methylene blue", *Materials Chemistry and Physics*, 223, 751–757 (2019). Doi:10.1016/j.matchemphys.2018.11.056.

22. Minkee Choi, Kyungsu Na, Jeongnam Kim, Yasuhiro Sakamoto, Osamu Terasaki and Ryong Ryoo, "Stable single-unit-cell nanosheets of zeolite MFI as active and long-lived catalysts", *Nature*, 461, 246–249 (2009). Doi: 10.1038/nature08288.

23. Chaorong Li, Q.T. Sun, Nianpeng Lu, B.Y. Chen and W.J. Dong, "A facile route for the fabrication of CeO$_2$ nanosheets via controlling the morphology of CeOHCO$_3$ precursors", *Journal of Crystal Growth*, 343, 1, 95–100 (2012). Doi:10.1016/j.jcrysgro.2012.01.041.

24. Wei Wang, Qin Zhu, Feng Qin, Qiguang Dai and Xingyi Wang, "Fe doped CeO$_2$ nanosheets as Fenton-like heterogeneous catalysts for degradation of salicylic acid", *Chemical Engineering Journal*, 333, 226–239 (2018). Doi: 10.1016/j.cej.2017.08.065.

25. Gülen Tekin, Gülin Ersöz and Süheyda Atalay, "Degradation of benzoic acid by advanced oxidation processes in the presence of Fe or Fe-TiO$_2$ loaded activated carbon derived from walnut shells: A comparative study", *Journal of Environmental Chemical Engineering*, 6, 1745–1759 (2018). Doi: 10.1016/j.jece.2018.01.067.

26. Fang-bai Li, Xiang Zhong Li, Xiaomin Li, Tongxu Liu and J. Dong, "Heterogeneous photodegradation of bisphenol A with iron oxides and oxalate in aqueous solution", *Journal of Colloid and Interface Science*, 311, 481–490 (2007). Doi: 10.1016/j.jcis.2007.03.067.

27. Tayyebeh Soltani and Mohammad H. Entezar, "Photolysis and photocatalysis of methylene blue by ferrite bismuth nanoparticles under sunlight irradiation", *Journal of Molecular Catalysis A: Chemical*, 377, 197–203 (2013). Doi: 10.1016/j.molcata.2013.05.004.

28. Rajasree Das and Kalyan Mandal, "Magnetic, ferroelectric and magnetoelectric properties of Ba-doped BiFeO$_3$", *Journal of Magnetism and Magnetic Materials*, 324, 1913–1918 (2012). Doi: 10.1016/j.jmmm.2012.01.022.

29. Tayyebeh Soltani, Ahmad Tayyebi and Byeong-Kyu Lee, "Quick and enhanced degradation of bisphenol A by activation of potassium peroxymonosulfate to SO$_4{}^{\cdot-}$with Mn-doped BiFeO$_3$ nanoparticles as a heterogeneous fenton-like catalyst", *Applied Surface Science*, 441, 853–861 (2018). Doi: 10.1016/j.apsusc.2018.02.063.

30. D. Chaara, Felipe Bruna, M.A. Ulibarri, Khalid Draoui, Cristobalina Barriga and I Pavlovic, "Organo/layered double hydroxide nanohybrids used to remove non ionic pesticides", *Journal of Hazardous Materials*, 196, 350–359 (2011). Doi: 10.1016/j.jhazmat.2011.09.034.

31. Divya S. Nair and Manju Kurian, "Catalytic peroxide oxidation of persistent chlorinated organics over nickel-zinc ferrite nanocomposites", *Journal of Water Process Engineering*, 16, 69–80 (2017). Doi: 10.1016/j.jwpe.2016.12.010.

32. Hong Xia, Zhen Zhang, Jia Liu, Yang Deng, Dongxu Zhang, Peiyao Du, Shouting Zhang and Xiaoquan Lua, "Novel Fe-Mn-O nanosheets/wood carbon hybrid with tunable surface properties as a superior catalyst for Fenton-like oxidation", *Applied Catalysis B: Environmental*, 259, 118058 (2019). Doi: 10.1016/j.apcatb.2019.118058.

33. E. Neyens and J. Baeyens, "A review of classic Fenton's peroxidation as an advanced oxidation technique", *Journal of Hazardous Materials*, 98, 33–50 (2003). Doi: 10.1016/S0304-3894(02)00282-0.

34. Jakelyne V. Coelho, Marina S. Guedes, Roberta G. Prado, Jairo Tronto, José D. Ardisson, Márcio C. Pereira and Luiz C.A. Oliveira, "Effect of iron precursor on the Fenton-like activity of Fe$_2$O$_3$/mesoporous silica catalysts prepared under mild conditions", *Applied Catalysis B: Environmental*, 144, 792–799 (2014). Doi: 10.1016/j.apcatb.2013.08.022.

35. Victor Augusto Araújo de Freitas, Samuel Moura Breder, Flávia Paulucci Cianga Silvas, Patrícia Radino Rouse and Luiz Carlos Alves de Oliveira, "Use of iron ore tailing from tailing dam as catalyst in a fenton-like process for methylene blue oxidation in continuous flow mode", *Chemosphere*, 219, (2018). Doi: 10.1016/j.chemosphere.2018.12.052.

36. Yu Mei, Jinchuan Zeng, Mengying Sun, Jianfeng Ma and Sridhar Komarnenid, "A novel Fenton-like system of Fe$_2$O$_3$ and NaHSO$_3$ for Orange II degradation", *Separation and Purification Technology*, 230, 115866 (2020). Doi: 10.1016/j.seppur.2019.115866.

37. Stefan Haase, M. Weiss, R. Langsch, T. Bauer and R. Lange, "Hydrodynamics and mass transfer in three-phase composite minichannel fixed-bed reactors", *Chemical Engineering Science*, 94, 224–236 (2013). Doi: 10.1016/j.ces.2013.01.050.

38. Fernando Martinez Castillejo, M. Isabel Pariente, Juan A. Melero, J.A. Botas and E. Gómez, "Catalytic wet peroxidation of phenol in a fixed bed reactor", *Water Science & Technology*, 55, 75–81 (2007). Doi: 10.2166/wst.2007.389.

39. Feiyan Liu, Huiping Zhang, Ying Yan and Haoxin Huang, "Graphene as efficient and robust catalysts for catalytic wet peroxide oxidation of phenol in a continuous fixed-bed reactor", *Science of the Total Environment*, 701, 134772 (2019). Doi: 10.1016/j.scitotenv.2019.134772.

40. Pegah Nazari, Omid Nouri, Zhiqun Xie, Shahrbanoo Rahman Setayesh and Zongsu Wei, "Delafossite-alumina nanocomposite for enhanced catalytic wet peroxide oxidation of anionic pollutants", *Journal of Hazardous Materials*, 417, 126015 (2021). Doi: 10.1016/j.jhazmat.2021.126015.

41. Asuncion Quintanilla, Jaime Carbajo, Jose Casas, P. Miranzo, Maria Osendi and M. Belmonte, "Graphene-based nanostructures as catalysts for wet peroxide oxidation treatments: From nanopowders to 3D printed porous monoliths", *Catalysis Today*, 356, (2019). Doi: 10.1016/j.cattod.2019.06.026.

42. Nawaz Shahid, Rashid Ehsan, Bagheri Ahmad, Aramesh Nahal, Bhatt Pankaj, Ali Nisar, Nguyen Tuan Anh and Bilal Muhammad, "Mitigation of environmentally hazardous pollutants by magnetically responsive composite materials", *Chemosphere*, 276, 130241 (2021). Doi: 10.1016/j.chemosphere.2021.130241.

43. Jumah Salmani M. Salmani, Sajid Asghar, Huixia Lv and Jianping Zhou, "Aqueous solubility and degradation kinetics of the phytochemical anticancer thymoquinone; Probing the effects of solvents, pH and light", *Molecules* (Basel, Switzerland), 19, 5925–39 (2014). Doi: 10.3390/molecules19055925.

44. Oliver Knoop, Marion Woermann, Holger Lutze, Bernd Sures and Torsten Schmidt, "Ecotoxicological effects prior to and after the ozonation of Tamoxifen", *Journal of Hazardous Materials*, 358, 286–293 (2018). Doi: 10.1016/j.jhazmat.2018.07.002.

45. Negin Nasseh, Tariq J. Al-Musawi, Mohammad Reza Miri, Susana Rodriguez-Couto and Ayat Hossein Panahi, "A comprehensive study on the application of FeNi₃/SiO₂/ZnO magnetic nanocomposites as a novel photocatalyst for degradation of tamoxifen in the presence of simulated sunlight", *Environmental Pollution*, 261, 114127 (2020). Doi: 10.1016/j.envpol.2020.114127.

46. Fatemeh Sadat Arghavan, Tariq J. Al-Musawi, Elaheh Allahyari, Mohammad Hadi Moslehi, Negin Nasseh and Ayat Hossein Panahi, "Complete degradation of tamoxifen using FeNi₃@SiO₂@ZnO as a photocatalyst with UV light irradiation: A study on the degradation process and sensitivity analysis using ANN tool", *Materials Science in Semiconductor Processing*, 128, 105725 (2021). Doi: 10.1016/j.mssp.2021.105725.

47. Shankar Subhash Kekade, Prashant Vijay Gaikwad, Suyog Asaram Raut, Ram Janay Choudhary, Vikas Laxman Mathe, Deodatta Phase, Anjali Kshirsagar and Shankar Ishwara Patil, "Electronic structure of visible light-driven photocatalyst δ-Bi₁₁VO₁₉ nanoparticles synthesized by thermal plasma", *ACS Omega*, 3, 5853–5864 (2018). Doi: 10.1021/acsomega.8b00564.

48. Wenlian William Lee, Chung-Shin Lu, Chung-Wei Chuang, Yen-Ju Chen, Jing-Ya Fu, Ciao-Wei Siao and Chiing-Chang Chen, "Synthesis of bismuth oxyiodides and their composites: Characterization, photocatalytic activity, and degradation mechanisms", *RSC Advances*, 5, 23450–23463 (2015). Doi: 10.1039/C4RA15072D.

49. Fatemeh Sadat Arghavan, Tariq J. Al-Musawi, Ghaida Abu Rumman, Rasool Pelalak, Alireza Khataee and Negin Nasseh, "Photocatalytic performance of a nickel ferrite/chitosan/bismuth(III) oxyiodide nanocomposite for metronidazole degradation under simulated sunlight illumination", 9, 105619 (2021). Doi: 10.1016/j.jece.2021.105619.

50. Mohammad Hossein Sayadim, Najmeh Ahmadpour and Shahin Homaeigohar, "Photocatalytic and antibacterial properties of Ag-CuFe₂O₄@WO₃ magnetic nanocomposite", *Nanomaterials*, 11, 298 (2021). Doi: 10.3390/nano11020298.

51. Michael De Volder, Sameh Tawfick, Ray Baughman and A. John Hart, "Carbon nanotubes: Present and future commercial applications", *Science* (New York, N.Y.), 339, 535–539 (2013). Doi: 10.1126/science.1222453.

52. Yanbiao Liu, Jianping Xie, Choon Nam Ong, Chad Vecitisc and Zhi Zhou, "Electrochemical wastewater treatment with carbon nanotube filters coupled with in situ generated H₂O₂", *Environmental Science: Water Research & Technology*, 1, 769–778 (2015). Doi: 10.1039/C5EW00128E.

53. Jéssica Martini, Carla A. Orge, Joaquim Faria Luis, M. Fernando R. Pereira and O. Salomé G. P. Soares, "Catalytic advanced oxidation processes for sulfamethoxazole degradation", *Applied Sciences*, 9, 2652 (2019). Doi: 10.3390/app9132652.

54. Giovanni De Filpo, Elvira Pantuso, Aleksander Mashin, Mariafrancesca Baratta and Fiore Pasquale Nicoletta, "WO₃/buckypaper membranes for advanced oxidation processes", 10, 157 (2020). Doi: 0.3390/membranes10070157.

55. G. Jeevitha, Abhinayaa Raja, Devanesan Mangalaraj and Ponpandian Nagamony, "Tungsten oxide-graphene oxide (WO₃-GO) nanocomposite as an efficient photocatalyst, antibacterial and anticancer agent", *Journal of Physics and Chemistry of Solids*, 116, 137–147 (2020). Doi: 10.1016/j.jpcs.2018.01.021.

56. Xiaohui Lu, Wei Qiu, Jiali Peng, Haodan Xu, Da Wang, Ye Cao, Wei Zhang and Jun Ma, "A review on additives-assisted ultrasound for organic pollutants degradation", *Journal of Hazardous Materials*, 403, 123915 (2021). Doi: 10.1016/j.jhazmat.2020.123915.

57. Nilsun H. Ince, "Ultrasound-assisted advanced oxidation processes for water decontamination", *Ultrasonics Sonochemistry*, 40, 97–107 (2018). Doi: 10.1016/j.ultsonch.2017.04.009.

58. Chih-Huang Weng and Kuen-Lung Tsai, "Ultrasound and heat enhanced persulfate oxidation activated with Fe⁰ aggregate for the decolorization of CI Direct Red 23", *Ultrasonics Sonochemistry*, 29, 11–18 (2016). Doi: 10.1016/j.ultsonch.2015.08.012.

59. Qun Wang, Ye Cao, Han Zeng, Youheng Lianga, Jun Ma and Xiaohui Lub, "Ultrasound-enhanced zero-valent copper activation of persulfate for the degradation of bisphenol AF", *Chemical Engineering Journal*, 378, 122143 (2019). Doi: 10.1016/j.cej.2019.122143.

60. Meimei Wang, Yunke Cui, Hongyang Cao, Ping Wei, Chen, Xiuyan Li and Juan Xu, "Activating peroxydisulfate with $Co_3O_4/NiCo_2O_4$ double-shelled nanocages to selectively degrade bisphenol A—A nonradical oxidation process", *Applied Catalysis B: Environmental*, 282, 119585 (2021). Doi: 10.1016/j.apcatb.2020.119585.

61. Houria Ghodbane and Oualid Hamdaoui, "Degradation of anthraquinonic dye in water by photoactivated periodate", *Desalination and Water Treatment*, 47, 1–10 (2014). Doi: 10.1080/19443994.2014.988657.

62. Xueming Tang and Linda K. Weavers, "Decomposition of hydrolysates of chemical warfare agents using photoactivated periodate", *Journal of Photochemistry and PhotobiologyA: Chemistry*, 187, 311–318 (2007). Doi: 10.1016/j.jphotochem.2006.10.029.

63. Chad D. Vecitis, Hyunwoong Park, Jie Cheng, Brian T. Mader and Michael R. Hoffmann, "Kinetics and mechanism of the sonolytic conversion of the aqueous perfluorinated surfactants, perfluorooctanoate (PFOA), and perfluorooctane sulfonate (PFOS) into inorganic products", *The Journal of Physical Chemistry. A*, 112, 4261–4270 (2008). Doi: 10.1021/jp801081y.

64. Oualid Hamdaoui and Slimane Merouani, "Improvement of sonochemical degradation of brilliant blue R in water using periodate ions: Implication of iodine radicals in the oxidation process", *Ultrasonics Sonochemistry*, 37, 344–350 (2017). Doi: 10.1016/j.ultsonch.2017.01.025.

65. Hui Lin, Junfeng Niu, Shiyuan Ding and Lilan Zhang, "Electrochemical degradation of perfluorooctanoic acid (PFOA) by Ti/SnO_2eSb, $Ti/SnO_2eSb/PbO_2$ and $Ti/SnO_2eSb/MnO_2$ anodes", *Water Research*, 46, 2281–2289 (2012). Doi: 10.1016/j.watres.2012.01.053.

66. Konstantinos Prevedouros, Ian T. Cousins, Robert C. Buck and Stephen H. Korzeniowski, "Sources, fate and transport of Perfluorocarboxylates", *Environmental Science & Technology*, 40, 32–44 (2006). Doi: 10.1021/es0512475.

67. Feifei Hao, Weilin Guo, Anqi Wang, Yanqiu Leng and Helian Li, "Intensification of sonochemical degradation of ammonium perfluorooctanoate by persulfate oxidant", *Ultrasonics Sonochemistry*, 21, 554–558 (2014). Doi:10.1016/j.ultsonch.2013.09.016.

68. Yu-Chi Lee, Meng-Jia Che, Chin-Pao Huang, Jeff Kuo and Shang-Lien Lo, "Efficient sonochemical degradation of perfluorooctanoic acid using periodate", *Ultrasonics Sonochemistry*, 31, 499–505 (2016). Doi: 10.1016/j.ultsonch.2016.01.030.

69. Xiaoyan Ma, Yongqing Cheng, Yongjian Ge, Huadan Wu, Qingsong Li, Naiyun Gao and Jing Deng, "Ultrasound-enhanced nanosized zero-valent copper activation of hydrogen peroxide for the degradation of norfloxacin", *Ultrasonics Sonochemistry*, 40, 1–34 (2017). Doi: 10.1016/j.ultsonch.2017.08.025.

70. Jing Zhang, Jing Guo, Yao Wu, Yeqing Lan and Ying Li, "Efficient activation of ozone by zero-valent copper for the degradation of aniline in aqueous solution", *Journal of the Taiwan Institute of Chemical Engineers*, 81, 335–342 (2017). Doi: 10.1016/j.jtice.2017.09.025.

71. Jing Deng, Mengyuan Xu, Yijing Chen, Jun Li, Chungen Qiu, Xueyan Li and Shiqing Zhou, "Highly-efficient removal of norfloxacin with nanoscale zero-valent copper activated persulfate at mild temperature", *Chemical Engineering Journal*, 366, 491–503 (2019). Doi: 10.1016/j.cej.2019.02.073.

72. Hailian Tang, Jiake Wei, Fei Liu, Botao Qiao, Xiaoli Pan, Lin Li, Jingyue Liu, Junhu Wang, Junhu Wang and Tao Zhang, "Strong metal–support interactions between gold nanoparticles and nonoxides", *Journal of the American Chemical Society*, 138, 56–59 (2016). Doi: 10.1021/jacs.5b11306.

73. Pan Li, Yuan Song, Shuai Wang, Zheng Tao, Shuili Yu and Yanan Liu, "Enhanced decolorization of methyl orange using zero-valent copper nanoparticles under assistance of hydrodynamic cavitation", *Ultrasonics Sonochemistry*, 22, 132–138 (2015). Doi: 10.1016/j.ultsonch.2014.05.025.
74. G. Kumaravel Dinesha, Malavika Pramodb and Sankar Chakmaa, "Sonochemical synthesis of amphoteric Cu^0-nanoparticles using hibiscus rosa-sinensis extract and their applications for degradation of 5-fluorouracil and lovastatin drugs", *Journal of Hazardous Materials*, 399, 123035 (2020). Doi: 10.1007/978-3-540-71095-0_4745.
75. Michael Naguib, Olha Mashtalir, Joshua Carle, Volker Presser, Jun Lu, Lars Hultman, Yury Gogotsi and Michel W. Barsoum, "Two-dimensional transition metal carbides", *ACS Nano*, 6, 1322–1331 (2012). Doi: 10.1021/nn204153h.
76. Junjie Wang, Tian-Nan Ye, Yutong Gong, Jiazhen Wu, Nanxi Miao, Tomofumi Tada and Hideo Hosono, "Discovery of hexagonal ternary phase Ti_2InB_2 and its evolution to layered boride TiB", *Nature Communications*, 10, 2248 (2019). Doi: 10.1038/s41467-019-10297-8.
77. Mohammadtaghi Vakili, Giovanni Cagnetta, Jun Huang, Gang Yu and Jing Yuan, "Synthesis and regeneration of a MXene-based pollutant adsorbent by mechanochemical methods", *Molecules*, 24, 2478 (20190). Doi: 10.3390/molecules24132478.
78. Junyu Chen, Qiang Huang, Hongye Huang, Liucheng Mao, Meiying Liu, Xiaoyong Zhang and Yen Wei, "Recent progress and advances in the environmental applications of MXene related materials", *Nanoscale*, 12, 3574–3592 (2020). Doi: 10.1039/C9NR08542D.
79. Yi Cuia, Dongao Zhang, Kelin Shen, Siqing Nie, Meiying Liu, Hongye Huang, Fengjie Deng, Naigen Zhou, Xiaoyong Zhang and Yen Wei, "Biomimetic anchoring of Fe_3O_4 onto Ti_3C_2 MXene for highly efficient removal of organic dyes by Fenton reaction", *Journal of Environmental Chemical Engineering*, 8, 104369 (2020). Doi: 10.1016/j.jece.2020.104369.
80. Yi Cui, Meiying Liu, Hongye Huang, Dongao Zhang, Junyu Chen, Liucheng Mao, Naigen Zhou, Fengjie Deng, Xiaoyong Zhang and Yen Wei, "A novel one-step strategy for preparation of Fe_3O_4-loaded Ti_3C_2 MXenes with high efficiency for removal organic dyes", *Ceramics International*, 46, 11593–11601 (2020). Doi: 10.1016/j.ceramint.2020.01.188.
81. Sehyeong Lim, Jin Hyung Kim, Hyunsu Park, Chaesu Kwak, Jeewon Yang, Jieun Kim, Seoung Young Ryu and Joohyung Lee, "Role of electrostatic interactions in the adsorption of dye molecules by Ti_3C_2-MXenes", RSC Advances, 11, 6201–6211 (2020). Doi: 10.1039/D0RA10876F.
82. Qui Thanh Hoai Ta, Nghe My Tran and Jin-Seo Noh, "Rice crust-like $ZnO/Ti_3C_2T_x$ MXene hybrid structures for Improved photocatalytic activity", *Catalysts*, 10, 1140 (2020). Doi: 10.3390/catal10101140.
83. Jiseon Janga, Asif Shahzadb, Seung Han Wooc and Dae Sung Lee, "Magnetic $Ti_3C_2T_x$ (Mxene) for diclofenac degradation via the ultraviolet/chlorine advanced oxidation process", *Environmental Research*, 182, 108990 (2020). Doi: 10.1016/j.envres.2019.108990.
84. Vahid avanbakht and Marzieh Mohammadian, "Photo-assisted advanced oxidation processes for efficient removal of anionic and cationic dyes using Bentonite/TiO$_2$ nanophotocatalyst immobilized with silver nanoparticles", *Journal of Molecular Structure*, 1239, 130496 (2021). Doi: 10.1016/j.molstruc.2021.130496.
85. Juhee Kim, Tianqi Zhang, Wen Liu, Penghui Du, Jordan T. Dobson and Ching-Hua Huang, "Advanced oxidation process with peracetic acid and Fe (II) for contaminant degradation", *Environmental Science & Technology*, 53, 13312–13322 (2019). Doi: 10.1021/acs.est.9b02991.

8 Fenton, Photo-Fenton, and Electro-Fenton Systems for Micropollutant Treatment Processes

Anjan Deb, Jannatul Rumky and Mika Sillanpää

CONTENTS

8.1 Introduction 158
8.2 Classical Fenton Process.. 159
8.3 Factors Affecting the Efficiency of the Fenton Process 160
    8.3.1 Operating pH 160
        8.3.1.1 Fe^{2+} Concentration 161
        8.3.1.2 H_2O_2 Concentration... 162
        8.3.1.3 Stoichiometric Ratio of Fe^{2+}/H_2O_2 162
        8.3.1.4 Operating Temperature 162
8.4 Catalysts for the Fenton Process.. 162
    8.4.1 Homogeneous versus Heterogeneous 162
    8.4.2 Heterogeneous Iron Catalysts... 164
        8.4.2.1 Zero-valent Iron 164
        8.4.2.2 Magnetite 164
        8.4.2.3 Goethite.. 165
        8.4.2.4 Hematite 165
        8.4.2.5 Ferrihydrite 165
        8.4.2.6 Pyrite.. 166
        8.4.2.7 Ferrite... 166
    8.4.3 Non-ferrous Heterogeneous Catalysts 166
        8.4.3.1 Aluminum... 167
        8.4.3.2 Cerium 167
        8.4.3.3 Chromium 168
        8.4.3.4 Manganese 168
        8.4.3.5 Copper... 169
        8.4.3.6 Ruthenium... 169
8.5 Fluidized Bed Fenton Process 169
8.6 Photo-Fenton.. 171

DOI: 10.1201/9781003247913-8

8.7   Electro-Fenton 172
8.8   Factors Affecting the EF Process 174
       8.8.1   Cell Configuration 174
       8.8.2   Current Density 175
       8.8.3   Electrode Materials... 176
       8.8.4   Electrolytes 177
       8.8.5   Oxygen-sparging Rate 177
8.9   Application of Fenton and Modified Fenton Processes for
       Micropollutant Treatment... 177
8.10   Mineralization Study of Phenol by Fenton Process................................. 179
8.11   Concluding Remarks and Future Perspective.. 181
References.. 181

8.1   INTRODUCTION

Ubiquitous existence of micropollutants in the aquatic environment has become a matter of serious concern in the last few decades. These micropollutants include a wide variety of industrial chemicals, personal care products, pharmaceuticals, pesticides, UV filters, flame retardants, surfactants, and other endocrine-disrupting substances [1, 2]. Most of these micropollutants are organic in nature, and their existence in the environment is detected in the range of µg/L to ng/L [3]. Even though the detrimental impact of micropollutants on human health, aquatic life, and the ecosystem is still under investigation or completely unknown in some cases, their existence has been repeatedly attributed to several negative consequences, including short-term and long-term toxicity, endocrine disruption, and antibiotic resistance of microbes [2].

Most of the wastewater treatment plants (WWTPs) currently operating can reduce the concentrations of a wide range of micropollutants but are not designed to eliminate them entirely. As a result, effluent from these WWTPs is one of the primary sources of micropollutants, which are typically discharged into surface waterways and eventually end up in the soil, groundwater, and the ocean [4]. Due to their ease of operation and simplicity, separation-based technologies, such as adsorption and membrane filtration (nanofiltration and reverse osmosis), are often adopted for water purification. However, such technologies are merely phased transfer methods, which do not destroy or mineralize the micropollutants. As a result, micropollutants may persist in the system and ultimately reach the surrounding environment if the contaminated adsorbents or membrane concentrates are not adequately handled.

Advanced oxidation processes (AOPs) are innovative treatment technologies intended to degrade and mineralize refractory organic micropollutants in the aqueous phase by reacting with hydroxyl radicals ($•OH$). These radicals can be produced in situ by employing a primary oxidant (e.g., hydrogen peroxide and ozone), UV light, electrochemical techniques, or a combination of these processes. The in-situ generated $•OH$ radicals have a very high oxidation potential (2.8V) and are therefore capable of oxidizing a wide range of organic molecules in water. Over 1,700 rate constants for $•OH$-mediated reactions with organic and inorganic compounds in aqueous solution have been listed in the last few decades, recognizing their relevance in the natural environment, biological systems, and beneficial chemical processes such as waste valorization.

A prominent example of a homogeneous AOP is Fenton's oxidation, in which Fe^{2+} ions act as the catalyst and H_2O_2 serves as the oxidant. The key benefits of the Fenton process over other oxidation techniques are the simplicity of equipment, ambient operating conditions, and ease of integration with existing water treatment facilities. Unlike other common oxidants, Fenton reagents (Fe^{2+} and H_2O_2) are affordable, safe, easy to handle, and do not pose any long-term environmental risks. Compared to all other AOPs, HO• production is relatively faster due to the rapid interaction between Fe^{2+} and H_2O_2. In this chapter, we endeavored to explain the complicated mechanisms of Fenton and modified Fenton processes and the important factors that influence their effectiveness in removing organic micropollutants from wastewater. A sincere attempt is made to provide the reader with information on state-of-the-art technology, the extent of potential applications, and remaining obstacles to developing these processes on a larger scale.

8.2 CLASSICAL FENTON PROCESS

Fenton chemistry began in 1876 when Henry J. Fenton, for the first time, observed the degradation of tartaric acid by hydrogen peroxide (H_2O_2) in the presence of ferrous ions (Fe^{2+}) [5, 6]. Later in the 1930s, Haber and Weiss postulated that the catalytic breakdown of H_2O_2 by iron salt (Fe^{2+}) generates hydroxyl radicals (•OH), one of the most powerful oxidants known ($E° = 2.73$ V) [7, 8]. Since then, Fenton and related reactions have attracted much attention because of their importance in the synthesis and functionalization of materials, hazardous waste minimization, biomass treatment, and biomedical application, particularly in cancer treatment and dental care [9].

Generally, the mixture of H_2O_2 and Fe^{2+} is known as Fenton's reagent, which generates •OH radical according to the following reaction [6, 10, 11]:

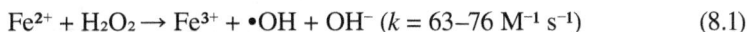

$$Fe^{2+} + H_2O_2 \rightarrow Fe^{3+} + \text{•OH} + OH^- \ (k = 63\text{–}76 \text{ M}^{-1} \text{ s}^{-1}) \tag{8.1}$$

The breakdown of H_2O_2 is initiated and catalyzed by Fe^{2+} ion in acidic media, which results in the formation of •OH. Chemical probes or spectroscopic techniques such as spin trapping can be used to confirm the formation of •OH in the Fenton process. The *in-situ* generated •OH interacts with the organic molecules via hydroxyl addition or hydrogen abstraction [12]. Organic molecules with an aromatic system or numerous carbon–carbon bonds experience hydroxyl addition (Eq. 8.2), whereas unsaturated organic compounds undergo hydrogen abstraction (Eq. 8.3).

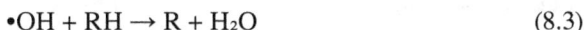

$$\text{•OH} + R \rightarrow R(OH)^{•} \tag{8.2}$$

$$\text{•OH} + RH \rightarrow R + H_2O \tag{8.3}$$

In the absence of competitive scavenging of either •OH or R•, a suitable concentration of Fe^{2+} and H_2O_2 should theoretically convert all organic molecules to carbon dioxide and water. However, the enhanced and nonspecific reactivity of •OH toward both organic and inorganic molecules result in several undesired and competing reactions (Eqs. 8.4–8.7) that can sabotage the oxidation efficiency of the Fenton process.

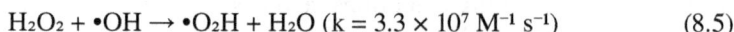

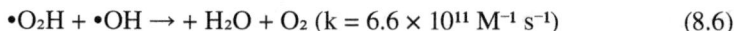

$$Fe^{2+} + \text{•OH} \rightarrow Fe^{3+} + OH^- \ (k = 3.2 \times 10^8 \ M^{-1} \ s^{-1}) \tag{8.4}$$

$$H_2O_2 + \text{•OH} \rightarrow \text{•}O_2H + H_2O \ (k = 3.3 \times 10^7 \ M^{-1} \ s^{-1}) \tag{8.5}$$

$$\text{•}O_2H + \text{•OH} \rightarrow + H_2O + O_2 \ (k = 6.6 \times 10^{11} \ M^{-1} \ s^{-1}) \tag{8.6}$$

$$\text{•OH} + \text{•OH} \rightarrow H_2O_2 \ (k = 6 \times 10^9 \ M^{-1} \ s^{-1}) \tag{8.7}$$

The regeneration of Fe^{2+} takes place by the reaction between H_2O_2 and the generated Fe^{3+} from the Fenton reaction. This reaction is known as Fenton-like reaction and is represented by the following reaction:

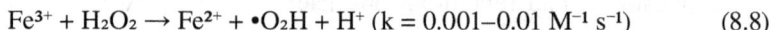

$$Fe^{3+} + H_2O_2 \rightarrow Fe^{2+} + \text{•}O_2H + H^+ \ (k = 0.001\text{–}0.01 \ M^{-1} \ s^{-1}) \tag{8.8}$$

As can be observed, the regeneration of Fe^{2+} by a Fenton-like reaction is much slower (~6,000 times) than the Fenton reaction (Eq. 8.1). Fortunately, the reduction of Fe^{3+} with •O_2H from reaction (8.9), a superoxide ion (O_2 $^-$) from reaction (8.10), or an organic radical $R^•$ from reaction (8.11) can regenerate Fe^{2+} more quickly.

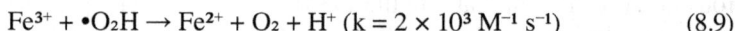

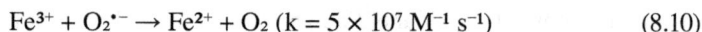

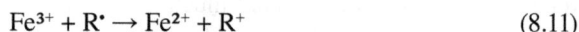

$$Fe^{3+} + \text{•}O_2H \rightarrow Fe^{2+} + O_2 + H^+ \ (k = 2 \times 10^3 \ M^{-1} \ s^{-1}) \tag{8.9}$$

$$Fe^{3+} + O_2^{•-} \rightarrow Fe^{2+} + O_2 \ (k = 5 \times 10^7 \ M^{-1} \ s^{-1}) \tag{8.10}$$

$$Fe^{3+} + R^• \rightarrow Fe^{2+} + R^+ \tag{8.11}$$

There are few rate constants available for reaction (8.11); however, the most reactive species are tertiary alkyl radicals and radicals α to -OH, -OR -CONH$_2$, which have k values of 10^7 to 10^8 M^{-1} s^{-1}.

8.3 FACTORS AFFECTING THE EFFICIENCY OF THE FENTON PROCESS

8.3.1 OPERATING pH

The pH of the reaction medium plays an essential role in the Fenton process as it affects the form of iron species and their catalytic activity to produce •OH. The speciation of iron species (Fe^{2+} and Fe^{3+}) as a function of pH is presented in Figure 8.1(a).

As shown in Figure 8.1(a), Fe^{2+} remains in the dissolved form up to neutral pH. Therefore, the fact that the optimum pH for the Fenton reaction is 3 cannot be explained by Fe^{2+} speciation. During the Fenton process, Fe^{3+} generates, and it can be seen in Figure 8.1(a) that Fe^{3+} vanishes at pH value of 4 and begins to precipitate as $Fe(OH)_3$. When both ions (Fe^{2+} and Fe^{3+}) are present together and pH is raised above 3, the Fe^{2+} tends to coprecipitate with ferric oxyhydroxides, reducing the amount of active catalyst in the solution. In addition, with increasing

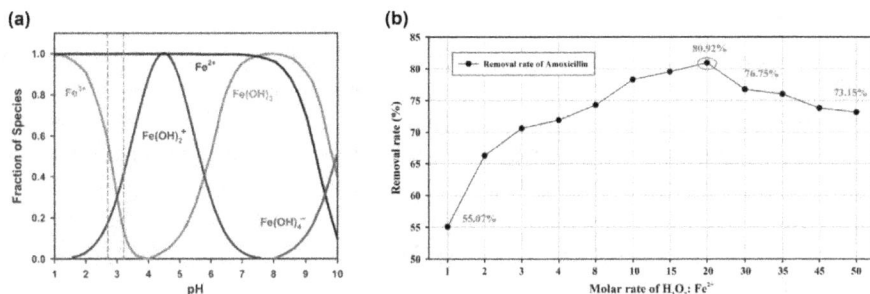

FIGURE 8.1 (a) Speciation of iron species as a function of solution pH. The dotted line indicates the optimum working range of pH for the Fenton reaction. Adapted from Reference [13] with permission. (b) Effect of H_2O_2/Fe^{2+} molar ratio on amoxicillin removal. Adapted from Reference [14] with permission.

pH, the oxidation potential of •OH diminishes. For example, at pH 0, the oxidation potential of the redox couple •OH/H_2O is 2.59 V versus NHE, while at pH 14, it is 1.64 V versus NHE [15].

Moreover, the auto decomposition of H_2O_2 into O_2 and H_2O occurs at higher pH value. On the other hand, a reduction in the degradation efficiency of the organic contaminants by the Fenton process was observed below pH 3. This is because at lower pH, Fe (II) exists as the hexaquo complex, $[Fe(H_2O)_6]^{2+}$, and it interacts with H_2O_2 more slowly than other species. Furthermore, in excess H^+ ions, the peroxide is solubilized to produce the stable oxonium ion, $[H_3O_2]^+$, which is more stable and hinders the reactivity of Fe^{2+} with H_2O_2 [16]. Therefore, under a suitable pH, usually in the range of 2.8–3.2, the production of •OH and subsequent breakdown of pollutants via Fenton oxidation are efficient. Therefore, to ensure that ferrous and ferric irons keep their catalytic activity, the pH of the solution must be kept within this range.

8.3.1.1 Fe^{2+} Concentration

In the Fenton process, Fe^{2+} catalyzes the breakdown of H_2O_2 to create highly oxidative •OH, which may destroy even the most recalcitrant organic contaminants. The degradation efficiency of the organic contaminants usually improves as Fe^{2+} concentration rises; however, the excess amount of Fe^{2+} not only raises the operating expenses and iron sludge formation but also raises the possibility of •OH scavenging, which results in a detrimental impact on Fenton reaction on organic contaminants degradation. Therefore, determining the optimum Fe^{2+} concentration is essential to achieve the maximum degradation of organic contaminants. The optimum concentration may differ for various organic compounds and must be established experimentally. For example, Kavitha and Palanivelu studied the Fenton process for nitrophenols degradation. They observed that the optimum Fe^{2+} concentration for the maximum degradation of mono nitrophenol was 0.45 mM, while for di- and tri-nitro phenol, it was 0.36 mM [17].

8.3.1.2 H_2O_2 Concentration

The initial concentration of H_2O_2, like Fe^{2+}, is a key factor in the degradation of organic contaminants by the Fenton process. When the concentration of H_2O_2 increases, organic contaminants' degradation efficiency usually improves. However, increasing the amount of H_2O_2 beyond the optimal limit has an adverse effect on the conversion efficiency, which might be due to the auto decomposition of H_2O_2 to O_2 and H_2O and the recombination of •OH and •O_2H (Eq. 8.6). Furthermore, an excess amount of H_2O_2 serves as a free radical scavenger, reducing the system's oxidation power (Eq. 8.5) [16].

8.3.1.3   Stoichiometric Ratio of Fe^{2+}/H_2O_2

Besides the initial concentration of Fenton reagents (Fe^{2+} and H_2O_2), their stoichiometric ratio is also an important parameter that determines the degradation efficiency of organic micropollutants. The optimal ratio of Fe^{2+} to H_2O_2 is usually determined from the laboratory scale study. A high Fe^{2+}/H_2O_2 ratio promotes the decomposition of H_2O_2 to •OH; however, it may not improve the mineralization degree due to scavenging reactions (Eqs. 8.4 and 8.7). Similarly, a lower ratio of Fe^{2+}/H_2O_2 means an excess amount of H_2O_2, which can also promote the scavenging of •OH (Eq. 8.5) and reduce the efficiency of the Fenton process. Guo et al. studied the amoxicillin removal using the Fenton oxidation process and showed that the maximum removal was obtained at a molar ratio of $Fe^{2+}/H_2O_2 = 1/20$ (Figure 8.1(b)) [14]. Beyond that optimum ratio up to 1/50, the removal efficiency of amoxicillin decreased from 80.92% to 73.15%.

8.3.1.4   Operating Temperature

In the Fenton process, increasing the temperature speeds up the reaction between H_2O_2 and Fe^{2+}, resulting in a high rate of •OH production. More •OH radicals could potentially encourage collisions with organic contaminants, improving degradation efficiency. For example, Zazo et al. [18] reported that when the reaction temperature increased from 25 °C to 90 °C in a homogeneous Fenton process, the mineralization efficiency of phenol increased from 28% to 80%. However, a further increase in temperature up to 130 °C did not improve the mineralization efficiency significantly. This is because the spontaneous breakdown of H_2O_2 and Fe^{2+} precipitation occurs at higher temperatures. In another study, it was reported that removal rate of amoxicillin improved from 74.02% to 80.02% as the temperature increased from 30 to 40 °C [14]. Therefore, maintaining an appropriate temperature should be evaluated on the basis of the target pollutants to be treated.

8.4   CATALYSTS FOR THE FENTON PROCESS

8.4.1   Homogeneous versus Heterogeneous

In the homogeneous Fenton process, both Fenton reagents (Fe^{2+} and H_2O_2) exist in the same phase. Therefore, the system shows negligible mass transfer restriction, which allows easily accessible iron ions in the reaction media to participate efficiently in the degradation process. However, there are some limitations in the

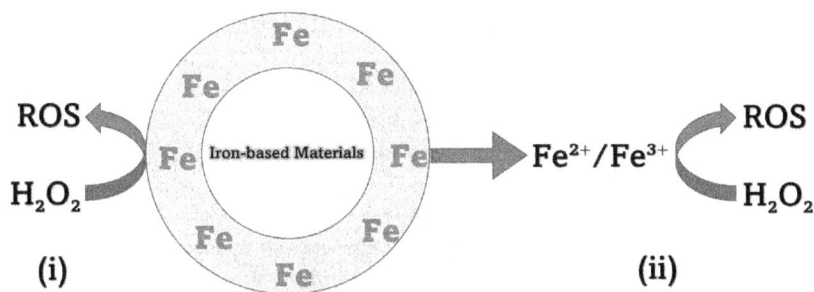

FIGURE 8.2 Schematic representation of the heterogeneous Fenton system: (i) Surface-leached iron-mediated homogeneous reaction mechanism and (ii) True heterogeneous catalytic reaction mechanism.

homogeneous Fenton process, including a narrow pH range, loss of Fe^{2+} due to poor regeneration kinetics, and the generation of iron sludge which further increases the operating costs of the Fenton process. These limitations can be overcome to some extent by employing heterogeneous iron catalysts, where iron is stabilized inside the interlayer space of the catalyst's structure and may efficiently generate •OH from H_2O_2 oxidation under noncontrolled pH environments.

Two possible reaction routes have been proposed to describe the reaction mechanism between heterogeneous Fe catalysts and H_2O_2 (Figure 8.2): (i) the homogeneous catalytic mechanism where Fe species leached out from the support material and the reaction takes place in the bulk solution and (ii) the actual heterogeneous catalytic mechanism where the reactions take place at the catalytic surface bounded on support. Especially at acidic pH, certain iron-based compounds can contribute to a homogeneous Fenton-reaction-dominated mechanism by serving as a source of continuous iron dissolution. However, leaching depletes the metal sites in the solid catalyst, resulting in the loss of catalytic activity over time. On the other hand, heterogeneous Fenton reactions are generally regulated by kinetic processes such as reactant adsorption, surface chemical reaction, and product desorption. Lin and Gurol [19] studied the reaction of H_2O_2 on solid iron oxide catalysts (goethite) in 1998 and proposed a widely accepted reaction mechanism for the heterogeneous catalytic breakdown of H_2O_2. In brief, H_2O_2 forms a primary surface complex, $(H_2O_2)_s$, with its interaction at the goethite ($\equiv Fe(III)$) surface (Eq. 8.12). Within the complex, charge transfer from ligand to metal creates a transitional state for the surface site ($\equiv Fe(II)•O_2H$), which is further deactivated by dissociation of the $•O_2H$ (Eqs. 8.13 and 8.14). Then, the formed $\equiv Fe(II)$ site catalyzes the formation of •OH (Eq. 8.15) and the $\equiv Fe(III)$ surface regenerates.

$$\equiv Fe(III)\text{-}OH + H_2O_2 \leftrightarrow (H_2O_2)_s \tag{8.12}$$

$$(H_2O_2)_s \leftrightarrow \equiv Fe(II)•O_2H + H_2O \tag{8.13}$$

$$\equiv Fe(II)•O_2H \rightarrow \equiv Fe(II) + •O_2H \tag{8.14}$$

$$\equiv Fe(II) + H_2O_2 \rightarrow \equiv Fe(III)\text{-}OH + •OH \tag{8.15}$$

A two-step degradation process is observed in several heterogeneous Fenton systems, with an initial sluggish induction period followed by a fast oxidation period. According to the researchers, the induction period is characterized by heterogeneous reactions such as surface iron leaching and heterogeneous catalysis. The second kinetic stage is characterized by homogeneous Fenton reactions caused by the leached iron [20, 21].

One of the primary advantages of the heterogeneous Fenton process over the homogeneous Fenton process is the overall stability and reusability of the iron catalysts. However, the iron concentration in the solution decreased with reaction time in the case of the homogeneous Fenton process owing to iron sludge formation. Furthermore, separating the remaining ferrous ions is extremely difficult since they are all homogeneously dissolved in water. On the other hand, all catalysts are solid in the case of the heterogeneous Fenton process, making separation from solution very simple.

8.4.2 HETEROGENEOUS IRON CATALYSTS

Many iron-based materials are stable and only experience little iron leaching when subjected to reactions. Iron-based materials have the benefits of having a high Fenton activity, being inexpensive, having minimal toxicity, and being simple to recover.

8.4.2.1 Zero-valent Iron

Zero-valent iron (ZVI) is a very effective reducing agent with a typical reduction potential of E^0 (Fe^{2+}/Fe^0) = -0.44 V. In the presence of H_2O_2 or O_2, ZVI can release two electrons, forming Fe^{2+} ions (Eqs. 16 and 17). The generated Fe^{2+} ions are responsible for effective organic pollutant degradation either through a homogeneous Fenton mechanism (at pH 2–3) or via a heterogeneous reaction-induced homogeneous process (at pH 4–6). ZVI continues to dissolve in the second stage of the heterogeneous reaction-induced homogeneous mechanism, largely by Fe^{3+} reduction (Eq. 8.18), which speeds up Fe^{2+} regeneration in the homogeneous Fenton reaction.

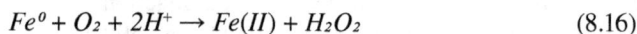

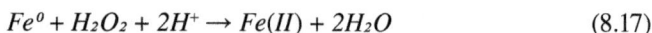

$$Fe^0 + O_2 + 2H^+ \rightarrow Fe(II) + H_2O_2 \tag{8.16}$$

$$Fe^0 + H_2O_2 + 2H^+ \rightarrow Fe(II) + 2H_2O \tag{8.17}$$

$$2Fe^{3+} + Fe^0 \rightarrow 3Fe^{2+} \tag{8.18}$$

8.4.2.2 Magnetite

Magnetite (Fe_3O_4) is a mixed-valence iron oxide that belongs to the spinel family and has special redox characteristics. Fe_3O_4 is the most magnetic of all the naturally occurring minerals on the planet. Fe_3O_4 has attracted a lot of attention in heterogeneous Fenton systems because of its structural ferrous ions, which are crucial in accelerating the production of •OH. Furthermore, magnetic separation may be used to quickly remove Fe_3O_4 particles from an aqueous media after their catalytic purification application. pH, on the other hand, may influence the interfacial mechanism of Fe_3O_4-catalyzed heterogeneous Fenton systems as a result of Fe_3O_4 dissolution.

Fe_3O_4 nanoparticles have high stability and reusability at neutral pH, owing to a heterogeneous catalytic reaction mechanism. However, at pH value of 3, organic pollutants are degraded by a heterogeneous reaction-induced homogeneous mechanism. Furthermore, the activity of Fe_3O_4 nanoparticles reduces gradually over several runs, most likely because of the leached iron and other factors such as active catalytic sites' poisoning by adsorbed organic species or surface oxidation.

8.4.2.3   Goethite

Goethite is an iron (III) oxide mineral with the chemical formula Fe^{III}-O(OH), found abundantly in soil and other low-temperature environments. It is one of the commonly used iron oxides in the Fenton process for the degradation of organic contaminants. Goethite has received considerable interest as a heterogeneous Fenton catalyst due to its (i) ability to operate in a wide pH range, (ii) higher thermodynamic stability, (iii) positive performance under sunlight, and (iv) low cost and environmental friendliness. Based on the experimental evidence, a goethite catalyst appeared to be effective in the photo Fenton process in the presence of sunlight [22].

8.4.2.4   Hematite

Hematite (α-Fe_2O_3) is the most stable Fe(III) oxide ore and is found in a variety of forms, including kidney ore, iron rose, martite, and specularite. Its color ranges from black to steel or silver-grey, brown to reddish-brown [23]. In recent years, there has been considerable interest in hematite as a heterogeneous Fenton catalyst because it is (i) an inexpensive and readily available mineral and (ii) nontoxic and environmentally benign. With its high surface area, distinct morphology, and well-defined topology, hematite exhibits relatively high catalytic performance and stability. Huang and co-workers [24] showed that the {001} facet exposed hematite nanoplates had higher Fenton catalytic performance than the {002} exposed facet hematite nano cubes. The development of iron-ascorbate complexes on the facets of the hematite can considerably limit the dissolution of surface bound Fe^{2+} ions, resulting in greater stability and increased catalytic efficiency of hematite. Moreover, hematite can absorb light up to 560 nm and capture about 40% of the solar spectrum energy due to its small bandgap (2.2 eV), making it a potential material for photo-Fenton catalytic applications [25].

8.4.2.5   Ferrihydrite

Ferrihydrite is a naturally occurring Fe(III) hydroxide with a large specific surface area (>200 m²/g) and numerous chemically active surface groups found in the earth's crust, soils and sediments [26]. The large surface area of ferrihydrite results in more interaction between the catalyst, H_2O_2 and the contaminants when employed as a heterogeneous Fenton catalyst, improving the effectiveness of H_2O_2 activation and, therefore, the oxidation of organic micropollutants. Huang et al. [27] conducted a comparative investigation of catalytic effectiveness in the decomposition of H_2O_2 using various iron(III) oxides. The maximum activity was obtained with amorphous ferrihydrite, which had the biggest surface area, followed by crystalline oxides such as needle-like goethite and plate-like hematite, which had the lowest surface area.

8.4.2.6   Pyrite

Pyrite (FeS_2), one of the most common sulfide minerals with high ferrous content, can be considered an ideal heterogeneous Fenton reagent for decomposing H_2O_2 to oxidize organic pollutants [28]. The use of FeS_2, being a natural source of Fe^{2+} and H^+ ions, brings several benefits to the Fenton process. In fact, dissolved oxygen in aqueous solutions oxidizes FeS_2, forming Fe^{2+} and H^+ ions (Eq. 19). The *in-situ* generated Fe^{2+} reacts with H_2O_2 according to the classic Fenton reaction. Meanwhile, FeS_2 interacts with H_2O_2 to form Fe^{3+} through reaction (Eq. 8.20), which is subsequently reduced to Fe^{2+} by reaction (Eq. 8.21). As a result, the production and regeneration of Fe^{2+} by FeS_2 result in a self-regulatory Fenton process [29, 30].

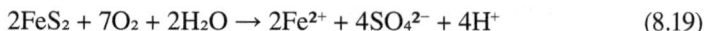

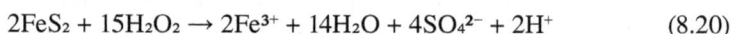

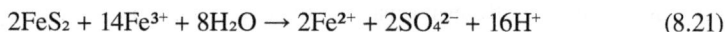

$$2FeS_2 + 7O_2 + 2H_2O \rightarrow 2Fe^{2+} + 4SO_4^{2-} + 4H^+ \qquad (8.19)$$

$$2FeS_2 + 15H_2O_2 \rightarrow 2Fe^{3+} + 14H_2O + 4SO_4^{2-} + 2H^+ \qquad (8.20)$$

$$2FeS_2 + 14Fe^{3+} + 8H_2O \rightarrow 2Fe^{2+} + 2SO_4^{2-} + 16H^+ \qquad (8.21)$$

8.4.2.7   Ferrite

The interaction of iron oxides with other transition metals results in the formation of ferrites, which are recognized as ceramic compounds. Ferrites are categorized as spinel, garnet, or hexagonal, depending on their crystal forms. Spinel ferrites have attracted considerable attention as a heterogeneous Fenton catalyst for eliminating different organic contaminants [23]. They have a face-centered cubic lattice with the general formula $M_xFe_{3-x}O_4$ (where M denotes bivalent metal ions such as Cu, Mn, Co, and Zn) in which oxide ions are organized in a face-centered cubic fashion. In contrast, metal ions are distributed in tetrahedral and octahedral positions. Wang et al. [31] developed mesoporous copper ferrite ($CuFe_2O_4$) for the degradation of imidacloprid using the heterogeneous Fenton process. Template-assisted fabrication of $CuFeS_2$ exhibited high surface area and porosity, which contributed to the high number of catalytically active sites and enhanced production of OH radicals. Moreover, the thermodynamic suitability of Fe^{3+} reduction by Cu^+ promoted the redox recycle of Fe^{2+}/Fe^{3+} and Cu^+/Cu^{2+} in mesoporous $CuFe_2O_4$.

8.4.3   Non-ferrous Heterogeneous Catalysts

Besides the heterogeneous Fe catalysts, several other materials with multiple oxidation states and redox stability are employed to decompose H_2O_2 via Fenton like reaction mechanisms. Zero-valent aluminum (ZVAl), cerium, chromium, manganese, copper, and ruthenium have recently gained significant interest as Fenton catalysts due to their ability to operate over a wider pH range and their high chemical stability. Although the specific method of activation is significantly reliant on the type of catalyst, it is largely controlled by the pH of the solution or the presence of metal–ligand complexes. Apart from that, the pH-dependent dual function of H_2O_2 as both oxidant and reductant makes the redox transformation of these non-ferrous metal species simple [13].

8.4.3.1 Aluminum

The aqueous solution of aluminum contains just one oxidation state, Al^{3+}. Consequently, unlike in the case of iron, which can exist in both the Fe^{2+} and Fe^{3+} forms, there is no possibility of an electron transfer process between Al^{3+} and H_2O_2. However, zero-valent aluminum (ZVAl or Al^0), a strong reducing agent with a reduction potential of $E^0 (Al^{3+}/Al^0) = -1.66$ V, offers a considerably higher thermodynamic driving force for the decomposition of H_2O_2. In 2009, Bokare and Choi [32] demonstrated the generation of $\bullet OH$ by zero-valent aluminum (ZVAl) in a Fenton-like process for the first time. Under acidic conditions, the native surface Al_2O_3 layer on ZVAl was dissolved to expose the bare Al^0 surface (Eq. 8.19). In the presence of dissolved O_2, electron transfer from ZVAl resulted in the *in-situ* production of H_2O_2 (Eqs. 8.20 and 8.21) and the subsequent breakdown into $\bullet OH$ (Eq. 8.22), which further triggered the oxidative mineralization of organic contaminants like phenol, 4-chlorophenol, nitrobenzene, and dichloroacetate.

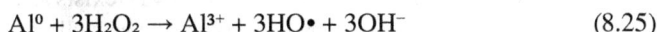

$$Al^0 \rightarrow Al^{3+} + 3e^- \tag{8.22}$$

$$O_2 + H^+ + e^- \rightarrow \bullet O_2H \tag{8.23}$$

$$2(\bullet O_2H) \rightarrow H_2O_2 + O_2 \tag{8.24}$$

$$Al^0 + 3H_2O_2 \rightarrow Al^{3+} + 3HO\bullet + 3OH^- \tag{8.25}$$

Several noteworthy features of ZVAl's usage to decompose H_2O_2 into $\bullet OH$ and subsequent contaminants' oxidation include its high natural availability and low weight. In addition, higher reduction potential and enhanced water solubility of Al^{3+}-species result in a considerably higher oxidative ability of ZVAl aerobic system compared to the Fe-based [32]. However, it is not feasible to employ ZVAl-based AOP systems to treat the organic contaminants in a non-acidic environment due to the inefficient removal of the Al_2O_3 surface layer under a neutral or near-neutral pH environment.

8.4.3.2 Cerium

Cerium is the only metal in the lanthanide or rare-earth group that exhibits both +3 and +4 oxidation states in the solution. In alkaline conditions, the cerous (Ce^{3+}) form is a powerful reducing agent, whereas, in acidic conditions, the ceric (Ce^{4+}) form is a powerful oxidant. Therefore, under an appropriate redox environment, cerium can effectively cycle between Ce^{3+} and Ce^{4+} oxidation states and is capable of activating H_2O_2 via a Fenton-like mechanism [33].

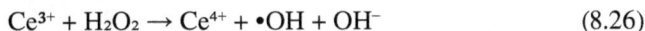

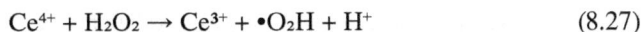

$$Ce^{3+} + H_2O_2 \rightarrow Ce^{4+} + \bullet OH + OH^- \tag{8.26}$$

$$Ce^{4+} + H_2O_2 \rightarrow Ce^{3+} + \bullet O_2H + H^+ \tag{8.27}$$

Among all cerium compounds, ceria (CeO_2) is the most studied material. Research has shown that the generation of $\bullet OH$ radicals, as well as the overall oxidation process, is inextricably linked to the surface characteristics of the CeO_2. In a study, Wang and co-workers [34] showed that the interaction of H_2O_2 and bare CeO_2 leads to the production of persistent peroxide-like species ($\equiv Ce^{III}-O_2H^-$), which do not readily degrade into •OH. However, the surface functionalization of CeO_2 with sulfuric acid accelerated the decomposition of peroxide species to •OH radicals via an acid-catalyzed intra-molecular electron transfer mechanism.

8.4.3.3   Chromium

Chromium (Cr) is a poisonous metal that may exist in various oxidation states, ranging from −2 to +6. However, only trivalent [Cr(III)] and hexavalent [Cr(VI)] species are widely found in water, and both react strongly with H_2O_2 to produce •OH in a sequence of Fenton-like reactions. The reaction between the Cr(III) and H_2O_2 strongly relies on the solution pH. Cr(III) exists in the form of $[Cr(H_2O)_6]^{3+}$ at pH ≤ 3 and is entirely unreactive toward H_2O_2, while at neutral pH, it exhibits the highest reactivity and yield of •OH generation [35]. A Fenton-like route leads to the formation of •OH (Eqs. 8.28, 8.29, and 8.30):

$$Cr(III) + H_2O_2 \rightarrow Cr(IV) + •OH + OH^- \tag{8.28}$$

$$Cr(IV) + H_2O_2 \rightarrow Cr(V) + •OH + OH^- \tag{8.29}$$

$$Cr(V) + H_2O_2 \rightarrow Cr(VI) + •OH + OH^- \tag{8.30}$$

On the other hand, the interaction between Cr(VI) (exists as $Cr^{VI}O_4^-$) and H_2O_2 causes the replacement of oxo ligands by peroxo groups as well as one electron reduction of the metal center, resulting in the formation of a $[Cr^V(O_2)_4]^{3-}$ complex, which decomposes at acidic pH to produce •OH and regenerates Cr(VI) [36].

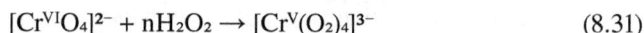

$$[Cr^{VI}O_4]^{2-} + nH_2O_2 \rightarrow [Cr^V(O_2)_4]^{3-} \tag{8.31}$$

$$[Cr^V(O_2)_4]^{3-} + H^+ \rightarrow [Cr^{VI}(O_2)_3(O)]^{2-} + •OH \tag{8.32}$$

Despite the fact that the dissociation of Cr(V) complex into •OH is significantly preferred at acidic pH, the formation of Cr(V) intermediate occurs throughout a wide pH range (3.0–9.0), which is a key benefit of employing Cr(VI) for Fenton-like activation of H_2O_2. Moreover, the aqueous solubility of Cr(VI) over the entire pH range is another advantage of employing Cr(VI)/H_2O_2 as a homogeneous AOP. Despite these realistic beneficial reaction conditions, the toxicity of Cr(VI) precludes its intentional inclusion in wastewater treatment. However, treating all Cr-contaminated wastewater using H_2O_2 as both oxidant and reductant is possible. Precise control of solution pH will allow the oxidation of Cr(III) to Cr(VI) and subsequent regeneration via Cr(VI) reduction [35].

8.4.3.4   Manganese

Manganese (Mn) may exist in a wide range of oxidation states, from 0 to +7; however, only the +2 and +4 oxidation states have significant environmental and catalytic implications. Generally, under aerobic conditions, Mn^{2+} is oxidized to Mn^{4+}, resulting in colloidal intermediates of Mn^{3+}-oxyhydroxides which are further converted to MnO_2. Simultaneously, chemical redox reactions readily convert the Mn^{4+} species to Mn^{2+} (Eq. 8.33).

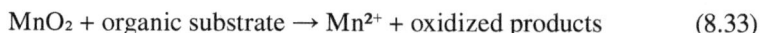

$$MnO_2 + organic\ substrate \rightarrow Mn^{2+} + oxidized\ products \tag{8.33}$$

To illustrate the possibility of manganese as a Fenton-like catalyst, a homogeneous Mn^{2+}/H_2O_2 system for the degradation of 1-hexanol was studied by Watts and co-workers and revealed that the system was capable of 99% degradation of 1-hexanol in acidic pH [37]. Moreover, several oxide polymorphs of Mn (MnO_2, MnOOH and Mn_3O_4) can react effectively with H_2O_2 and generate various kinds of reactive oxygen species, including $\bullet OH$ and $O_2^{\bullet -}$ in the pH range of 3.5–7.0.

8.4.3.5 Copper

Copper (Cu) is another metal that has been utilized effectively as a Fenton-like catalyst. It exists in two oxidation states: monovalent (Cu^+) and divalent (Cu^{2+}). The redox reaction of Cu with H_2O_2 is analogous to that of the Fe^{2+}/Fe^{3+} Fenton system:

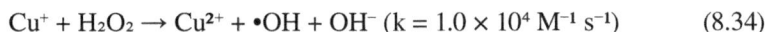

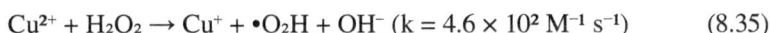

$$Cu^+ + H_2O_2 \rightarrow Cu^{2+} + \bullet OH + OH^- \ (k = 1.0 \times 10^4 \ M^{-1} \ s^{-1}) \tag{8.34}$$

$$Cu^{2+} + H_2O_2 \rightarrow Cu^+ + \bullet O_2H + OH^- \ (k = 4.6 \times 10^2 \ M^{-1} \ s^{-1}) \tag{8.35}$$

Although the more stable hydrolyzed complex of Cu ($Cu(OH)_2$ or Cu^{2+}) is a Fenton active catalyst similar to the iron Fenton system, the aqueous solubilities of Cu^{2+} and Fe^{3+} are quite different. While the iron aquo complex $[Fe(H_2O)_6]^{3+}$ is insoluble above pH value of 5, the copper aquo complex $[Cu(H_2O)_6]^{2+}$ is soluble at neutral pH, allowing Cu^{2+}/H_2O_2 system to operate across a wider pH range than the Fe^{3+}/H_2O_2 redox system [38].

8.4.3.6 Ruthenium

Ruthenium (Ru) is the sole member of the platinum group metals, which shows Fenton-like activity in the presence of H_2O_2. Hu et al. oxidized bisphenol A using the Ru^{3+}/Ru^{2+} ($E°(Ru^{3+}/Ru^{2+})$ = 1.29 V) redox couple. In the pH range 4.0–8.0, the interaction between Ru^{2+} and H_2O_2 effectively produced $\bullet OH$, and higher pH showed increasing oxidation efficiency [39]. The optimal degradation efficiency at neutral or near-alkaline pH is advantageous for Ru catalysts in practical application. Moreover, the high stability of Ru catalysts prevents metal leaching, which is beneficial for multiple catalytic cycles.

8.5 FLUIDIZED BED FENTON PROCESS

A Fluidized Bed Fenton (FBF) process is an innovative technology where a fluidized bed reactor is coupled with the Fenton process to increase the overall process performance and minimize iron sludge formation. The fluidized materials, specified as carriers in the FBF process, can be inert solid or iron-based materials. Fe^{3+} ions generated from the Fenton reaction can precipitate and crystallize on the surface of the carriers. Fe-O(OH) is the most common form of crystallized iron oxide, and it can further function as a heterogeneous catalyst for the decomposition of H_2O_2. However, the crystallization of ferric ions might be hampered by the pH of the reaction medium. The Fe (III) species are insoluble above pH 5.0 and precipitate in the hydroxide form ($Fe(OH)_3$), which can prevent Fe-O(OH) crystallization onto the carriers. A heterogeneous iron catalyst, such as goethite or iron oxide wastes, can be employed directly as a fluidization carrier to solve this problem.

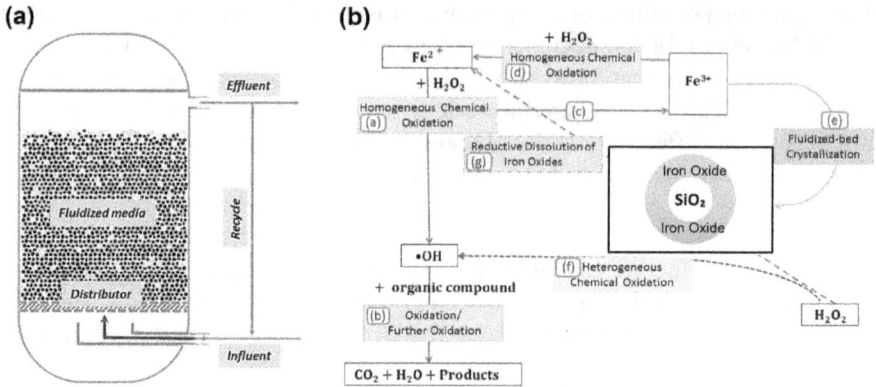

FIGURE 8.3 Schematics of (a) basic fluidized bed reactor and (b) reactions that take place in the Fluidized Bed Fenton process. Adapted from Reference [41] with permission.

It is possible to have several reactions occurring simultaneously in the FBF process. These reactions include [40]:

    i. Homogeneous chemical oxidation (H_2O_2/Fe^{2+})
    ii. Heterogeneous chemical oxidation (H_2O_2/iron oxide)
    iii. Fluidized-bed crystallization, and
    iv. Reductive dissolution of iron oxides

The schematic representation of a basic fluidized bed reactor and the possible reaction mechanism in a fluidized-bed Fenton process is depicted in Figure 8.3. The homogeneous reactions occur in both the fluidized-bed and conventional Fenton processes, and they include: (a) homogeneous •OH production from Fe^{2+} and H_2O_2, (b) reaction of organic matter with the •OH, (c) conversion of Fe^{2+} to Fe^{3+}, and (d) slower conversion of Fe^{3+} back to Fe^{2+}. Additionally, the presence of a carrier causes the following heterogeneous reactions in the fluidized-bed Fenton process: (e) crystallization of Fe^{3+} in iron oxide forms, which is initiated by the carrier, (f) production of •OH from H_2O_2, which is catalyzed by iron oxides, and (g) redissolution of iron oxide to Fe^{2+}.

During the FBF process, the carrier materials, as well as their surface properties and size distribution, all play an important role, with the following effects in (i) fluid velocity for the starting fluidization of carrier material, (ii) crystallization thermodynamics by modifying the Gibbs energy of iron oxide nucleation and crystal growth, (iii) iron oxide crystallization kinetics, and (iv) final weight of the carrier. As a result, the choice of carriers is an essential part that impacts the efficiency of the FBF process. In this context, various carriers, including pure SiO_2, Al_2O_3, goethite, iron oxide waste, chitosan nanoparticles, quartz sand, and even activated-fly ash containing Fe_2O_3, have been utilized in the FBF process.

8.6 PHOTO-FENTON

During the classical Fenton process, Fe^{3+} generates and accumulates in the system due to the slow reaction kinetics of the Fenton-like reaction (Eq. 8.7), through which the regeneration of Fe^{2+} takes place from Fe^{3+}. As a result, the oxidation efficiency of the Fenton process decreases gradually, and the system generates iron sludges. It has been reported that the organic compounds' mineralization efficiency of the Fenton process ranges from 40 to 60%, depending on the materials used. The oxidation efficiency of the traditional Fenton process can be improved by combining it with a light source, such as UV or visible light, and this is known as the photo-Fenton process.

Fenton process generates Fe^{3+}, which mainly exists in the process as $[Fe(OH)]^{2+}$ at pH 2.8–3.5 [42]. Under light irradiation ($\lambda < 580$ nm), $[Fe(OH)]^{2+}$ undergoes photo-reduction, regenerating Fe^{2+} and creating more •OH (Eq. 8.36). The regenerated Fe^{2+} recombines with H_2O_2 to produce •OH radicals and Fe^{3+}, and the cycle continues:

$$Fe(OH)^{2+} + hv \rightarrow Fe^{2+} + \bullet OH \qquad (8.36)$$

In addition, direct photolysis of H_2O_2 takes place upon light irradiation ($\lambda < 310$ nm), which generates additional •OH in the system according to Eq (8.37):

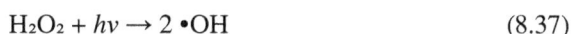

$$H_2O_2 + hv \rightarrow 2 \bullet OH \qquad (8.37)$$

Therefore, in comparison to the Fenton reaction, the photo-Fenton reaction has a much lower total iron consumption and sludge production. Apart from that, the photo-Fenton process allows using Fe(III) compounds instead of Fe(II) to initiate the Fenton reaction. A schematic reaction mechanism of the photo-Fenton process for the decomposition of organic micropollutants is depicted in Figure 8.4.

Similar to the Fenton process, photo-Fenton reactions can be categorized as homogeneous and heterogeneous depending on the catalysts employed in the system (free

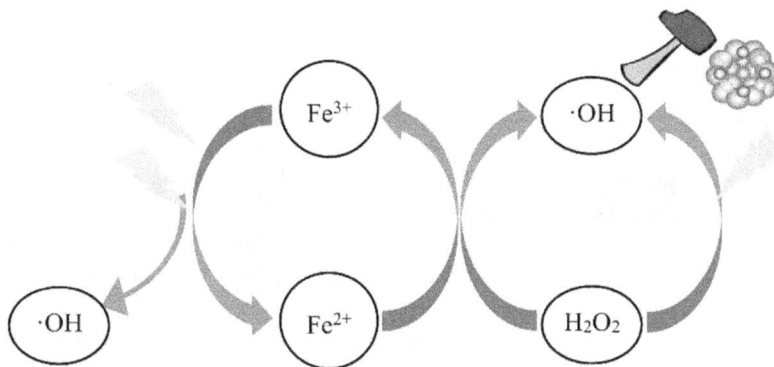

FIGURE 8.4   Reaction mechanism for the photo-Fenton process. Adapted from Reference [43] with permission.

Fe-ions or Fe-based solid catalysts). Compared to a homogeneous system, the heterogeneous Fenton process with light irradiation may generate more •OH. Combining iron oxides with semiconductors makes it easier to create a heterojunction structure separating photogenerated carriers, resulting in increased photo-Fenton degradation efficiency. Hernández-Uresti and co-workers [44] developed Fe-impregnated $BiVO_4$ catalyst by directly hydrolyzing the Fe salt onto $BiVO_4$ nanoparticles. The synergistic impact of photocatalysis and photo-Fenton was shown to be responsible for the enhanced and complete degradation of 10 mg/L ciprofloxacin (CIP) by 1 wt% Fe-doped $BiVO_4$ catalysts under UV–Vis irradiation for 30 min. Based on the experimental results, the photogenerated hole (h^+) and H_2O_2 played an important role in CIP degradation. The following reactions took place in the 1 wt% Fe-$BiVO_4$-catalyzed photo-Fenton system:

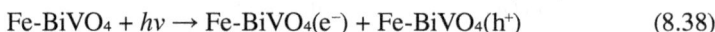

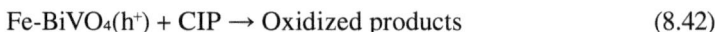

$$Fe\text{-}BiVO_4 + hv \rightarrow Fe\text{-}BiVO_4(e^-) + Fe\text{-}BiVO_4(h^+) \tag{8.38}$$

$$Fe\text{-}BiVO_4(e^-) + H_2O_2 \rightarrow \bullet OH + OH^- \tag{8.39}$$

$$Fe\text{-}BiVO_4(e^-) + Fe^{3+} \rightarrow Fe^{2+} \tag{8.40}$$

$$Fe\text{-}BiVO_4(h^+) + Fe^{2+} \rightarrow Fe^{3+} \tag{8.41}$$

$$Fe\text{-}BiVO_4(h^+) + CIP \rightarrow \text{Oxidized products} \tag{8.42}$$

$$\bullet OH + CIP \rightarrow \text{Oxidized products} \tag{8.43}$$

Upon light irradiation, photogenerated charge carriers (Fe-$BiVO_4(e^-)$ and Fe-$BiVO_4(h^+)$) are developed due to the excitation of an electron from the valence band to the conduction band of Fe-$BiVO_4$. The reduction of H_2O_2 with the photogenerated electrons leads to the generation of •OH (Eq. 8.34), which can degrade the CIP. Simultaneously, H_2O_2 interacts with the surface-bound Fe according to the Fenton reaction and generates •OH. Apart from that, the redox conversion of Fe^{3+}/Fe^{2+} takes place on the catalyst's surface with the photogenerated charge carriers (Eqs. 8.35 and 8.36), inhibiting the recombination of charge carriers.

8.7   ELECTRO-FENTON

The efficiency of a Fenton process can be improved by coupling the Fenton reaction with electrochemical processes, known as electro-Fenton (EF) process. In the EF process, at least one of the Fenton reagents is either in-situ generated (H_2O_2 and Fe^{2+}) or regenerated (Fe^{2+}) electrochemically. The earliest investigations on EF-like reactions took place in the late 1970s and early 1980s, when hydroxylation reactions of organic molecules such as cyclohexane, benzene, toluene, and phenol were carried out in acidic conditions with the addition of various catalysts, including Fe^{2+} [45, 46]. However, in 1986, Sudoh et al. studied the EF process for water treatment for the first time. Using platinum and graphite electrodes in the presence of Fe^{2+} catalysts, they investigated the degradation of phenol at a cathodic potential of −0.6 V [47].

H_2O_2 can be generated continuously at the cathode of an electrochemical cell by directly injecting pure O_2 gas or air in an acidic aqueous solution. Injected O_2 gas is first dissolved in the aqueous phase and then transported from the bulk to the cathodic surface. Finally, it is converted to H_2O_2 by a two-electron oxygen reduction reaction according to Eq. (8.14).

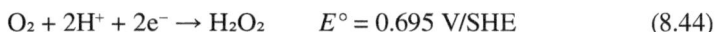

$$O_2 + 2H^+ + 2e^- \rightarrow H_2O_2 \qquad E^\circ = 0.695 \text{ V/SHE} \tag{8.44}$$

In the same acidic condition, dissolved oxygen can convert into H_2O by a four-electron oxygen reduction reaction (Eq. 8.15) and compete with the reaction Eq. (8.14).

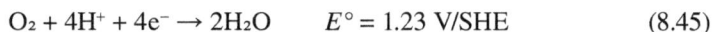

$$O_2 + 4H^+ + 4e^- \rightarrow 2H_2O \qquad E^\circ = 1.23 \text{ V/SHE} \tag{8.45}$$

On the other hand, Fe^{2+} can be generated electrochemically either by the oxidative dissolution of the sacrificial iron anode (Eq. 8.16) or cathodic reduction of Fe^{3+} (Eq. 8.17)

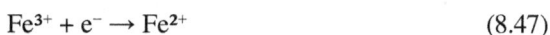

$$Fe^\circ \rightarrow Fe^{2+} + 2e^- \tag{8.46}$$

$$Fe^{3+} + e^- \rightarrow Fe^{2+} \tag{8.47}$$

The cell configuration, cathode materials, and operating environment affect the generation and stability of H_2O_2. For example, in an undivided cell, H_2O_2 can undergo anodic oxidation via the formation of hydroperoxyl radicals as intermediates according to the following equations:

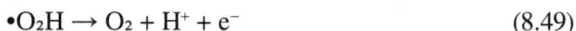

$$H_2O_2 \rightarrow {}^\bullet O_2H + H^+ + e^- \tag{8.48}$$

$${}^\bullet O_2H \rightarrow O_2 + H^+ + e^- \tag{8.49}$$

In addition, electrogenerated H_2O_2 can be reduced further at the cathode surface or disproportioned to O_2 gas and H_2O according to the following reactions:

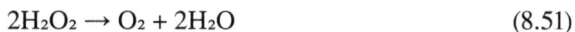

$$H_2O_2 + 2H^+ + 2e^- \rightarrow 2H_2O \tag{8.50}$$

$$2H_2O_2 \rightarrow O_2 + 2H_2O \tag{8.51}$$

Depending on the addition or in-situ electro-generation of Fenton reagents, the EF process can be divided into four types, as shown in Table 8.1 [46, 43]. In type I, Fe^{2+} is supplied externally while H_2O_2 is produced in situ by electrochemical reduction of O_2 at the cathode, avoiding the costs and dangers of handling, shipping, and storage of H_2O_2. In type II, both Fenton reagents, Fe^{2+} and H_2O_2, are supplied externally, and then Fe^{2+} is regenerated at the cathode from Fe^{3+}, which is generated from the Fenton reaction (Eq. 8.1). This process can reduce the initial Fe^{2+} input and iron sludge production. H_2O_2 is supplied externally in type III, but Fe^{2+} is electro-generated using a sacrificial anode. This method has certain drawbacks, such as a high anode consumption rate and considerable iron sludge generation. Finally, in type IV, both Fenton reagents, H_2O_2 and Fe^{2+}, are in situ generated electrochemically using an air-sparging cathode and sacrificial anode, respectively.

TABLE 8.1

Different Strategies of EF Process and Their Corresponding Electrochemical Reactions.

Type	Fe²⁺	H₂O₂	Electrochemical Reactions
I.	External addition	Cathodic in situ electro-generation	Anode: $2H_2O \rightarrow 4H^+ + O_2 + 2e^-$ Cathode: $O_2 + 2H^+ + 2e^- \rightarrow H_2O_2$
II.	Cathodic in-situ electro-regeneration	External addition	Anode: $2H_2O \rightarrow 4H^+ + O_2 + 2e^-$ Cathode: $Fe^{3+} + e^- \rightarrow Fe^{2+}$
III.	Anodic in-situ electro-generation	External addition	Anode: $Fe^0 \rightarrow Fe^{2+} + 2e^-$ Cathode: $2H_2O + 2e^- \rightarrow H_2 + 2OH^-$
IV.	Anodic in-situ electro-generation	Cathodic in-situ electro-generation	Anode: $Fe^0 \rightarrow Fe^{2+} + 2e^-$ Cathode: $O_2 + 2H^+ + 2e^- \rightarrow H_2O_2$; $Fe^{3+} + e^- \rightarrow Fe^{2+}$

EF process offers several advantages over the classical Fenton process, such as

(i) The ability to generate Fenton's reagents in situ eliminates the need for storage and dosing facilities for those compounds.
(ii) Cathodic regeneration of Fe³⁺ to Fe²⁺ accelerates the Fenton reaction and minimizes iron sludge production.
(iii) Because of the additional electro-generation of •OH at the anode, organic contaminants mineralization rates are higher than with the traditional Fenton process.

8.8 FACTORS AFFECTING THE EF PROCESS

8.8.1 Cell Configuration

Fenton's reagent can be electrogenerated in either a divided or an undivided cell using two- or three-electrode systems. Divided cell systems consist of two solutions, anolyte and catholyte, which are generally divided by an ion exchange membrane or salt bridge that enables ions to flow between the cells in order to maintain electroneutrality in both solutions. H₂O₂ is electrogenerated in the catholyte from the reduction of dissolved oxygen (Eq. 8.39) and participates in pollutant degradation via the Fenton process. Depending on whether a reference electrode is used or not, the system can function with three or two electrodes. Three-electrode systems typically work under potentiostatic conditions, while two-electrode systems operate under galvanostatic conditions. In the potentiostatic condition, a constant potential is applied to the cathode against a reference electrode (usually Ag/AgCl or SCE), resulting in a current

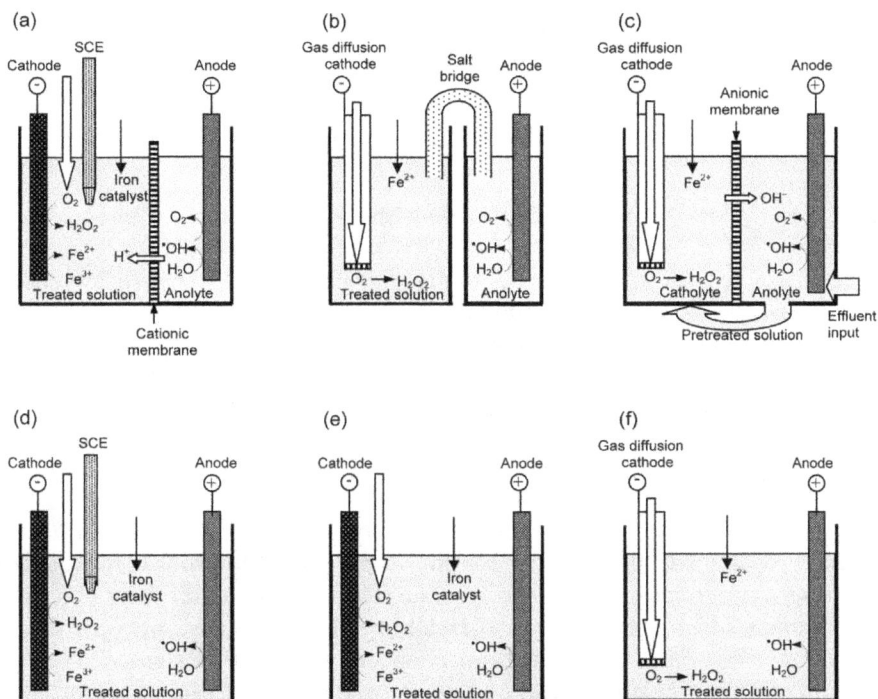

FIGURE 8.5 Schematic representation of divided (a to c) and undivided cell (d to f) configurations used in EF treatment of organic micropollutants with their corresponding electrode reactions. Adapted from Reference [11] with permission.

flow between the anode and cathode, while in the galvanostatic condition, a constant current or current density is directly applied to the cell.

On the other hand, in an undivided cell, there is no separator between the anode and cathode which provides an advantage of lower cell voltage requirement for electrolysis. However, this type of system is not appropriate for obtaining a large amount of H_2O_2 due to the anodic oxidation of H_2O_2 to O_2 (Eqs. 8.43 and 8.44). A schematic representation of divided and undivided EF cell configurations with their corresponding electrode reactions is depicted in Figure 8.5.

8.8.2 Current Density

Current density, which is the driving force of H_2O_2 electro-generation, plays a significant role in the efficiency of the EF process. Higher current density increases the production of H_2O_2, which in turn enhances the production of •OH. A higher current density also leads to rapid Fe^{2+} regeneration (Eq. 8.42) and improves the Fenton process efficiency. However, the higher current density may hasten the development of side reactions such as anodic oxygen evolution (Eq. 8.47), cathodic hydrogen reduction (Eq. 8.48), and parasitic reactions of •OH (Eq. 8.49). Apart from that, higher current density necessitates an increased energy usage.

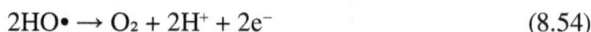

$$2H_2O \rightarrow 4H^+ + O_2 + 4e^- \tag{8.52}$$

$$2H^+ + 2e^- \rightarrow H_2 \tag{8.53}$$

$$2HO• \rightarrow O_2 + 2H^+ + 2e^- \tag{8.54}$$

Therefore, increasing current density has the advantages of faster pollutant degradation rates and shorter reaction times, but it also has the disadvantages of higher energy consumption with worse current efficiency. Therefore, it is important to adjust the applied current density in order to achieve a satisfactory balance between the intended process efficiency and the associated energy expenditures [48, 49].

8.8.3 ELECTRODE MATERIALS

Selecting the appropriate electrode materials is an essential step in the electro-Fenton process. Proper anode materials are essential in preventing electrode corrosion, and high oxygen overvoltage anodes can generate •OH to increase treatment efficiency [50]. Platinum's high conductivity and excellent stability make it a good choice for functioning as an anode in electrochemical operations [48, 49]. Extensively studied platinum anodes include Pt flake, Pt gauze, Pt grid, Pt mesh, Pt plate, and Pt sheet. However, the high costs of Pt reduce its economic feasibility. A better strategy for dealing with this issue is to utilize platinized anodes, which have a layer of Pt plating applied on a suitable metal. Similar results to Pt anodes have been achieved. Apart from Pt, other noble metals like gold and silver can also work as an anode material, but their exorbitant prices do not make them viable for practical use.

Boron-doped diamond (BDD) is another efficient and suitable electrode material for the EF process. In many electrochemical processes, BDD has been used as either an anode or both an anode and cathode because of the broader potential window and smaller background currents in aqueous solutions. Compared to Pt electrodes, BDD aids in producing a significantly higher amount of •OH and may fully eliminate unsaturated and aromatic contaminants [51]. The performance of the BDD electrode compared with the Pt electrode was monitored by studying the degradation of 2-(2,4-dichlorophenoxy)-propionic acid, and the findings revealed that the BDD electrode provided higher degradation efficiency due to its higher electrochemical activity and enhanced production of •OH on the surface [52]. Dimensionally stable anodes (e.g., Ti rod coated with RuO_2/IrO_2), carbon felt, graphite and carbon nanotubes (CNTs) are other forms of anodes utilized in EF processes.

Carbon-based materials are commonly employed as cathodes for EF systems due to their effectiveness in DO reduction, lower H_2O_2 decomposition capability, and enhanced hydrogen evolution overpotential. The most common carbon-based cathode materials used for H_2O_2 electro-generation are graphite rods, felts, paper and plates, carbon felts, carbon sponge, CNTs, activated carbon fiber, and reticulated vitreous carbon. A further improvement of H_2O_2 electro-generation can be achieved by modifying the cathode materials to generate more active sites at the cathode surface.

Porous carbon, porous CNT sponge, and carbon aerogel are some prominent examples of modified carbon cathodes [11].

8.8.4 ELECTROLYTES

Electrolytes play an important role in the efficiency of an electro-Fenton process by improving the solution conductivity. Sodium sulphate (Na_2SO_4) is the most widely used electrolyte due to its high ionic strength and little interference with aqueous solutions in the EF process. However, other electrolytes have also been used, including NaCl, KCl, $NaClO_4$, and $NaNO_3$. When chloro-salts are utilized as the electrolyte, the presence of Cl^- is expected to improve degradation efficiency and reduce reaction time by generating chlorine (Cl_2(aq)) or hypochlorous acid (HClO), both of which are powerful oxidizing agents in aqueous solutions. However, the presence of these active Cl species might lead to the generation of toxic organochlorinated compounds as intermediates. For example, electrochemical oxidation of phenol generated nonvolatile chlorinated aliphatic acids and volatile trichloromethane when NaCl was used as an electrolyte. The formation of toxic intermediate compounds can be prevented only when organic molecules are fully mineralized.

8.8.5 OXYGEN-SPARGING RATE

The rate of oxygen sparging is a key limiting factor in EF process efficiency. A higher oxygen sparging rate can increase the DO level and mass transfer rate in the system, promoting H_2O_2 generation on which the removal efficiency of the organic contaminants is dependent [49]. However, the generation of H_2O_2 is not proportional to the amount of oxygen flowing through the system. In fact, the increasing O_2 sparging rate does not increase the DO concentration in the system once the DO concentration of the system reaches saturation (8.28 mg/L at 25 °C), and, therefore, H_2O_2 production remains steady.

8.9 APPLICATION OF FENTON AND MODIFIED FENTON PROCESSES FOR MICROPOLLUTANT TREATMENT

Fenton and modified Fenton processes have been extensively studied for the degradation and mineralization of various emerging micropollutants. In a study, Höfl et al. [53] evaluated the effectiveness of three AOPs (UV/H_2O_2, UV/O_3, and Fenton process) for the degradation of adsorbable organic halogen (AOX) and COD from pharmaceutical wastewater. They pointed out the benefits of the Fenton process over other AOPs. The findings revealed that all three techniques effectively degraded AOX and COD. When comparing AOX degradation to COD degradation, UV irradiation showed strong selectivity for AOX. Processes relying on •OH radicals, on the other hand, were less selective but more efficient in COD removal. COD and AOX could be eliminated almost entirely using the Fenton process, with a very short reaction time compared to the other two AOPs.

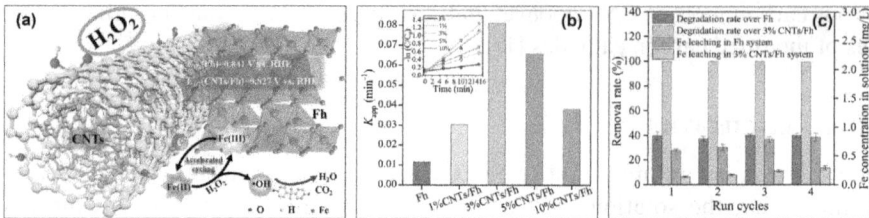

FIGURE 8.6 Degradation of bisphenol-A by CNT/Fh composite catalysts: (a) reaction mechanism, (b) apparent degradation rate constants, and (c) stability test of CNT/Fh catalysts. Adapted with permission from Reference [54].

Zhu et al. [54] studied the degradation of bisphenol A (BPA) using a composite material of oxidized multi-walled carbon nanotubes and ferrihydrite (CNTs/Fh) as a heterogeneous Fenton catalyst. They synthesized a series of CNTs/Fh composites by simply swirling CNTs and Fh together. Compared to simple Fh, a 3% CNTs/Fh system degraded the pollutant seven times faster. Furthermore, after four cycles of reuse, CNTs/Fh maintained strong Fenton activity, and XPS findings revealed that the graphitic structure of the used CNTs is highly robust. DFT calculations and cyclic voltammetry characterization of the CNTs/Fh composite were useful in revealing the catalytic mechanism of the possible degradation process (Figure 8.6). The increased electron conductivity of CNTs/Fh composites facilitated Fe(III)/Fe(II) redox cycling, promoting the breakdown of H_2O_2 and the formation of •OH.

Zhong et al. [55] used a heterogeneous photo-Fenton process catalyzed by titano-magnetite to study the degradation of tetrabromobisphenol A (TBBPA), a brominated flame retardant. Within 240 minutes of UV irradiation, an almost complete break-down of TBBPA was accomplished using the optimal conditions of 0.125g/L catalyst dosage, 10 mmol/L H_2O_2, and pH 6.5. The addition of Ti^{4+} to magnetite significantly enhanced its catalytic activity. TBBPA degradation was triggered by debromina-tion and the ß-scission reaction. The produced intermediates, which mostly com-prise mono-, di-, and tri-BBPA, were identified by GC-MS, and their concentration approached undetectable levels during the experiment. In another research, Ayoub et al. [56] studied the heterogeneous photo-Fenton process for removing 21 micro-pollutants from the water samples collected from the Meurthe river in France. The identified micropollutants belong to the pharmaceuticals, personal care products, endocrine disruptors, and perfluorinated compounds with different initial con-centrations ranging from 0.8 to 88 ng/L. The heterogeneous catalyst used for the photo-Fenton process was iron (III)-impregnated zeolite. Under optimized reaction conditions, (20 wt% Fe (III) impregnated zeolite dosage, 0.007 mol/L H_2O_2 concen-tration, pH 5.5, and UV light irradiation), 19 out of 21 micropollutant concentrations reached below the detection limit within 30 minutes to 6 hours depending on their initial concentration.

Barhoumi et al. [57] studied the degradation of levofloxacin using heterogeneous pyrite-catalyzed electro-Fenton process using BDD/carbon felt cell. Pyrite serves as a source of Fe^{2+} ions and regulates the solution pH. Another Fenton reagent, H_2O_2, was *in-situ* electrogenerated within the system under oxygen bubbling. The major oxidizing agent •OH was produced from anodic water oxidation and the Fenton process. A 95% elimination of total organic carbon was possible when a levofloxacin solution with a 0.23 mM concentration was degraded at 300 mA for 8 hours. An analysis of mineralization of levofloxacin by ion chromatography revealed the presence of F^-, NO_3^-, and NH_4^+ in the solution.

Zhao et al. [58] investigated the electro-Fenton degradation of emerging pesticide imidacloprid using iron oxide modified activated carbon aerogel ($Fe_3O_4@Fe_2O_3/ACA$) composite cathode. The system operated within a wide pH range, from 3 to 9, without compromising the ability to degrade 90% of imidacloprid and TOC within 30 and 60 minutes, respectively. A comprehensive analysis of surface characteristics and produced oxidants was used to suggest the oxidation process of the E-Fenton process with the new composite cathode at varied pH ranges. In acidic pH, electro-generated H_2O_2 interacts with dissolved iron ions and surface Fe (II) sites to produce •OH through a Haber–Weiss process. In contrast, in basic pH, it interacts with Fe (II) and Fe (III) sites on the surface and produces superoxide anion ($•O_2^-/•O_2H$) and •OH.

8.10   MINERALIZATION STUDY OF PHENOL BY FENTON PROCESS

Among the various toxic and persistent organic micropollutants, phenol and phenolic compounds are found in high concentrations in wastewater discharge of various industries, including pharmaceuticals, foods, resins, dyes, papermaking, and petrochemicals. The existence of phenol in water bodies causes severe pollution, and both human and aquatic life are severely affected by its carcinogenic and mutagenic consequences [59]. Brillas and Gracia-Segura proposed a Fenton-based reaction mechanism (Figure 8.7) for phenol degradation and mineralization in water on the basis of experimental evidence [60–62]. The information on the production of various oxidation products can be obtained using high-performance liquid chromatography (HPLC) and gas chromatography-mass spectrometry (GC-MS) analysis of treated wastewater. Hydroxylation and charge transfer processes between the oxidation derivatives are involved in the mineralization process of phenol. When phenol is first hydroxylated, it produces hydroquinone (HQ), catechol (CT), and resorcinol (RC) as intermediate products. The activation energies (E_a) for the formation of HQ, CT, and RC by phenol hydroxylation are 0.855, 1.174, and 2.534 kcal/mol, respectively [61]. As a result, the detection of HQ is prevalent over other hydroxylated phenol intermediates. It's worth noting that HQ and CT have redox equilibrium with their respective quinones (4-benzoquinone and 2-benzoquinone), even though these species are also classified as intermediates. Simultaneously, 1,2,3-trihydroxybenzene is produced by further hydroxylation of CT and RC. Following the ring opening of these aromatic intermediates, a mixture of carboxylic acids is produced, which is then converted to oxalic and formic acids. The final transformations of these two acids end up with CO_2 and H_2O.

Although the mineralization pathway illustrated in Figure 8.7 is apparently simpler, the complete mineralization of phenol may be impeded by several species in the aqueous matrix. The presence of iron ions, which either naturally occur in water

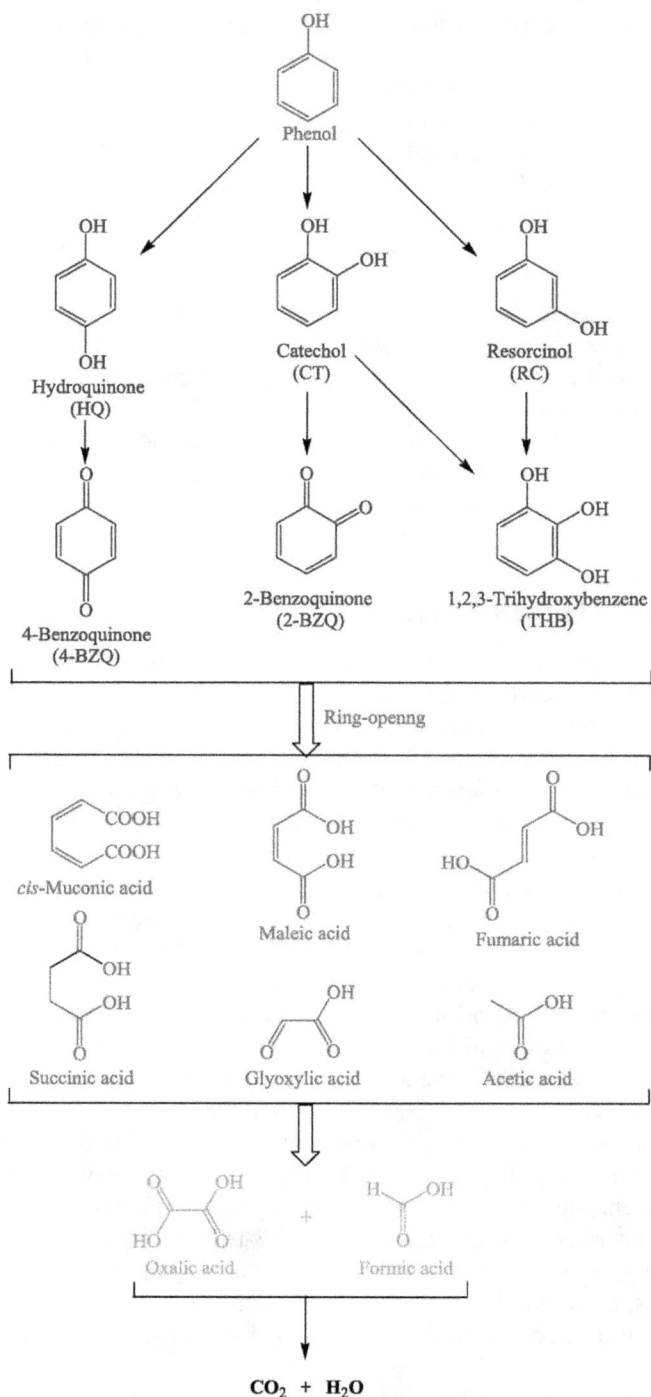

FIGURE 8.7 Degradation and mineralization pathway of phenol by •OH generated from Fenton process. Adapted from Reference [60] with permission.

or is introduced as Fenton's catalyst, is one of the most unwanted consequences. For example, Fe^{3+}, generated by the Fenton reaction, is well recognized for forming stable complexes with final carboxylic acids. However, these Fe (III)-carboxylate complexes are seldom degraded by the •OH and induce a poor degree of mineralization. This situation can be overcome by introducing EF or PF process. In EF, Fe^{3+} ions are effectively reduced to Fe^{2+} and generate Fe (II) carboxylate complexes by interacting with carboxylic acids. Fe (II) carboxylate complexes are more easily degraded than Fe (III)-carboxylate complexes, and more phenol mineralization is obtained. In the case of the PF process, photolysis of Fe (III) carboxylate complexes improves the mineralization efficiency.

8.11 CONCLUDING REMARKS AND FUTURE PERSPECTIVE

Fenton oxidation is a promising AOP to overcome the challenges of treating emerging and persistent micropollutants from an aqueous environment. Both Fenton reagents (Fe^{2+} and H_2O_2) are easily available, inexpensive, and environment friendly. The homogeneous Fenton process exhibits minimal limitation on mass transfer and therefore exhibits faster reaction kinetics. However, a limited working pH range and an enormous amount of iron sludge production limit its application on an industrial scale. The development of heterogeneous Fenton catalysts is a promising solution to overcome the limitations of the conventional Fenton process. Recent studies on various kinds of ferrous and non-ferrous heterogeneous Fenton catalysts showed an enormous potential for large-scale application of the Fenton process. However, the commercial use of heterogeneous Fenton processes may be hindered primarily by the long-term deactivation of the catalyst's active sites. Therefore, more research is necessary to develop novel catalysts with a wider workable pH range, high stability, minimal deactivation, and leaching.

FBF process is a promising strategy to implement the Fenton process on a bigger scale. A synergistic combination of homogeneous and heterogeneous Fenton processes can be achieved in an FBF process which further promotes the removal efficiency of emerging contaminants. Moreover, the development of modified Fenton processes, such as photo-Fenton and electro-Fenton, shows enormous potential for application on a broader scale. Several research groups worldwide are currently developing the photo-Fenton and electro-Fenton processes, particularly developing novel reactor configurations, electrode materials, and operational parameters. Integrating the FBF process with light irradiation or an electric field may provide an excellent solution for large-scale applications.

Finally, there should be more focus on identifying reaction intermediates, developing rate expression for complex contaminants transformation and mineralization, identifying scale-up factors, and establishing cost-effectiveness criteria.

REFERENCES

1. Y. Luo, W. Guo, H.H. Ngo, L.D. Nghiem, F.I. Hai, J. Zhang et al., *A review on the occurrence of micropollutants in the aquatic environment and their fate and removal during wastewater treatment*, Sci. Total Environ. 473–474 (2014), pp. 619–641.

2. P.R. Rout, T.C. Zhang, P. Bhunia and R.Y. Surampalli, *Treatment technologies for emerging contaminants in wastewater treatment plants: A review*, Sci. Total Environ. 753 (2021), pp. 141990.

3. J.-Q. Jiang, Z. Zhou and V.K. Sharma, *Occurrence, transportation, monitoring and treatment of emerging micro-pollutants in waste water—A review from global views*, Microchem. J. 110 (2013), pp. 292–300.

4. L. Prieto-Rodríguez, I. Oller, N. Klamerth, A. Agüera, E.M. Rodríguez and S. Malato, *Application of solar AOPs and ozonation for elimination of micropollutants in municipal wastewater treatment plant effluents*, Water Res. 47 (2013), pp. 1521–1528.

5. H.J.H. Fenton, *On a new reaction of tartaric acid*, Chem News. 33 (1876), pp. 190.

6. H.J.H. Fenton, *Oxidation of tartaric acid in presence of iron*, J. Chem. Soc., Trans. 65 (1894), pp. 899–910.

7. F. Haber and J. Weiss, *On the catalysis hydroperoxydes*, Naturwissenschaften. 20 (1932), pp. 948–950.

8. F. Haber and J. Weiss, *The catalytic decomposition of hydrogen peroxide by iron salts*, Proc. R. Soc. London. Ser. A—Math. Phys. Sci. 147 (1934), pp. 332–351.

9. S. Giannakis, *A review of the concepts, recent advances and niche applications of the (photo) Fenton process, beyond water/wastewater treatment: Surface functionalization, biomass treatment, combatting cancer and other medical uses*, Appl. Catal. B Environ. 248 (2019), pp. 309–319.

10. J.J. Pignatello, E. Oliveros and A. MacKay, *Advanced oxidation processes for organic contaminant destruction based on the Fenton reaction and related chemistry*, Crit. Rev. Environ. Sci. Technol. 36 (2006), pp. 1–84.

11. E. Brillas, I. Sirés and M.A. Oturan, *Electro-Fenton process and related electrochemical technologies based on Fenton's reaction chemistry*, Chem. Rev. 109 (2009), pp. 6570–6631.

12. C.P. Huang, C. Dong and Z. Tang, *Advanced chemical oxidation: Its present role and potential future in hazardous waste treatment*, Waste Manag. 13 (1993), pp. 361–377.

13. A.D. Bokare and W. Choi, *Review of iron-free Fenton-like systems for activating H_2O_2 in advanced oxidation processes*, J. Hazard. Mater. 275 (2014), pp. 121–135.

14. R. Guo, X. Xie and J. Chen, The degradation of antibiotic amoxicillin in the Fenton-activated sludge combined system, Environ. Technol. 36 (2015), pp. 844–851.

15. S.H. Bossmann, E. Oliveros, S. Göb, S. Siegwart, E.P. Dahlen, L. Payawan et al., *New evidence against hydroxyl radicals as reactive intermediates in the thermal and photochemically enhanced Fenton reactions*, J. Phys. Chem. A. 102 (1998), pp. 5542–5550.

16. M.M. Bello, A.A. Abdul Raman and A. Asghar, *A review on approaches for addressing the limitations of Fenton oxidation for recalcitrant wastewater treatment*, Process Saf. Environ. Prot. 126 (2019), pp. 119–140.

17. V. Kavitha and K. Palanivelu, *Degradation of nitrophenols by Fenton and photo-Fenton processes*, J. Photochem. Photobiol. A Chem. 170 (2005), pp. 83–95.

18. J.A. Zazo, G. Pliego, S. Blasco, J.A. Casas and J.J. Rodriguez, *Intensification of the Fenton process by increasing the temperature*, Ind. Eng. Chem. Res. 50 (2011), pp. 866–870.

19. S.-S. Lin and M.D. Gurol, *Catalytic decomposition of hydrogen peroxide on iron oxide: Kinetics, mechanism, and implications*, Environ. Sci. Technol. 32 (1998), pp. 1417–1423.

20. J. He, X. Yang, B. Men and D. Wang, *Interfacial mechanisms of heterogeneous Fenton reactions catalyzed by iron-based materials: A review*, J. Environ. Sci. 39 (2016), pp. 97–109.

21. L. Xu and J. Wang, *Fenton-like degradation of 2,4-dichlorophenol using Fe_3O_4 magnetic nanoparticles*, Appl. Catal. B Environ. 123–124 (2012), pp. 117–126.

22. G.B. Ortiz de la Plata, O.M. Alfano and A.E. Cassano, *The heterogeneous photo-Fenton reaction using goethite as catalyst*, Water Sci. Technol. 61 (2010), pp. 3109–3116.

23. P.V. Nidheesh, *Heterogeneous Fenton catalysts for the abatement of organic pollutants from aqueous solution: A review*, RSC Adv. 5 (2015), pp. 40552–40577.

24. X. Huang, X. Hou, F. Jia, F. Song, J. Zhao and L. Zhang, *Ascorbate-promoted surface iron cycle for efficient heterogeneous fenton alachlor degradation with hematite nanocrystals*, ACS Appl. Mater. Interfaces. 9 (2017), pp. 8751–8758.

25. Z. Zhang, M.F. Hossain and T. Takahashi, *Fabrication of shape-controlled α-Fe$_2$O$_3$ nanostructures by sonoelectrochemical anodization for visible light photocatalytic application*, Mater. Lett. 64 (2010), pp. 435–438.

26. Y. Zhu, R. Zhu, L. Yan, H. Fu, Y. Xi, H. Zhou et al., *Visible-light Ag/AgBr/ferrihydrite catalyst with enhanced heterogeneous photo-Fenton reactivity via electron transfer from Ag/AgBr to ferrihydrite*, Appl. Catal. B Environ. 239 (2018), pp. 280–289.

27. H.-H. Huang, M.-C. Lu and J.-N. Chen, *Catalytic decomposition of hydrogen peroxide and 2-chlorophenol with iron oxides*, Water Res. 35 (2001), pp. 2291–2299.

28. W. Liu, Y. Wang, Z. Ai and L. Zhang, *Hydrothermal synthesis of FeS 2 as a high-efficiency fenton reagent to degrade alachlor via superoxide-mediated Fe(II)/Fe(III) cycle*, ACS Appl. Mater. Interfaces. 7 (2015), pp. 28534–28544.

29. S. Ammar, M.A. Oturan, L. Labiadh, A. Guersalli, R. Abdelhedi, N. Oturan et al., *Degradation of tyrosol by a novel electro-Fenton process using pyrite as heterogeneous source of iron catalyst*, Water Res. 74 (2015), pp. 77–87.

30. M. Arienzo, *Oxidizing 2,4,6-trinitrotoluene with pyrite-H$_2$O$_2$ suspensions*, Chemosphere. 39 (1999), pp. 1629–1638.

31. Y. Wang, H. Zhao, M. Li, J. Fan and G. Zhao, *Magnetic ordered mesoporous copper ferrite as a heterogeneous Fenton catalyst for the degradation of imidacloprid*, Appl. Catal. B Environ. 147 (2014), pp. 534–545.

32. A.D. Bokare and W. Choi, *Zero-valent aluminum for oxidative degradation of aqueous organic pollutants*, Environ. Sci. Technol. 43 (2009), pp. 7130–7135.

33. E.G. Heckert, S. Seal and W.T. Self, *Fenton-Like Reaction Catalyzed by the Rare Earth Inner Transition Metal Cerium*, Environ. Sci. Technol. 42 (2008), pp. 5014–5019.

34. Y. Wang, X. Shen and F. Chen, *Improving the catalytic activity of CeO$_2$/H$_2$O$_2$ system by sulfation pretreatment of CeO$_2$*, J. Mol. Catal. A Chem. 381 (2014), pp. 38–45.

35. A.D. Bokare and W. Choi, *Advanced oxidation process based on the Cr(III)/Cr(VI) redox cycle*, Environ. Sci. Technol. 45 (2011), pp. 9332–9338.

36. A.D. Bokare and W. Choi, *Chromate-induced activation of hydrogen peroxide for oxidative degradation of aqueous organic pollutants*, Environ. Sci. Technol. 44 (2010), pp. 7232–7237.

37. R.J. Watts, J. Sarasa, F.J. Loge and A.L. Teel, *Oxidative and Reductive Pathways in Manganese-Catalyzed Fenton's Reactions*, J. Environ. Eng. 131 (2005), pp. 158–164.

38. J.I. Nieto-Juarez, K. Pierzchła, A. Sienkiewicz and T. Kohn, *Inactivation of MS2 coliphage in Fenton and Fenton-like systems: Role of transition metals, hydrogen peroxide and sunlight*, Environ. Sci. Technol. 44 (2010), pp. 3351–3356.

39. Z. Hu, C.-F. Leung, Y.-K. Tsang, H. Du, H. Liang, Y. Qiu et al., *A recyclable polymer-supported ruthenium catalyst for the oxidative degradation of bisphenol A in water using hydrogen peroxide*, New J. Chem. 35 (2011), pp. 149–155.

40. C. Ratanatamskul, S. Chintitanun, N. Masomboon and M.-C. Lu, *Inhibitory effect of inorganic ions on nitrobenzene oxidation by fluidized-bed Fenton process*, J. Mol. Catal. A Chem. 331 (2010), pp. 101–105.

41. E.M. Matira, T.-C. Chen, M.-C. Lu and M.L.P. Dalida, *Degradation of dimethyl sulfoxide through fluidized-bed Fenton process*, J. Hazard. Mater. 300 (2015), pp. 218–226.

42. C. Barrera-Díaz, F. Ureña-Nuñez, E. Campos, M. Palomar-Pardavé and M. Romero-Romo, *A combined electrochemical-irradiation treatment of highly colored and polluted industrial wastewater*, Radiat. Phys. Chem. 67 (2003), pp. 657–663.

43. M. Zhang, H. Dong, L. Zhao, D. Wang and D. Meng, *A review on Fenton process for organic wastewater treatment based on optimization perspective*, Sci. Total Environ. 670 (2019), pp. 110–121.

44. D.B. Hernández-Uresti, C. Alanis-Moreno and D. Sanchez-Martinez, *Novel and stable Fe-BiVO4 nanocatalyst by efficient dual process in the ciprofloxacin degradation*, Mater. Sci. Semicond. Process. 102 (2019), pp. 104585.

45. R. Tomat and A. Rigo, *Electrochemical production of OH. Radicals and their reaction with toluene*, J. Appl. Electrochem. 6 (1976), pp. 257–261.

46. M. Sillanpää and M. Shestakova, Emerging and combined electrochemical methods, in *Electrochemical water treatment methods*, Elsevier, 2017, pp. 131–225.

47. M. Sudoh, T. Kodera, K. Sakai, J.Q. Zhang and K. Koide, *Oxidative degradation of aqueous phenol effluent with electrogenerated Fenton's reagent*, J. Chem. Eng. Japan. 19 (1986), pp. 513–518.

48. P.V. Nidheesh and R. Gandhimathi, *Trends in electro-Fenton process for water and wastewater treatment: An overview*, Desalination. 299 (2012), pp. 1–15.

49. H. He and Z. Zhou, *Electro-Fenton process for water and wastewater treatment*, Crit. Rev. Environ. Sci. Technol. 47 (2017), pp. 2100–2131.

50. H. Zhang, C. Fei, D. Zhang and F. Tang, *Degradation of 4-nitrophenol in aqueous medium by electro-Fenton method*, J. Hazard. Mater. 145 (2007), pp. 227–232.

51. G. Pliego, J.A. Zazo, P. Garcia-Muñoz, M. Munoz, J.A. Casas and J.J. Rodriguez, *Trends in the intensification of the fenton process for wastewater treatment: An overview*, Crit. Rev. Environ. Sci. Technol. 45 (2015), pp. 2611–2692.

52. E. Brillas, M.Á. Baños, M. Skoumal, P.L. Cabot, J.A. Garrido and R.M. Rodríguez, *Degradation of the herbicide 2,4-DP by anodic oxidation, electro-Fenton and photo electro-Fenton using platinum and boron-doped diamond anodes*, Chemosphere. 68 (2007), pp. 199–209.

53. C. Höfl, G. Sigl, O. Specht, I. Wurdack and D. Wabner, *Oxidative degradation of aox and cod by different advanced oxidation processes: A comparative study with two samples of a pharmaceutical wastewater*, Water Sci. Technol. 35 (1997), pp. 257–264.

54. R. Zhu, Y. Zhu, H. Xian, L. Yan, H. Fu, G. Zhu et al., *CNTs/ferrihydrite as a highly efficient heterogeneous Fenton catalyst for the degradation of bisphenol A: The important role of CNTs in accelerating Fe(III)/Fe(II) cycling*, Appl. Catal. B Environ. 270 (2020), p. 118891.

55. Y. Zhong, X. Liang, Y. Zhong, J. Zhu, S. Zhu, P. Yuan et al., *Heterogeneous UV/Fenton degradation of TBBPA catalyzed by titanomagnetite: Catalyst characterization, performance and degradation products*, Water Res. 46 (2012), pp. 4633–4644.

56. H. Ayoub, T. Roques-Carmes, O. Potier, B. Koubaissy, S. Pontvianne, A. Lenouvel et al., *Iron-impregnated zeolite catalyst for efficient removal of micropollutants at very low concentration from Meurthe river*, Environ. Sci. Pollut. Res. 25 (2018), pp. 34950–34967.

57. N. Barhoumi, L. Labiadh, M.A. Oturan, N. Oturan, A. Gadri, S. Ammar et al., *Electrochemical mineralization of the antibiotic levofloxacin by electro-Fenton-pyrite process*, Chemosphere. 141 (2015), pp. 250–257.

58. H. Zhao, Y. Wang, Y. Wang, T. Cao and G. Zhao, *Electro-Fenton oxidation of pesticides with a novel Fe3O4@Fe2O3/activated carbon aerogel cathode: High activity, wide pH range and catalytic mechanism*, Appl. Catal. B Environ. 125 (2012), pp. 120–127.

59. H. Kusic, N. Koprivanac, A. Bozic and I. Selanec, *Photo-assisted Fenton type processes for the degradation of phenol: A kinetic study*, J. Hazard. Mater. 136 (2006), pp. 632–644.

60. E. Brillas and S. Garcia-Segura, *Benchmarking recent advances and innovative technology approaches of Fenton, photo-Fenton, electro-Fenton, and related processes: A review on the relevance of phenol as model molecule*, Sep. Purif. Technol. 237 (2020), p. 116337.

61. C. Valdés, J. Alzate-Morales, E. Osorio, J. Villaseñor and C. Navarro-Retamal, *A characterization of the two-step reaction mechanism of phenol decomposition by a Fenton reaction*, Chem. Phys. Lett. 640 (2015), pp. 16–22.

62. Y.-H. Huang, Y.-J. Huang, H.-C. Tsai and H.-T. Chen, *Degradation of phenol using low concentration of ferric ions by the photo-Fenton process*, J. Taiwan Inst. Chem. Eng. 41 (2010), pp. 699–704.

9 Techno-Economic Assessment of the Application of Advanced Oxidation Processes for the Removal of Seasonal and Year-Round Contaminants from Drinking Water

Reece Lima-Thompson and Stephanie Leah Gora

CONTENTS

9.1 Introduction 188
9.2 Techno-economic Assessment 189
9.3 Advanced Oxidation Processes 190
9.4 AOPs for Drinking Water Treatment 190
9.5 TEA for AOPs 191
9.6 Methods 192
    9.6.1 Cost Estimates and Cost Curves 192
    9.6.2 Treatment Conditions 194
    9.6.3 Inflation 195
    9.6.4 Error Bounds 195
    9.6.5 Data Analysis and Visualization 195
9.7 Results 196
    9.7.1 Ozone 196
    9.7.2 UV/H_2O_2 200
    9.7.3 Key Costing Factors for AOPs 204
    9.7.4 Implications of Influent Water Quality on System
            Operation and Costs 204
    9.7.5 Integrating AOPs into Water Treatment Plants 204
    9.7.6 Upstream and Downstream Considerations 206

DOI: 10.1201/9781003247913-9

9.8   Summary and Conclusion.. 210
    9.8.1   Limitations of Analysis 210
    9.8.2   Future Research Directions 211
        9.8.2.1   Positioning Statement 212
References..212

9.1   INTRODUCTION

Design engineers and system owners require accurate cost predictions that account for these costing factors in order to choose between different design options. Cost curves are widely used in the water and wastewater treatment industry to develop conceptual cost estimates allowing design engineers and their clients to compare different technical solutions. In this desktop study, we used cost estimates published in government and academic studies to build and update cost curves for ozone and UV/H_2O_2. Costing information for ozone was obtained from the vendor-validated USEPA document: *Technologies and Costs Document for the Final Long Term 2 Enhanced Surface Water Treatment Rule and Final Stage 2 Disinfectants and Disinfection Byproducts Rule*. Costing data for UV/H_2O_2 is very limited in the public domain, and cost curves for these AOPs were developed by combining curves for their component unit processes (UV, H_2O_2, and HOCl for quenching) originally published by the USEPA. These curves were compared to cost estimates, cost curves, and cost functions in the academic literature and to the final constructed costs of a small set of full-scale North American installations.

Our analysis indicates that the accuracy and comparability of cost estimates, cost curves, and costing functions available in the public domain are varied and depend on the number and type of costing elements included. Most of the cost estimates available in the public domain do not specify the UV dose, H_2O_2 dose, or other design assumptions used to cost out the equipment, making it difficult to compare them to one another. Seasonal operation of AOPs, which would be expected in drinking water plants using AOPs for T&O and cyanotoxin removal, will substantially decrease operation and maintenance (O&M) costs but will have less of an impact on capital costs as the process must be powerful enough to achieve the design goals at maximum flow rates and influent contaminant concentrations. In UV/H_2O_2 treatment processes, UV equipment drives capital costs and H_2O_2-related costing factors, especially quenching and drive operating costs. Granular activated carbon (GAC) contactors may be a viable alternative to hypochlorite for H_2O_2 quenching. Still, their feasibility will depend on the plant capacity, the H_2O_2 dose applied, whether the target contaminants are seasonal or non-seasonal, existing infrastructure, and local funding paradigms.

In this study, we reviewed the existing costing data and models for ozone and UV/H_2O_2 and developed costing curves for each process based on data and methods published by the USEPA and academic researchers. The cost curves and subsequent analyses were used to identify important capital and operational cost drivers for each process and fruitful areas for future research. The UV/H_2O_2 cost curves presented here should not be used for cost estimation as they were built by combining cost estimates originally developed for disinfection applications and have not been validated by external vendors or system designers.

The results of this analysis were not rigorously validated against real systems and should not be used in place of professional conceptual cost estimates. We recommend that a comprehensive project be initiated to solicit and analyse cost estimates developed by multiple equipment vendors and system designers to accurately characterize the contributions of different cost elements to total capital and O&M costs, and develop realistic, reliable, and vendor-validated cost curves for common AOPs.

9.2 TECHNO-ECONOMIC ASSESSMENT

Techno-economic assessment encompasses many approaches to predicting the cost of applying a technological solution to solve an engineering problem. These approaches range from cost estimates for the component parts of a single product to long-ranging analyses of the annualized costs of building, operating, and maintaining a piece of infrastructure throughout its entire lifetime. In this study, we have chosen a techno-economic assessment approach based on cost estimates for full-scale drinking water treatment unit processes and the development and analysis of preliminary cost curves for ozonation and UV/H_2O_2, an advanced oxidation process (AOP) that has been well characterized at the laboratory scale but is only now beginning to be used at the full scale for drinking water treatment.

Cost estimates are usually developed by soliciting and comparing quotes from multiple vendors to design a system that meets the stated treatment goals at the chosen plant capacities. Estimates can also be "built-up" component by component, based on unit costs for different equipment and infrastructure[1,2] or be based on professional experience with similar treatment processes and total costs for similar projects that have been completed in the past.[3] The most useful cost estimates are site-specific and include local considerations (e.g. treatment standards, source water quality, environmental requirements, cost of purchasing land, local labour rates). Accurate and site-specific cost estimates allow system designers and owners to decide which unit process(es) should be built to achieve a treatment goal.

In aggregate, cost estimates can also be used to create more generalizable costing curves (cost capacity curves) and costing functions. Cost curves plot the estimated cost to construct or operate a unit process against capacity (daily flow rate). Cost curves are available for conventional water treatment unit processes in government or industry publications. When they exist, publicly accessible cost curves for less conventional technologies, including AOPs, are usually found in the academic literature.[4] Some of the most useful cost curves have been developed by soliciting cost estimates for a hypothetical unit process or its component parts from multiple expert design engineers (process designers and/or vendors). The USEPA has used this approach to develop cost curves for processes ranging from coagulation to UV disinfection to nanofiltration.[2]

A costing function is an equation that relates plant capacity to total cost and thus describes the trend of a cost curve. It may be a linear, polynomial, or power function depending on the specifics of the unit process and the volume and quality of the data used to create the cost curve.[3] Like a cost curve, a costing function can be used to estimate the cost of a unit process on the basis of its flow capacity. They are useful for programming and modelling exercises but are no more accurate than visually examining a cost curve.

Even the most general cost estimates quickly become outdated and must be updated to account for inflation and other cost increases. For example, the Construction Cost Index (CCI) is published monthly by the Engineering News-Record (ENR). It is widely used to convert a cost estimate developed in one year into an equivalent cost in a different year.[3,5] For example, a cost estimate prepared in 2008 can be updated to 2020 dollars by multiplying the 2008 estimate by the ratio of the 2020 and 2008 CCI values, as shown in Eq. (9.1).

$$Cost_{2020} = \frac{CCI_{2020}}{CCI_{2008}} \times Cost_{2008} \qquad (9.1)$$

The CCI is a useful metric to account for the overall impacts of inflation on the materials and labour costs of building a new structure. Still, it was developed specifically for construction projects. As a result, it did not explicitly account for many factors that might impact the cost of water treatment plant (WTP) equipment or ongoing O&M costs,[1] and a more complex approach to cost updating may be required for the more detailed cost estimates that are required further along in the design process.

9.3　ADVANCED OXIDATION PROCESSES

AOPs are drinking water and wastewater treatment applications used to degrade recalcitrant and potentially toxic micropollutants that are otherwise difficult to remove in conventional drinking water treatment processes. AOPs can degrade recalcitrant micropollutants because they produce hydroxyl radicals and other highly oxidative radical species,[6] which can break down micropollutants into products that are less harmful to human health and the environment.[7]

The effectiveness of AOPs is heavily influenced by the water matrix.[8] Depending on the water matrix being treated, different AOPs may be more efficient and, hence, more economical at producing safe drinking water quality that abides by the governing drinking water standards. In addition to treatment efficacy, the composition of the water matrix has implications for the formation of undesirable intermediate oxidation products,[9,10] increased disinfection byproduct (DBP) formation,[9,11–13] and microbial regrowth in the distribution system.[12–16]

9.4　AOPS FOR DRINKING WATER TREATMENT

Conventional water treatment processes like coagulation, flocculation, clarification, filtration, and disinfection are optimized to remove pathogens, particulates, natural organic matter (NOM), and other common health and aesthetic contaminants from surface or groundwater before it is sent to users. Existing processes are proven, robust, and usually sufficient to remove the most regulated contaminants. However, they are not always effective for removing emerging contaminants increasingly detected in source water and regulated in treated drinking water, such as cyanobacteria-related cyanotoxins, taste and odour compounds (T&O compounds), and industrial pollutants like 1,4-dioxane.

Cyanotoxins are emerging contaminants increasingly being detected in surface water bodies across the globe due to climate change and high nutrient loadings of nitrogen and phosphorus from anthropogenic agricultural activities.[17] Cyanotoxins are produced by harmful blooms of cyanobacteria and have the potential to cause serious problems to the environment and human health. When cyanotoxins are consumed in drinking water, they can affect the liver or nervous system and cause gastrointestinal illness, fever, and headaches. Microcystin-LR, a cyanotoxin, has also been identified as a possible carcinogen.[18] Conventional drinking WTPs are not specifically designed to remove cyanotoxins or cyanobacterial cells, with the decentralized water treatment systems at most risk as they have less extensive treatment.[17]

People often perceive good drinking water quality as water with good taste and no smell. Therefore, it is critical for drinking water treatment engineers to design treatment processes capable of providing good aesthetic water quality. T&O compounds impact drinking water quality aesthetics. T&O compounds may not have significant health impacts when consumed in drinking water; however, most people will not drink it if it does not taste good and/or has an unpleasant smell.[19] AOPs like ozone, UV/H_2O_2, and UV/Cl have been applied for T&O control at the laboratory scale.[13, 16, 19, 20]

Another emerging contaminant of concern in drinking water sources in North America is 1,4-dioxane. 1,4-Dioxane is a versatile synthetic compound with applications in fabric cleaning, pharmaceuticals, electronics, metal finishing, antifreeze, paper manufacturing, and more.[22] Wastewater discharge, accidental spills, and improper disposal practices have increased the presence of 1,4-dioxane in drinking water sources.[14] 1,4-Dioxane is classified as a probable human carcinogen highlighting the importance and need for treatment methods to degrade the compounds.[14] 1,4-Dioxane is not biodegradable, and traditional remediation methods have shown inadequate degradation of the compound.[22] UV/H_2O_2, one of the AOPs discussed in this chapter, has shown a promising ability to degrade 1,4-dioxane either completely or into partial degradation products that are easier to remove from water.[14]

Some drinking WTPs in North America have already installed full-scale AOPs to control seasonal taste and odour concerns[21, 23] and cyanotoxins[24] in surface water or 1,4-dioxane from groundwater.[25, 26] As climate change drives temperature increases across the world, the incidence of seasonal contaminants in surface water supplies impacted by nutrient (nitrogen and phosphorus) pollution is likely to increase,[27] resulting in a greater demand for technologies that can remove them from drinking water. Increasing urbanization and industrialization worldwide will also increase the demand for treatment processes that can remove industrial pollutants from potential drinking water supplies and other water bodies. AOPs are strong contenders for these applications.

9.5　TEA FOR AOPs

Existing TEAs for AOPs in the academic literature are based on general scientific feasibility,[28, 29] focused on energy costs,[30] or related to small-scale or highly specific applications.[31, 32] The exception is Plumlee et al.,[4] who developed and published costing curves for ozone, peroxone, and UV/H$_2$O$_2$, among other advanced processes, in the context of water reclamation. Their analysis was based on full process costing estimates developed by engineering design practitioners and equipment manufacturers. Unfortunately, the researchers only had access to limited costing data and validation opportunities for UV/H$_2$O$_2$ because there were few full-scale installations of this technology at the time of writing.[4] Sharma et al.[1] is another useful reference that includes costing functions for the individual components of conventional water treatment processes as well as some more advanced processes like ion exchange and ozonation. Sharma et al.'s costing functions were validated through comparison with real bids prepared by reputable engineering firms.

9.6   METHODS

9.6.1   Cost Estimates and Cost Curves

Direct capital cost and O&M cost curves for ozone were prepared using costing data from USEPA's *Technologies and Costs Document for the Final Long Term 2 Enhanced Surface Water Treatment Rule and Final Stage 2 Disinfectants and Disinfection Byproducts Rule.*[2] These cost curves were compared to curves developed using the approaches and data described by Plumlee et al. (2014) and Sharma et al. (2013). Additional details about these three references are provided in the following sections. The total final capital cost for three full-scale UV/H$_2$O$_2$ installations in North America was gathered from industry magazines[33, 34] and seminar proceedings[35] and compared to the capital cost curves built in this study. Additional single-point conceptual cost estimates for UV/H$_2$O$_2$ were collected from various industry and academic literature sources for comparison purposes.[36–39]

The USEPA document includes in-depth cost estimates for many drinking water treatment processes that are used to remove natural organic matter (NOM) and disinfection byproduct (DBP) precursors, as well as technologies like UV and ozone that can be used as alternatives or add-ons to chlorine disinfection, thus minimizing the formation of regulated chlorinated DBPs. The ozone and UV cost estimates were prepared using a cost build-up approach where: (i) Preliminary designs were developed for individual components with standard engineering design calculations and based on the chosen chemical and/or UV doses and at various plant capacities (daily flow rates); (ii) costing information was obtained from vendors and chemical suppliers based on the preliminary designs; (iii) costs for general equipment like electrical, instrumentation, piping, and valves were determined using cost estimating guides; (iv) professional engineering judgment was used to cost out any remaining items. The USEPA's GAC cost estimates were created using a combination of existing costing models. All of the USEPA capital cost estimates included redundant process equipment; process monitoring; installation of electrical and instrumentation (E&I) equipment; and any additional or adjusted piping, storage, and pumping. Operations and maintenance costs included electricity, chemical inputs, labour, and replacement parts. Additional process-specific capital and O&M costs like ozone gas destruction (ozone), reactor housing (UV), and media regeneration (GAC) are also included in the USEPA estimates as appropriate. Finally, each estimate consists of a capital cost multiplier for design engineering and equipment installation. The USEPA data set also includes estimates for indirect capital costs for items like piloting, training, and permitting. These were omitted from our analysis because these indirect cost items are often jurisdiction-specific and were not included in the other studies we drew from.

The UV/H_2O_2 cost estimates and cost functions in this study were developed by combining the USEPA's direct capital cost estimates for UV disinfection with direct capital costs for H_2O_2 storage and dosing as described in the USEPA's ozone direct capital cost estimates. O&M cost estimates for the UV/H_2O_2 were prepared using the same approach, adding some more recent chemical costing data.[20] Quenching is an important cost driver for UV/H_2O_2. The cost estimates in this study assume that HOCl is used for quenching and that the equipment required to store, monitor, and apply it to the water is already installed on-site.

The Plumlee et al. study aimed to develop and validate cost curves and costing functions for advanced water treatment processes, including but not limited to ozone and UV/H_2O_2, in the context of water reclamation. The authors solicited conceptual cost estimates from equipment vendors and design engineers and used this data to create capital and O&M cost curves for each process. Their ozone cost curves were based on cost estimates from a single vendor. They assumed that the ozone treatment system was located ahead of reverse osmosis membrane filtration in the water reclamation process train. The O&M estimates did not consider maintenance, labour, or the ongoing costs associated with producing oxygen on-site or shipping it from offsite. Plumlee et al. consulted three vendors for cost estimates for UV/H_2O_2 to remove organic contaminants from reclaimed water that had undergone reverse osmosis pretreatment and was assumed to have a UV transmittance (UVT) of 95%. Two vendors provided similar capital cost estimates, and these were combined to create the curves presented in the Plumlee et al. paper. The third vendor provided a capital cost estimate that was substantially higher than the other two. This higher estimate was not included in the UV/H_2O_2 capital cost curve presented by Plumlee et al. All three vendor estimates were developed to achieve 1.2 log removal of NDMA and 0.5 log removal of 1,4-dioxane on the basis of draft regulations for these contaminants in California. The vendors did not disclose their choice of UV dose or H_2O_2 dose. All the capital cost estimates in the Plumlee study included percent-based multipliers to account for yard piping, site work and landscaping, electrical and instrumentation installation costs, contingency costs, and fees for contractors, engineering, legal, and administration. The multipliers for these costing factors were derived from McGivney and Kawamura.[3,4] The O&M cost curves in Plumlee et al. were based on estimates provided by all three vendors. They included H_2O_2 supply, hypochlorite quenching, energy, and lamp replacement. Labour costs were assumed to be negligible.

Sharma et al.'s ozone costing functions are updated equations from curves originally developed by Gumerman et al. on behalf of the USEPA[40] updated to 2011 costs using the ENR CCI and other cost indexes. Sharma et al. validated their costing functions through comparison with real bids prepared by reputable engineering firms. However, the ozone generation costing function presented by Sharma is only valid for an ozone demand of 3,500 lb/day. It thus could not be applied to some of the larger plant capacities and ozone doses explored in our analysis.

9.6.2 Treatment Conditions

In general, the choice of chemical and UV doses (for UV-based AOPs) and contact time for an AOP will be determined by:

- The vulnerability of the target contaminant(s) to the oxidative species produced by the AOP
- The concentrations of non-target oxidant-demanding substances in the water matrix
- The presence of UV-light-attenuating substances in the water matrix
- Other water quality parameters (e.g. pH) that can impact the rate of oxidation reactions

This study restricted our analysis to two design ozone doses used in USEPA (2005), 4.50 mg/L and 10.88 mg/L. These doses were originally chosen for the USEPA study because they correspond to CT conditions required to achieve 0.5 log and 2 log reduction of *Cryptosporidium* at the design flow rates and water quality conditions that were used to develop the cost estimates for ozone treatment systems.[2] Although these estimates were originally developed for disinfection rather than chemical oxidation, the 4.50 mg/L dose falls within the range that has been reported for cyanotoxin and T&O compound oxidation in laboratory studies[30, 41] and reported for 1,4-dioxane removal in industry publications.[25] The 10.88 mg/L dose is a useful comparator that allowed us to explore the potential impacts of ozone dose on direct capital and O&M costs. Operation and maintenance costs for ozone were calculated at lower average doses (2.43 and 5.88 mg/L) to reflect that the system is unlikely to operate at its design dose regularly. The capital costs were based on the design doses, as the ozone equipment and infrastructure must be large enough to meet these design doses if required.

Ozone cost curves were also built on the basis of cost estimates presented by Plumlee et al. (2014) and costing functions from Sharma et al. (2013) and compared to the USEPA curves. Plumlee et al.'s cost curve relied on an estimate provided by a single vendor who calculated costs for plant capacities ranging from 10 to 535 MGD, assuming an ozone dose of 3 mg/L and a hydraulic residence time of 5 minutes. These capital and O&M cost estimates were adjusted based on a methodology presented by Plumlee et al. to account for ozone doses other than 3 mg/L. The Sharma et al. costing functions are based on the ozone generator capacity and ozone contactor volume. The former was calculated on the basis of the USEPA ozone design doses (4.50 mg/L and 10.88 mg/L), and the latter was determined on the basis of the plant capacity and assuming a 5-minute ozone contact time.

The UV/H_2O_2 cost estimates in this study were developed on the basis of a 750 mJ/cm^2 UV dose and H_2O_2 doses ranging from 0.5 mg/L to 20 mg/L, which are in line with previous academic studies.[4, 13, 20] These conditions also agree with full-scale UV/H_2O_2 installations that oxidize seasonal T&O compounds and/or cyanotoxins in drinking water.[23, 24] Like UV disinfection in full-scale drinking water plants, UV/H_2O_2 is usually installed after coagulation, flocculation, and filtration but ahead of chlorination. As a result, the influent water to the process generally contains low levels of UV-attenuating substances (e.g. NOM, turbidity) and non-target oxidant-demanding substances (e.g. NOM, metals). Depending on source water quality, upstream processes, and downstream treatment and water quality goals, the influent water to the UV/H_2O_2 process may contain moderate concentrations of carbonate ions, commonly reported as alkalinity, which can scavenge some of the oxidants produced by AOPs (e.g. hydroxyl radicals). Accounting for these factors is complex and outside of the scope of this preliminary study. In the interests of simplicity, the UV/H_2O_2 cost curves were developed using the same assumptions as the USEPA UV disinfection cost estimates, except for the H_2O_2 and HOCl costs. The former was based on a line item in the ozone cost estimates and thus assumed slightly different water quality conditions than the UV disinfection estimates (89% UVT, turbidity = 0.1 NTU, alkalinity = 60 mg/L as $CaCO_3$). The HOCl costs were based on stoichiometric dosing of H_2O_2 and HOCl and costing information from more recent studies.[20,42] No allowances were made in our analysis to account for design versus average UV and H_2O_2 doses, as we did not have enough information to predict these. The UV/H_2O_2 capital cost estimate and O&M cost curve prepared by Plumlee et al. were also visualized for comparison with the UV/H_2O_2 curves developed on the basis of the USEPA costing data. Plumlee et al.'s data is based on estimates from two vendors who did not disclose the UV doses used to prepare their estimates and gave a range of H_2O_2 doses between 2.5 and 3.5 mg/L.

9.6.3 Inflation

The estimates published by the USEPA in 2005 were based on vendor estimates and other cost data from 2003, while the Sharma and Plumlee studies both relied on costing information from 2011. All costing information collected from these government and academic sources was updated to account for inflation between the year of preparation and 2020 using the ENR CCI.[5]

9.6.4 Error Bounds

Error bounds for direct capital costs and O&M costs were calculated as +50% and -30% of the calculated total cost on the basis of the recommendation for conceptual cost estimates.[3] This is the least detailed type of construction estimate and is intended to help designers and project proponents choose between multiple conceptual treatment options (e.g. ozone versus UV for *Cryptosporidium* removal).

9.6.5 Data Analysis and Visualization

Data analysis and data visualization were done in RStudio using the dplyr and ggplot2 packages.

9.7   RESULTS

9.7.1   OZONE

Figure 9.1 shows cost curves built from the estimates of the direct capital costs for ozone water treatment systems from the USEPA,[2] Plumlee et al.,[4] and Sharma et al.[1] Estimates for ozone doses of 4.50 mg/L and 10.88 mg/L and design flow rates from 0.091 MGD (0.344 MLD) to 430 MGD (1,628 MLD) were examined. In Figure 9.1, the estimates have been broken into small systems (≤1 MGD) and large systems (>1 MGD) to better visualize the trends and relationships between the estimates prepared

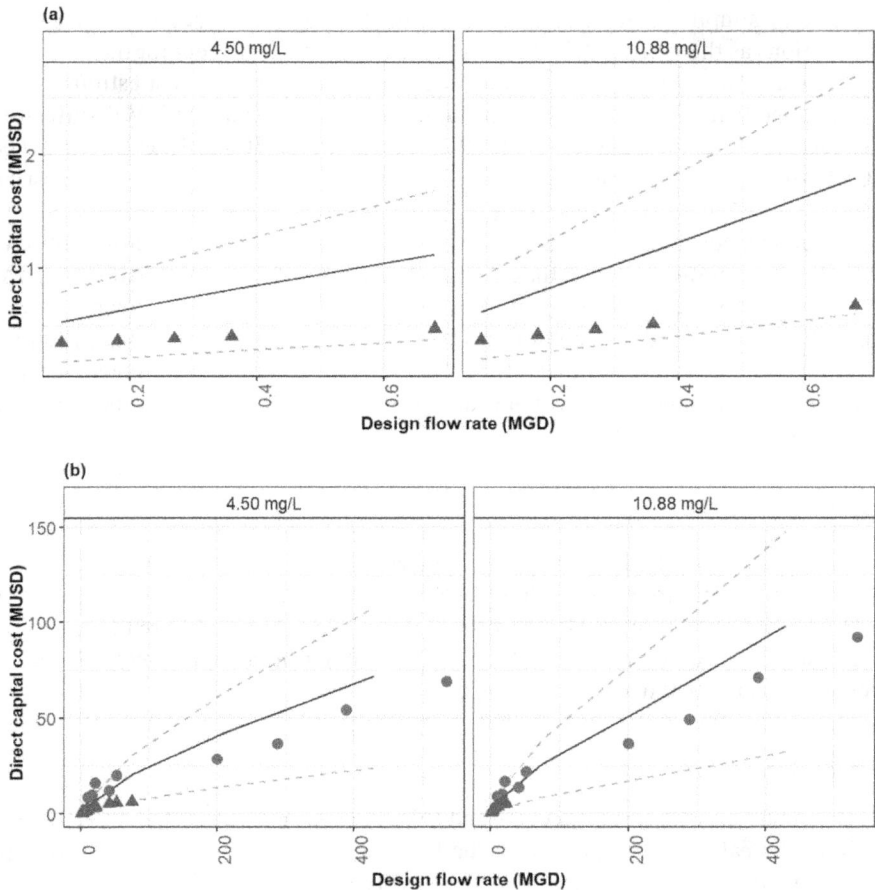

FIGURE 9.1   Estimated direct capital costs (millions of US dollars—MUSD) for ozone process equipment and ancillary monitoring and pumping in small (a) and large (b) water treatment plants applying a design dose of 4.5 mg/L or 10.88 mg/L of ozone (corresponding to an average dose of 2.43 mg/L and 5.88 mg/L, respectively). Data from the USEPA, 2005[2] (line), Plumlee et al., 2014[4] (circles), and Sharma et al. 2013[1] (triangles). The dotted lines represent the standard range of error for initial cost estimates.[3] All costs have been updated to December 2020 using the ENR CCI.[5]

with the assumptions and data from the three sources. As expected, at both the small and large scale, estimated capital costs were greater for the higher ozone dose (10.88 mg/L) than for the lower ozone dose (4.50 mg/L).

The direct capital cost curves based on the data and/or equations from Plumlee et al. and Sharma et al. were within the standard error bounds of the USEPA estimates. However, the estimates by Sharma et al. were always substantially lower than those reported by the USEPA and Plumlee et al. (Figure 9.1). Only two elements were considered for the Sharma et al. estimates: the ozone generator and the ozone contactor. In contrast, the USEPA and Plumlee estimates included ozonation equipment (generator, contractor, injector, etc.), ancillary piping and pumping, monitoring, and other cost factors described in the respective sources in the methods section of this chapter. Note that the curve based on the Sharma et al. costing functions is only valid at ozone generation demands below 3,500 lb/day and thus only includes points at lower plant capacities and/or lower ozone doses.

Cost curves for ozone O&M costs are presented in Figure 9.2. Each plot includes two sets of curves—one based on constant system operation and one that assumes that the process is only operated for eight weeks of the year to manage seasonal T&O and potential seasonal cyanotoxins. Naturally, the seasonal cost curves are substantially lower than the curves for constant operation—highlighting the importance of taking frequency of operation into account when developing cost estimates and the

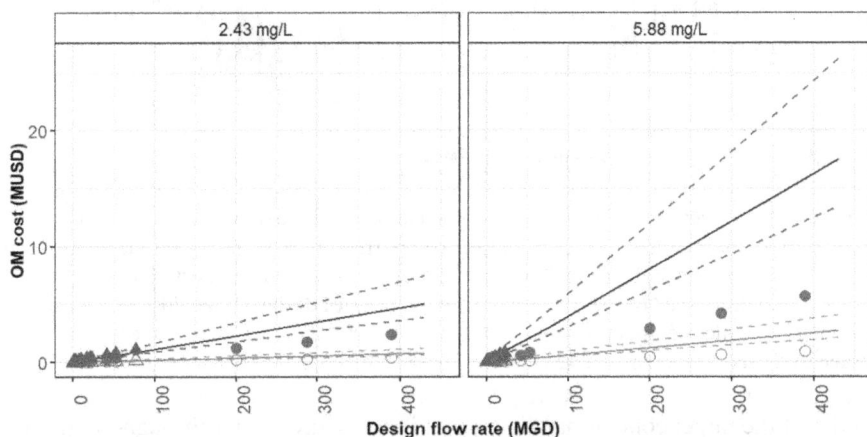

FIGURE 9.2 Operations and maintenance costs in millions of US dollars (MUSD) for ozone treatment in water treatment plants applying an average ozone dose of 2.43 mg/L and 5.88 mg/L of ozone (design dose of 4.5 mg/L or 10.88 mg/L) continuously (black line, filled shapes) or seasonally (grey line, hollow shapes) for T&O and/or cyanotoxin removal. Data from the USEPA, 2005[2] (solid lines), Plumlee et al., 2014[4] (solid circles), and Sharma et al. 2013[1] (triangles). The dotted lines represent the standard error range for initial predesign cost estimates (+50%, −30%) under each set of conditions. All costs have been updated to December 2020 cost values using the ENR CCI.[5] Seasonal O&M cost estimates assume that the system is operated for eight weeks a year to manage seasonal outbreaks of taste and odour compounds and/or cyanotoxins.

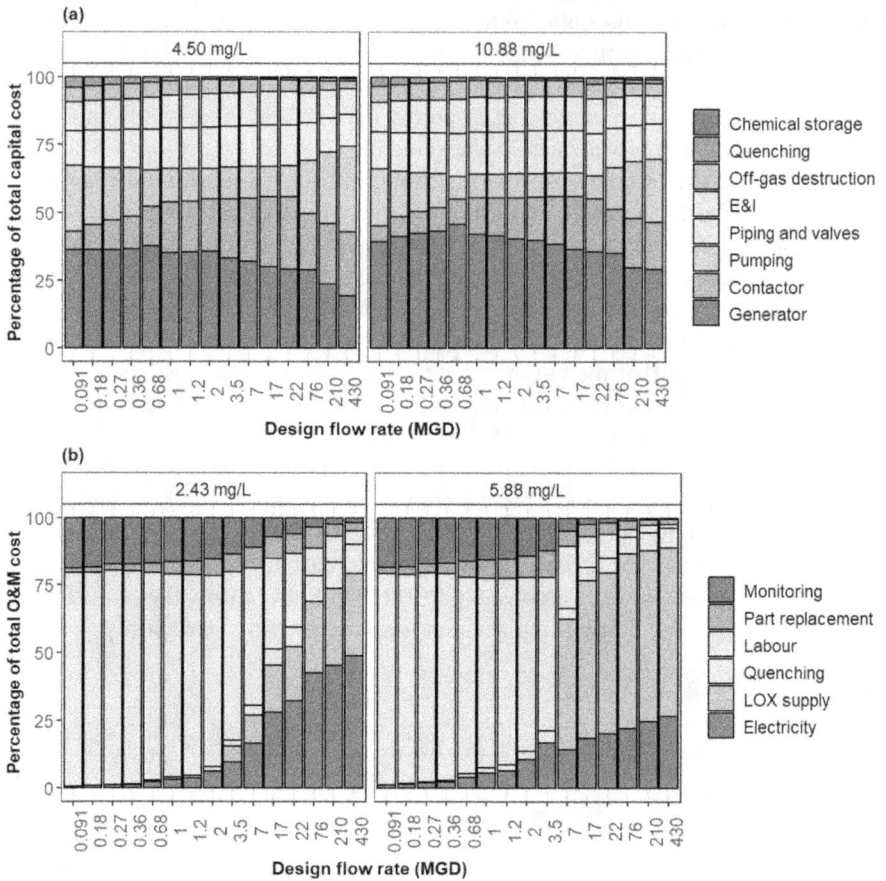

FIGURE 9.3 Percent contribution of different elements to the overall direct capital costs (a) and operation and maintenance costs (b) for ozone treatment processes on the basis of cost information presented in USEPA, 2005.[2] Note that the direct capital cost values provided in USEPA, 2005, are based on estimates made on the basis of design ozone dose (4.50 mg/L and 10.88 mg/L), while the operation and maintenance cost values are based on estimates at the average expected ozone dose (2.43 mg/L and 5.88 mg/L).

fact that if the target contaminant is present in the source water throughout the year, O&M costs will be substantially higher.

The curves in Figure 9.2 show that ozone dose had a clear and strong impact on overall O&M costs for ozonation, especially at higher plant capacities. This impact is more than that for direct capital costs in Figure 9.1. As presented later in Figure 9.3(b), at high plant capacities, O&M costs are dominated by electricity and LOX costs, which would be expected to scale directly with ozone dose and plant capacity. The equipment costs for ozone generators, contactors, and ancillary items may be less likely to increase in such a direct way.

The ozone O&M cost curves developed using the Plumlee et al. approach were consistently lower than those developed on the basis of USEPA cost estimates. Unlike the USEPA estimates, the Plumlee et al. O&M estimates do not include labour or LOX supply costing. These two cost factors account for a large proportion of the total O&M costs in the USEPA cost estimates, with labour costs dominating at lower plant capacities and LOX supply costs becoming more important at higher plant capacities (see Figure 9.3(b)). This explains why there is such a clear discrepancy between the curve based on the Plumlee et al. approach and that from the USEPA estimate. The cost curves based on the Sharma et al. costing functions were limited to lower ozone doses and capacities, but within this range, they were a good match to the USEPA-derived cost curves. This is likely because, unlike Plumlee et al., Sharma et al. accounted implicitly for labour in their costing functions. As a result, the cost curve developed on the basis of the Sharma et al. cost functions closely matches the USEPA-based cost curve.

The data presented in Figure 9.3(a) indicates that the ozone-generation equipment was the most important driver of capital costs in an ozone treatment system under almost all conditions. As expected, the generator made up a greater proportion of the total direct capital cost at the higher ozone dose (10.88 mg/L) than the lower ozone dose (4.50 mg/L). The relative importance of the generator was higher at lower flow rates than at higher flow rates because of important differences in the design of ozone generators for small versus large applications. For example, at the 4.50 mg/L ozone dose, the generator accounted for 35–37% of the total direct capital cost for systems between 0.091 MGD and 3.5 MGD, but this decreased to 19% at the highest flow rate (430 MGD). This change reflects the fact that smaller systems often generate both oxygen and ozone on-site.

In contrast, systems that require more than 100 lbs/day of ozone usually rely on shipments of liquid oxygen (LOX) to supply their ozone generators, reducing their direct capital costs while increasing ongoing chemical costs. The importance of the ozone contactor was higher at higher flow rates, especially at the lower ozone dose, likely reflecting the declining importance of the generator under these conditions. The relative contribution of the ancillary pumping did not clearly relate to flow rate or ozone dose. Off-gas destruction, electrical and instrumentation (E&I) equipment, and pipes and valves made smaller contributions, and their importance relative to other elements of the system remained essentially constant at both ozone doses and through a large range of flow rates. Chemical storage and the equipment required to supply the quenching agent (H_2O_2) to remove any residual ozone from the treated water were responsible for only small fractions of the direct capital costs at all ozone doses and flow rates.

Figure 9.3(b) shows the relative importance of different O&M cost elements in the ozone cost estimates. Labour and performance monitoring was responsible for most of the estimated OM costs at lower flow rates. In comparison, costs related to liquid oxygen (LOX) supply and electricity became dominant at higher flow rates. The USEPA cost estimates assume that ozone systems that require less than 100 pounds (45.5 kg) of ozone per day use an on-site oxygen generator ahead of the ozone generator. In contrast, those that require more than this rely on LOX shipped from elsewhere. The 100 lb cutoff occurs between 2 and 3.5 MGD for the 4.50 mg/L design ozone dose (2.43 mg/L operating dose). The cutoff occurs at 1 MGD for the 10.88 mg/L ozone design dose (5.88 mg/L operating dose), but no LOX cost data is provided in the USEPA for this dose until 7.5 MGD. The decreasing importance of labour costs reflects the fact that the USEPA estimates assume that plants treating between 0.091 MGD and 76 MGD can be operated and managed by the same number of people and thus have nearly identical estimated labour costs. The USEPA cost estimates assume that plants with capacities above 100 MGD require twice as much technical labour and four times as much managerial labour. Still, at these higher flow rates, the costs for LOX supply and electricity are well above labour costs and thus continue to dominate in terms of percentage cost.

9.7.2   UV/H_2O_2

The cost estimates built for UV/H_2O_2 in this study suggest that the direct capital cost for a UV/H_2O_2 system is likely to be driven almost entirely by the cost of the UV equipment rather than by the H_2O_2 storage and dosing or ancillary equipment (e.g. pumping, piping). Figure 9.4 shows results for plant capacities between 0.091 and 80 MGD, but the trend was identical at higher plant capacities (data not shown). The percent contribution of the UV equipment ranged from 95% at 10.88 mg/L ozone

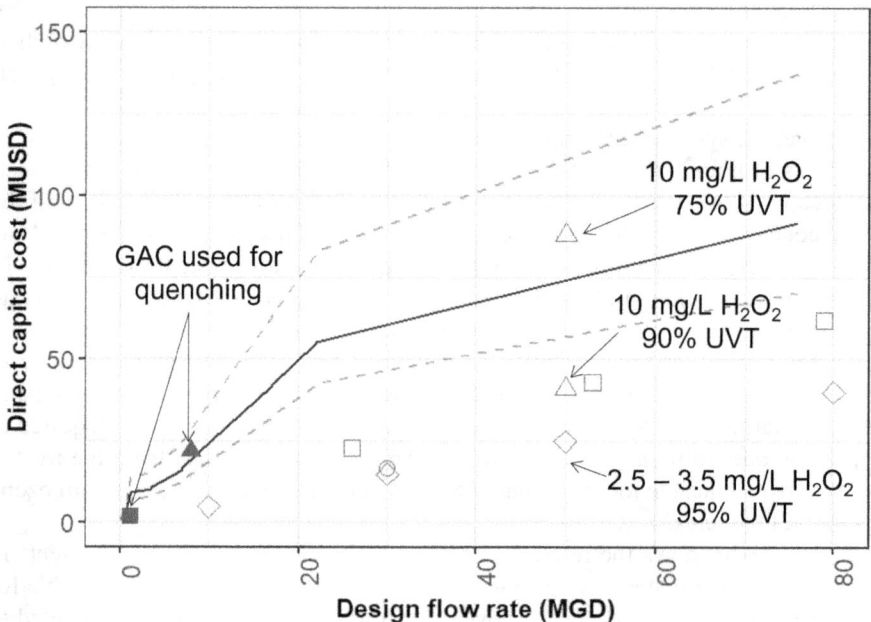

FIGURE 9.4   Estimated direct capital costs for UV/H_2O_2 at flow rates below 80 MGD based on 2005 cost estimates from the USEPA[2] updated to 2020 US dollars (UV dose = 750 mJ/cm^2, H_2O_2 dose 0.5 mg/L to 20 mg/L). The curve is compared to (solid points) final capital costs reported for full-scale UV/H_2O_2 water systems in North America as listed in industry magazines, conference proceedings, and other online sources[33–37] and (hollow points) conceptual cost estimates prepared by researchers based on information provided by equipment vendors.[4,38,39]

and a plant capacity of 2 MGD to greater than 99% at most of the other treatment conditions considered in this analysis. As a result, increasing the H_2O_2 dose from 0.5 mg/L to 20 mg/L had almost no impact on the cost curves that describe the total estimated direct capital cost.

The only publicly accessible UV/H_2O_2 cost curves found during this study's preparation were from Plumlee et al.[4] The Plumlee et al. curve was developed using equipment cost estimates provided by two reputable vendors and thus is likely a better representation of real-world costs for UV/H_2O_2 than the estimate prepared in the current study. Additional single-point conceptual capital cost estimates for UV/H_2O_2 drinking water treatment were collected from other academic sources,[38,39] government documents,[37] and press releases[36] and are shown in Figure 9.4 as solid points. Note that the updated Plumlee et al. conceptual capital cost estimates are shown as hollow diamonds in Figure 9.4. Finally, final capital costs reported for three full-scale UV/H_2O_2 treatment systems in North America were collected from industry publications[33,34] and conference proceedings,[35] updated to 2020 US dollars, and added to Figure 9.4 (solid points). Unfortunately, all of the cost estimates available that we could access in the public domain lack important context like influent water quality, specific treatment goals, UV dose, and H_2O_2 dose, among others. Therefore, any direct comparison between them and the UV/H_2O_2 cost curves built in this study is fraught.

Nonetheless, some interesting trends are apparent in the data presented in Figure 9.4. Nearly all the conceptual cost estimates are below the UV/H_2O_2 curve built in this study. There are two likely and potentially overlapping reasons for this. First, the curves developed in this study relied on estimates originally developed for UV disinfection and H_2O_2 for ozone quenching and, because of a lack of information related to larger UV reactors, unrealistically assumed that large banks of 40 mJ/cm² UV reactors were used to provide the UV dose required to drive the treatment. A multiplier was developed to estimate the number of UV lamps required. This multiplier was then applied to all other direct capital costs, including ancillary and monitoring equipment, likely inflated estimates of the total capital costs.

Additionally, however, the conceptual estimates from Plumlee et al. and other researchers either explicitly left out certain costing elements or provided only limited information about UV and H_2O_2 dosing—it may be that some or all of the conceptual estimates that fall below the new cost curve assumed lower UV doses and H_2O_2 doses compared to those that we assumed in this study (UV dose = 750 mJ/cm², H_2O_2 dose = 0.5–20 mg/L). Two estimates by Cotton and Collins (2006), shown as hollow triangles in Figure 9.4, demonstrate the importance of considering influent water quality. One hollow triangle is within the error bounds of the UV/H_2O_2 curve built in the current study. This estimate was based on a UVT of 75%. The second hollow triangle, based on a UVT of 90%, is well below the capital cost of the current study. Finally, there is a substantial disagreement between the cost estimates, whether drawn from the comprehensive USEPA document, academic studies like Plumlee et al., or real full-scale installations. One data point representing the final capital cost for a full-scale system was an order of magnitude greater than the rest of the estimates and is not shown in Figure 9.4. In addition, Plumlee et al. chose to omit a third vendor estimate that was three times higher than the two that were ultimately used to build the curve in Figure 9.4. Between 20 MGD and 80 MGD, this third estimate would have been in line with the UV/H$_2$O$_2$ cost curves depicted in Figure 9.4. These discrepancies indicate that, at this point, it remains difficult to accurately assess costs for UV/H$_2$O$_2$ because there are few existing installations available for cost benchmarking.

No matter its underlying reasons, the discrepancy among the capital cost curves built in this study, those published by Plumlee et al., and the other single-point cost estimates and final built cost points to the urgent need for new, detailed, and vendor-informed cost estimates for UV/H$_2$O$_2$ that can be validated against real full-scale drinking water installations.

Unlike the direct capital costs, the O&M cost estimates for UV/H$_2$O$_2$ built in this study (Figure 9.5) increased with the H$_2$O$_2$ dose applied. Figure 9.6 provides some insight into why O&M costs were more strongly impacted by H$_2$O$_2$ dose than direct capital costs: As H$_2$O$_2$ dose increases, so does the demand for quenching. As the flow rate, and thus the total demand for H$_2$O$_2$ and HOCl, increases, chemical costs begin to represent a more important fraction of the total estimated OM costs for UV/H$_2$O$_2$. At lower flow rates, H$_2$O$_2$ O&M costs are dominated by the cost of replacement parts for the UV reactors. Interestingly, the UV/H$_2$O$_2$ O&M cost curve for 1 mg/L H$_2$O$_2$ and 750 mJ/cm^2 developed in this study is nearly identical but slightly below the UV/H$_2$O$_2$ O&M cost curve developed by Plumlee et al. (data not shown). The Plumtree et al. curve was informed by three real vendor estimates within 20% of one another. This suggests that the O&M cost estimates and curves developed in the current study are within the range expected by vendors. This statement is tempered by the fact that the specific UV doses assumed by the vendors were not provided to Plumlee et al.

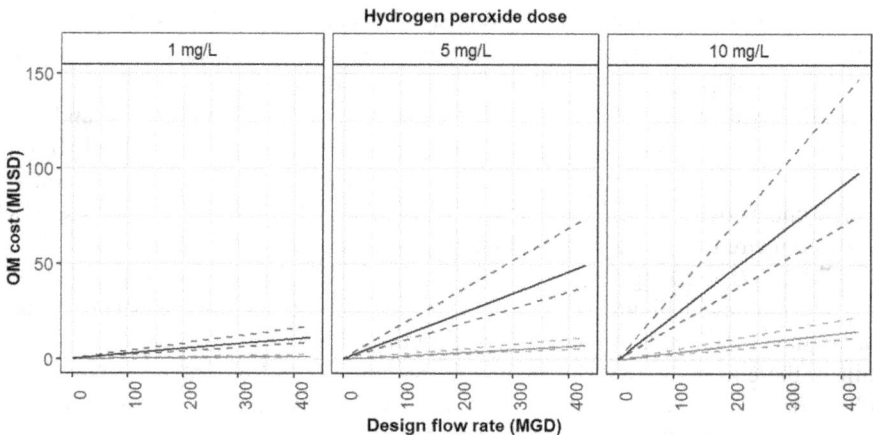

FIGURE 9.5   Preliminary O&M cost curves for UV/H$_2$O$_2$ operated continuously (black) or seasonally (grey). Cost curves were built by combining cost estimates for UV disinfection and H$_2$O$_2$ dosing from the USEPA[2] updated to 2020 US dollars using the ENR CCI[5]. The curves assume a UV dose of 750 mJ/cm^2. The seasonal operation curves assume 8 weeks of operation at the design UV dose and H$_2$O$_2$ dose.

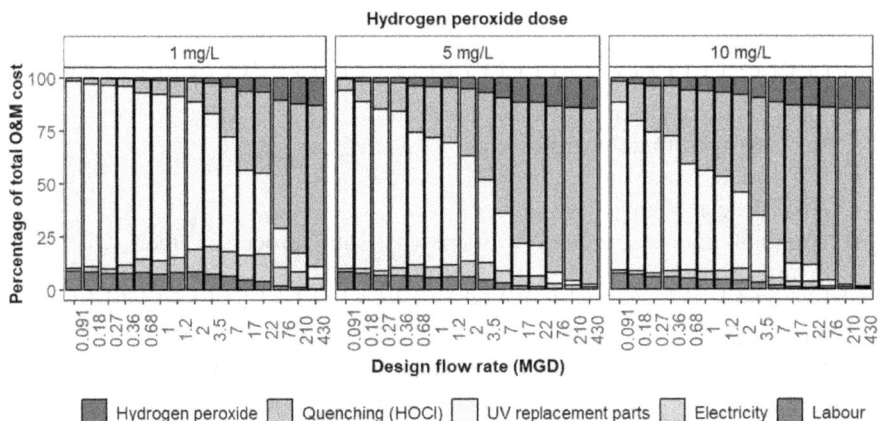

FIGURE 9.6 Percentage contribution of different cost elements to overall O&M costs for UV/H_2O_2 treatment in cost estimates built for UV/H_2O_2 in this study based on updated cost estimates for UV disinfection and H_2O_2 for ozone quenching from the USEPA (UV dose = 750 mJ/cm^2).

Unsurprisingly, the O&M cost estimates for seasonal operations are substantially lower than those for year-round operations, particularly at the higher H_2O_2 dose.

Not all of the H_2O_2 added for UV/H_2O_2 treatment will be converted to hydroxyl radicals within the UV reactor. The remaining H_2O_2 must be quenched before the water travels to the distribution system. In drinking water applications, hypochlorite (HOCl) is the quenching agent of choice because it is widely used for disinfection. Thus, most plants already have the required storage, dosing, and monitoring equipment. As a result, the capital cost for installing HOCl quenching is often assumed to be negligible.[2,4] However, as shown in Figure 9.6, O&M costs for HOCl quenching can be substantial, particularly at higher H_2O_2 doses and large plant capacities.

GAC contactors have also been used for H_2O_2 quenching in drinking water applications,[23,24,33–35,43] including at one of the longest-running UV/H_2O_2 installations in the world, the Andjik WTP in the Netherlands.[24,43] GAC contactors are essentially media filters and, as such, represent a large capital investment.

The USEPA estimates for the direct capital costs and O&M costs for a GAC contactor with an EBCT of 10 minutes and an annual regeneration frequency were updated using the ENR CCI (data not shown). This EBCT is within the range that has been reported for GAC contactors used for H_2O_2 quenching after UV/H_2O_2 treatment.[43] In practice, the media in GAC contactors used for H_2O_2 quenching can be used for many years without regenerating or replacing the media.[43] The updated estimates were used to examine the impacts of flow rate and H_2O_2 dose on the feasibility of GAC versus HOCl for H_2O_2 residual quenching.

The choice of GAC versus HOCl for quenching will come from trade-offs between capital and O&M costs. These will be driven by:

- Plant capacity (design operating flow rate)
- Average operating flow rate
- UV dose applied
- H_2O_2 dose applied
- Potential for integration with downstream processes (e.g. chlorine disinfection)
- Potential for synergistic contaminant removal (e.g. adsorption of T&O compounds by GAC)
- Frequency of system operation (seasonal versus constant)
- Ability to accommodate new process equipment within the WTP footprint

For example based on the cost curves built in this study, at 22 MGD and 1 mg/L H_2O_2, the capital cost of using GAC instead of HOCl will be approximately 21% higher, and the O&M costs will be approximately 13% higher. On the other hand, at 210 MGD and 10 mg/L H_2O_2, the capital cost increase associated with installing new GAC contactors is 25%, but the predicted annual O&M costs are decreased by approximately 58%. As is true throughout this chapter, the numbers underlying these calculations are based on conceptual estimates and should not be used in place of site-specific estimates.

9.7.3 Key Costing Factors for AOPs

Figures 9.7 and 9.8 summarize the main cost factors that drive direct capital costs, indirect capital costs, and O&M costs for ozone and UV/H_2O_2, respectively. These figures were developed on the basis of the costing elements included in USEPA 2005, Plumlee 2014, Sharma 2013, and Watts 2012 and the new analyses presented earlier in this chapter.

9.7.4 Implications of Influent Water Quality on System Operation and Costs

Water quality affects how UV light travels through water and the amount of oxidant required to successfully degrade the target compound(s). Tables 9.1 and 9.2 summarize the most relevant impacts of water quality on ozonation and UV/H_2O_2 and related cost implications.

9.7.5 Integrating AOPs into Water Treatment Plants

A full-scale WTP is made up of multiple interrelated unit processes. A traditional conventional water treatment process includes coagulation, flocculation, sedimentation, media filtration, and chlorination. Modern WTPs often incorporate more advanced processes like membrane filtration, dissolved air flotation, granular

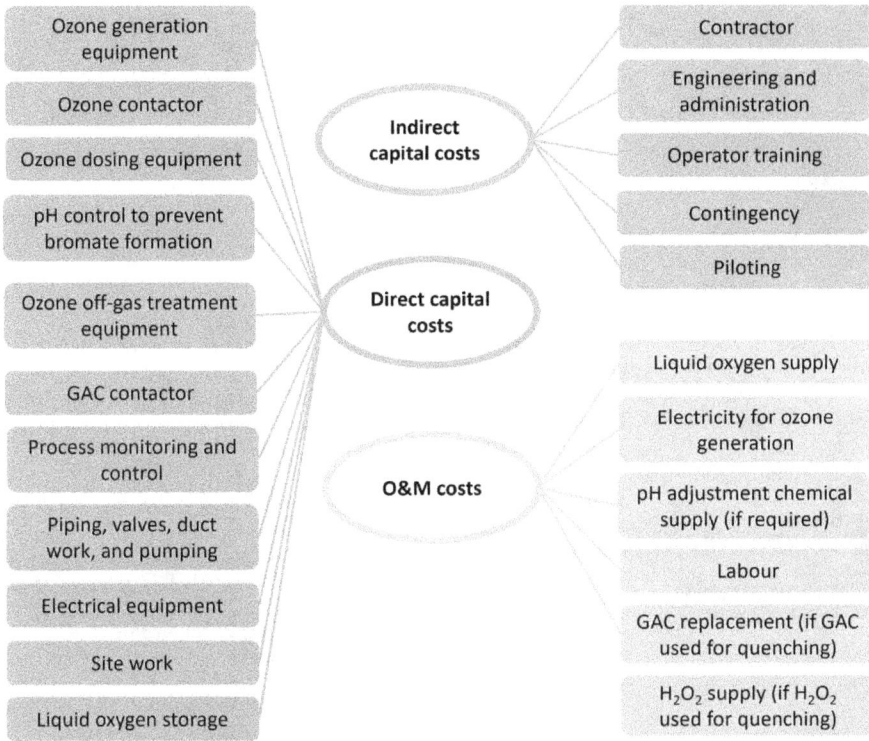

FIGURE 9.7 Direct, indirect, and operation and maintenance costs that contribute to the overall cost of ozone treatment based on insights from government[2] and academic literature.[1,4]

activated carbon (GAC) contactors, ozone, and biofiltration to improve treated water quality. Upstream and downstream water quality and treatment considerations, which are discussed in detail in the following section, often dictate where an AOP should be installed in a WTP.

AOPs are strongly impacted by water matrix components, especially natural organic matter (NOM).[44,47] As a result, they are usually included near the end of the treatment train after coagulation and filtration and before disinfection. This ensures that there is little NOM left in the water to interfere with the AOP treatment, thus maximizing the removal of target compounds.

AOPs can also be integrated earlier into the treatment plant. Ozone has also long been used as a pretreatment for biofiltration at the full scale because it breaks down larger, more recalcitrant organic compounds into smaller compounds that are more easily assimilated by the microorganisms in the biofilter.[48] Ozone's high oxidation potential also makes it a potent disinfectant, capable of quickly destroying chlorine-resistant pathogens like *Cryptosporidium* and *Giardia*.[48] As such, it has been used instead of or in addition to chlorine to enhance a WTP's disinfection capability without increasing the formation of regulated disinfection by-products.

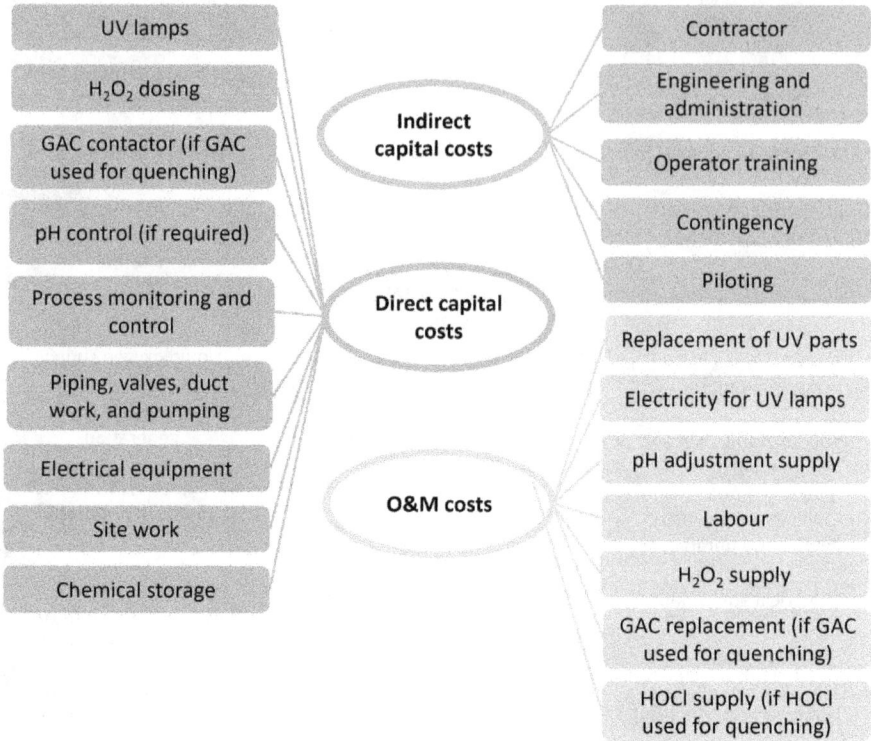

FIGURE 9.8 Direct and indirect capital costs and operation and maintenance costs that contribute to the overall cost of UV/H$_2$O$_2$ treatment based on insights from government[2] and academic literature.[4,20,43]

Finally, UV-AOPs can be directly integrated with a UV disinfection step. The latter approach has been taken at two WTPs in Ontario, Canada. In both cases, banks of UV lamps are continuously operated at a lower UV fluence (40 mJ/cm²) for disinfection and ramped up to provide a higher UV fluence when the AOP system is brought online to address seasonal T&O events.

9.7.6 Upstream and Downstream Considerations

Most AOPs are indiscriminate oxidants that will oxidize target contaminants, NOM, and other matrix constituents. As a result, these non-target contaminants can substantially increase the overall oxidant demand and the associated capital and O&M costs. This effect can be mitigated by optimizing the removal of NOM and other non-target contaminants in upstream processes; however, this optimization may incur costs related to increased coagulant dosing, increased mixing, increased reactor size, and the addition of settling aids, and/or more frequent filter backwashing.

Major downstream concerns for AOPs include the quenching of oxidant residuals and the formation of undesirable intermediate products when the target compounds

TABLE 9.1

Impacts of Water Quality on O_3 Treatment of Emerging Contaminants in Drinking Water and Potential Implications for Capital and O&M Costs.

Parameter	O_3	
	Known Effects	**Potential Cost Implications**
pH	High pH values favour O_3 decomposition to hydroxyl radicals and, therefore, higher efficiency at higher pH values[8]	Increased capital and O&M costs if pH adjustment is required
Natural organic matter (NOM)	Can react to form disinfection by-products such as aldehydes and organic acids[44]	Increased O&M cost to optimize coagulation for NOM removal ahead of AOP treatment
	Can reduce the efficiency of the degradation of targeted micropollutants through radical scavenging[8]	Higher ozone dose required to overcome demand exerted by NOM results in increased capital and O&M costs
Carbonates	Can reduce the efficiency of the degradation of targeted micropollutants through radical scavenging[8]	Higher ozone dose required to overcome the demand exerted by carbonates results in increased capital and O&M costs
Inorganic ions	Can reduce the efficiency of the degradation of targeted micropollutants through radical scavenging[8]	The higher ozone dose required to overcome demand exerted by inorganic ions results in increased capital and O&M costs
	React with hydroxyl radicals to form more selective radicals, hindering the efficiency of degradation of targeted micropollutants[8]	Bromide removal ahead of ozonation adds upstream costs, and bromate removal after ozonation adds downstream costs
	Bromide in the water can react with ozone to form bromate, a potential carcinogen[44]	
Nitrogen compounds	Nitrate can reduce the efficiency of the degradation of targeted micropollutants through radical scavenging[8]	

are not fully mineralized. The latter can lead to increased biodegradability of organic matter in treated water or increased formation of disinfection by-products in subsequent disinfection steps.

The oxidants added to the water in AOPs are not fully consumed in the process, and the residual oxidant must be "quenched" or otherwise removed from the treated water.[20,42,43] Ozone can be quenched through the addition of H_2O_2. H_2O_2 can be quenched with hypochlorite ("free chlorine"), sodium bisulfite, or using a GAC contactor. Hypochlorite used for quenching can be allowed to remain in the water to contribute to disinfection and the maintenance of a secondary disinfectant residual.[20] Adding GAC contactors for quenching can result in a substantial increase

TABLE 9.2

Impacts of Water Quality on UV/H$_2$O$_2$ Treatment of Emerging Contaminants in Drinking Water and Potential Implications for Capital and O&M Costs.

Parameter	UV/H$_2$O$_2$	
	Known Effects	Potential Cost Implications
pH	UV/H$_2$O$_2$ has a high pKa value (11.7), so pH is not a major driver of its activity13	Direct costs are difficult to assess because the effects reported in the literature are minor and, in some cases, contradictory
	pH 5–7 is more effective than pH 8[14]	Capital and operating costs will increase if post-AOP pH changes are required to achieve downstream disinfection and corrosion control targets
	Acidic and basic pH values more effective than neutral pH[45]	
Natural organic matter (NOM)	Reduces the efficiency of degradation of micropollutants by absorbing UV light meant for the photolysis of H$_2$O$_2$ and reacting with hydroxyl radicals[15,44]	Increased UV dose required to account for light attenuation will increase capital and O&M costs
	Can be broken down into more biodegradable compounds or DBP precursors such as aldehydes and organic acids[14]	Upstream and downstream processes must be optimized to reduce DBP formation, increasing O&M costs
	Depending on the composition of NOM, reaction with hydroxyl radicals may produce other reactive species that can contribute to the degradation of micropollutants[8]	If the production of additional reactive species can be predicted, this could reduce the amount of H$_2$O$_2$ required to achieve treatment goals and thus reduce O&M costs
Carbonates	React with hydroxyl radicals meant for the degradation of targeted micropollutants, reducing the efficiency of performance[46]	The higher H$_2$O$_2$ dose required to account for oxidant demand exerted by carbonates will increase O&M costs
	Form carbonate radicals when reacting with hydroxyl radicals which are a more selective oxidant than hydroxyl radicals[8]	Higher UV dose required to ensure that sufficient light reaches H$_2$O$_2$ molecules results in increased capital and O&M costs
	Attenuate light is meant for the photolysis of oxidants leading to a reduced yield of hydroxyl radicals[8]	
Inorganic ions	React with hydroxyl radicals meant to degrade targeted micropollutants reducing the efficiency of performance[46]	The higher H$_2$O$_2$ dose required to account for oxidant demand exerted by inorganic ions will increase O&M costs
	Attenuate light is meant for the photolysis of oxidants leading to a reduced yield of hydroxyl radicals[8]	Increased UV dose required to account for light attenuation will increase capital and O&M costs

Nitrate	Photolysis of nitrate results in nitrite formation, which is a potential carcinogen[44] Reduces the efficiency of degradation by absorbing UV light meant for the photolysis of oxidants and reacting with hydroxyl radicals[8] Photolysis of nitrate produces hydroxyl radicals[8]	Removal of nitrate ahead of AOP adds upstream expenses, and removal of nitrite after AOP adds downstream expenses Increased UV dose required to account for light attenuation will increase capital and O&M costs If the production of hydroxyl radicals through nitrate photolysis can be predicted, this could reduce the amount of H_2O_2 required to achieve treatment goals and thus reduce O&M costs
Other	Residual H_2O_2 will scavenge hydroxyl radicals meant for the degradation of targeted pollutants[46]	Cost implications will depend on the quenching agent being used

in capital cost but require little in the way of regular O&M cost inputs. H_2O_2 and hypochlorite dosing require less capital input but can substantially increase overall chemical operating costs. The relative importance of O&M versus capital cost for a utility will depend on local funding paradigms as well as whether the AOP is operated to remove seasonal contaminants like T&O compounds or non-seasonal contaminants like 1,4-dioxane. These considerations will also factor into the decision to use HOCl, GAC, or a different quenching agent after UV/H_2O_2 treatment.

AOP treatment can increase the biodegradability of NOM and other organic contaminants. In a pilot-scale study, Metz et al.[15] demonstrated that water treated with UV/H_2O_2 ahead of granular activated carbon (GAC) contained more assimilable organic carbon (AOC) than water treated with GAC alone. In the absence of a downstream biofiltration process, increased biodegradability can contribute to microbial regrowth and biofilm formation downstream of the WTP. Pilot and full-scale studies have demonstrated that the inclusion of an AOP in the treatment train can increase the formation of chlorinated organic DBPs when chlorine is added to the treated water for disinfection or quench H_2O_2 residuals.[12,13,16] AOPs can also interact directly with water matrix components to form undesirable oxidation products, many of which are classified as DBPs in regulatory literature. Additionally, ozone is known to react with bromide to form bromate, a regulated compound in many North American jurisdictions.

9.8 SUMMARY AND CONCLUSION

This book chapter summarizes existing cost curves for two AOPs that have been installed at the full scale in North America as of the time of writing (February 2022). Cost curves for ozone treatment were drawn from government and academic sources. New UV/H_2O_2 cost curves were built by combining cost estimates for the component parts (UV and H_2O_2 supply) and adjusting as required to achieve a UV dose of 750 mJ/cm^2 and H_2O_2 doses from 0.5 to 20 mg/L.

9.8.1 Limitations of Analysis

This analysis was squarely focused on the costs of using AOPs to remove cyanotoxins and T&O compounds from drinking water. It thus made use of data and cost curves from government and academic sources focused on drinking water treatment. In the future, AOPs will also likely find a full-scale niche for removing pharmaceuticals and other emerging organic contaminants from wastewater effluent before disposal to receiving water bodies or in water reuse applications. Wastewater effluent generally contains substantially higher concentrations of water matrix constituents that can interfere with AOP treatment. These will need to be considered when cost estimates and cost curves are developed for these applications.

The analysis presented in this chapter relies predominantly on data and models from the early 2000s. The overall cost curves for each process were updated to 2020 US dollars using the ENR CCI, a general index that accounts broadly for inflation but does not specifically address some of the individual cost drivers that will inevitably impact operational costs for AOPs (e.g. electricity costs, supply chain changes and disruptions due to the increased globalization of the economy, substantially increased labour costs in some jurisdictions)

Finally, the UV/H_2O_2 cost estimates and cost curves built in this study were predicated on the use of large numbers of 40 mJ/cm² UV reactors to achieve the specified 750 mJ/cm² UV dose, but it is more likely that 200 mJ/cm² reactors would be used in a full-scale UV/H_2O_2 application. This approach was necessary because the USEPA document did not provide information about the larger 200 mJ/cm² reactors for daily flow rates greater than 2 MGD. Since the number of 40 mJ/cm² reactors required to achieve the target UV dose is much higher than the number of 200 mJ/cm² reactors that would be required to achieve the same dose (e.g. 38 × 40 mJ/cm² reactors vs. 8 × 200 mJ/cm² at 1 MGD), the reported direct capital costs for UV/H_2O_2 are undoubtedly overestimated. As a result, the UV/H_2O_2 cost curves in this chapter should not be used for costing real projects, nor should they be directly compared to the ozone cost curves presented in this study. Rather, they should be viewed as the first step towards developing more realistic vendor-informed and practitioner-validated cost curves for UV/H_2O_2 and other AOPs.

9.8.2 Future Research Directions

The cost curves in this study require an in-depth update. We recommend that a diverse group of equipment vendors and design engineers be engaged to develop, compare, and integrate cost estimates for the components of UV/H_2O_2 and other AOPs that can be easily scaled up to municipal scale (e.g. UV/Cl, O_3/H_2O_2). These cost estimates can be used to create generalized cost curves and costing functions that will enable future design engineers to develop quick and accurate conceptual cost estimates for these systems. This exercise should be accompanied by rigorous validation of the model results based on installed systems.

A more nuanced understanding of the impacts of different water quality parameters on contaminant removal and full-scale AOP operation would improve the accuracy of cost estimates and accelerate the adoption of AOPs for drinking water treatment. Accounting for the myriad effects of water matrix composition at the full scale is challenging. Still, it could be made easier using alternative control parameters like overall oxidant demand[49] or machine learning.

Finally, although the definition of techno-economic assessment is open to interpretation, it is generally a limited tool that does not explicitly account for environmental and human health risk factors; the culture and history of an area and the influence of these factors on water management approaches; jurisdiction-specific treatment and water quality requirements; and the complexity of funding programs. In practice, techno-economic considerations must be balanced against local design standards, water quality regulations, and environmental conditions; the preferences of water system owners and water consumers; and a community's capacity to build, operate, and maintain a highly complex water treatment process. Holistic assessment methods like Triple Bottom Line Analysis, Quantitative Human Health Risk Assessment, and Water Safety Planning would help better understand the suitability of AOPs for different applications, communities, and regions.

9.8.2.1 Positioning Statement

The first author is recently graduated from a professionally accredited civil engineering program in Canada, who is now pursuing graduate studies in drinking water management. The second author is an assistant professor of civil engineering, holds a PhD in civil engineering, and is licensed as professional engineer in two Canadian provinces. The findings and recommendations described in this chapter are informed by North American water management practices and water quality and operating data.

REFERENCES

1. Sharma, J. R.; Najafi, M.; Qasim, S. R. Preliminary Cost Estimation Models for Construction, Operation, and Maintenance of Water Treatment Plants. *J. Infrastruct. Syst.* **2013**, *19* (4), 451–464.
2. United States Environmental Protection Agency. *Technologies and Costs Document for the Final Long Term 2 Enhanced Surface Water Treatment Rule and Final Stage 2 Disinfectants and Disinfection Byproducts Rule*; EPA 815-R-05–013; EPA Office of Water, **2015**.
3. McGivney, W.; Kawamura, S. *Cost Estimating Manual for Water Treatment Facilities*; John Wiley & Sons, Inc., 2008.
4. Plumlee, M. H.; Stanford, B. D.; Debroux, J.-F.; Hopkins, D. C.; Snyder, S. A. Costs of Advanced Treatment in Water Reclamation. *Ozone Sci. Eng.* **2014**, *36*, 485–495. https://doi.org/10.1080/01919512.2014.921565.
5. Zevin, A. Using ENR's Cost Indexes. *Engineering News-Record*, **2021**.
6. Yin, R.; Ling, L.; Shang, C. Wavelength-Dependent Chlorine Photolysis and Subsequent Radical Production Using UV-LEDs as Light Sources. *Water Res.* **2018**, *142*, 452–458. https://doi.org/10.1016/j.watres.2018.06.018.
7. Schulte, P.; Bayer, A.; Kuhn, F.; Luy, T.; Volkmer, M. H_2O_2/O_3, H_2O_2/UV and H_2O_2/Fe^{2+} Processes for the Oxidation of Hazardous Wastes. *Ozone Sci. Eng.* **1995**, *17* (2), 119–134.
8. Lado Ribeiro, A. R.; Moreira, N. F. F.; Li Puma, G.; Silva, A. M. T. Impact of Water Matrix on the Removal of Micropollutants by Advanced Oxidation Technologies. *Chem. Eng. J.* **2019**, *363*, 155–173. https://doi.org/10.1016/j.cej.2019.01.080.
9. Agbaba, J.; Jazić, J. M.; Tubić, A.; Watson, M.; Maletić, S.; Isakovski, M. K.; Dalmacija, B. Oxidation of Natural Organic Matter with Processes Involving O_3, H_2O_2 and UV Light: Formation of Oxidation and Disinfection by-Products. *RSC Adv.* **2016**, *6* (89), 86212–86219. https://doi.org/10.1039/C6RA18072H.
10. Sarathy, S. R.; Mohseni, M. The Impact of UV/H_2O_2 Advanced Oxidation on Molecular Size Distribution of Chromophoric Natural Organic Matter. *Environ. Sci. Technol.* **2007**, *41* (24), 8315–8320. https://doi.org/10.1021/es071602m.
11. Gora, S. L.; Liang, R.; Zhou, Y. N.; Andrews, S. A. Photocatalysis with Easily Recoverable Linear Engineered TiO_2 Nanomaterials to Prevent the Formation of Disinfection Byproducts in Drinking Water. *J. Environ. Chem. Eng.* **2018**, *6* (1), 197–207. https://doi.org/10.1016/j.jece.2017.11.068.
12. Dotson, A. D.; Keen, V. (Olya) S.; Metz, D.; Linden, K. G. UV/H_2O_2 Treatment of Drinking Water Increases Post-Chlorination DBP Formation. *Water Res.* **2010**, *44* (12), 3703–3713. https://doi.org/10.1016/j.watres.2010.04.006.
13. Wang, C.; Moore, N.; Bircher, K.; Andrews, S.; Hofmann, R. Full-Scale Comparison of UV/H_2O_2 and UV/Cl2 Advanced Oxidation: The Degradation of Micropollutant Surrogates and the Formation of Disinfection Byproducts. *Water Res.* **2019**, *161*, 448–458. https://doi.org/10.1016/j.watres.2019.06.033.

14. Lee, C.-S.; Venkatesan, A. K.; Walker, H. W.; Gobler, C. J. Impact of Groundwater Quality and Associated Byproduct Formation During UV/Hydrogen Peroxide Treatment of 1,4-Dioxane. *Water Res.* **2020**, *173*, 115534. https://doi.org/10.1016/j.watres. 2020.115534.

15. Metz, D. H.; Reynolds, K.; Meyer, M.; Dionysiou, D. D. The Effect of UV/H_2O_2 Treatment on Biofilm Formation Potential. *Water Res.* **2011**, *45* (2), 497–508. https:// doi.org/10.1016/j.watres.2010.09.007.

16. Wang, W.-L.; Wu, Q.-Y.; Huang, N.; Wang, T.; Hu, H.-Y. Synergistic Effect Between UV and Chlorine (UV/Chlorine) on the Degradation of Carbamazepine: Influence Factors and Radical Species. *Water Res.* **2016**, *98*, 190–198. https://doi.org/10.1016/j. watres.2016.04.015.

17. Svrcek, C.; Smith, D. W. Cyanobacteria Toxins and the Current State of Knowledge on Water Treatment Options: A Review. *J. Environ. Eng. Sci.* **2004**, *3* (3), 155–185. https:// doi.org/10.1139/s04-010.

18. Health Canada. *Guidelines for Canadian Drinking Water Quality Guideline Technical Document—Cyanobacterial Toxins*; Water and Air Quality Bureau, Healthy Environments and Consumer Safety Branch, Health Canada, **2021**.

19. Fakioglu, M.; Gulhan, H.; Ozgun, H.; Ersahin, M. E.; Ozturk, I. Removal of Taste and Odor Causing Compounds from Drinking Water Sources by Peroxone Process: Laboratory and Pilot Scale Studies. *Ozone Sci. Eng.* **2021**, *43* (6), 527–537. https://doi. org/10.1080/01919512.2020.1856641.

20. Watts, M. J.; Hofmann, R.; Rcdsenfeldt, E. J. Low-Pressure UV/Cl_2 for Advanced Oxidation of Taste and Odor. *J.—Am. Water Works Assoc.* **2012**, *104* (1), E58–E65. https://doi.org/10.5942/jawwa.2012.104.0006.

21. Sarathy, S. R.; Mohseni, M. An Overview of UV-Based Advanced Oxidation Processes for Drinking Water Treatment. *IUVA News*, **2006**.

22. Chitra, S.; Paramasivan, K.; Cheralathan, M.; Sinha, P. K. Degradation of 1,4-Dioxane Using Advanced Oxidation Processes. *Environ. Sci. Pollut. Res.* **2012**, *19* (3), 871–878. https://doi.org/10.1007/s11356-011-0619-9.

23. Corporation of the City of Cornwall, D. of I. and M. W., Environmental Services Division. *2019 Drinking Water Quality Report*; Compliance Report, **2020**.

24. Kruithof, J. C.; Kamp, P. C.; Martjin, B. J. UV H_2O_2 Treatment: A Practical Solution for Organic Contaminant Control and Primary Disinfection. *Ozone Sci. Eng.* **2007**, *29*, 273–280.

25. Flis, K. Advanced Treatment Solutions for 1,4-dioxane. *J Am Water Works Assoc.* **2021**, 34–39.

26. Region of Waterloo, W. S. *2019 Water Quality Report for Integrated Urban System and Rural Water Systems*; Region of Waterloo: Waterloo Region, **2020**.

27. Chorus, I.; Fastner, J.; Welker, M. Cyanobacteria and Cyanotoxins in a Changing Environment—Concepts, Controversies, Challenges. *Water.* **2021**, *13*, 2463.

28. Loeb, S. K.; Alvarez, P. J. J.; Brame, J. A.; Cates, E. L.; Choi, W.; Crittenden, J.; Dionysiou, D. D.; Li, Q.; Li-Puma, G.; Quan, X.; Sedlak, D. L.; David Waite, T.; Westerhoff, P.; Kim, J.-H. The Technology Horizon for Photocatalytic Water Treatment: Sunrise or Sunset? *Environ. Sci. Technol.* **2019**, *53* (6), 2937–2947. https://doi.org/10.1021/acs. est.8b05041.

29. Cates, E. L. Photocatalytic Water Treatment: So Where Are We Going with This? *Environ. Sci. Technol.* **2017**, *51* (2), 757–758. https://doi.org/10.1021/acs.est.6b06035.

30. Miklos, D. B.; Remy, C.; Jekel, M.; Linden, K. G.; Drewes, J. E.; Hübner, U. Evaluation of Advanced Oxidation Processes for Water and Wastewater Treatment—A Critical Review. *Water Res.* **2018**, *139*, 118–131. https://doi.org/10.1016/j.watres.2018.03.042.

31. Stirling, R.; Walker, W. S.; Westerhoff, P.; Garcia-Segura, S. Techno-Economic Analysis to Identify Key Innovations Required for Electrochemical Oxidation as Point of Use Treatment Systems. *Electrochimica Acta.* **2020**, *338*, 135874.

32. Ontario Ministry of the Environment and Climate Change. *Showcasing Water Innovation— 2015—Guidance Document for Integrating UV-Based Advanced Oxidation Processes into Municipal Wastewater Treatment Plants.Pdf*; Showcasing Water Innovation Program, 2015.

33. Advanced Oxidation Process for 1,4-Dioxane Removal. *Water and Wastes Digest.* **2018**.

34. Diebel, J.; Catalano, L. Providing for Aurora. *Civil Engineering Magazine*, September **2011**.

35. Biggs, J. Tucson Water's AOP Treatment Facility: Transformation Oif Tucson Water's CERCLA-To_Drinking Water Program After 25 Years, 2020.

36. Regional Development Corporation, P. of N. B. Investment in Moncton's Drinking Water System. February 22, 2022.

37. City of Moncton. *2020 Water Quality Annual Report: Moncton/Riverview/Dieppe*; City of Moncton, **2021**.

38. Cotton, C. A.; Collins, J. R. *Dual Purpose UV Light*; Using UV Light for Disinfection and for Taste and Odour Oxidation, **2006**.

39. Dore, M. *Global Drinking Water Management and Conservation: Optimal Decision Making*; Global Drinking Water Management and Conservation; Springer, **2014**.

40. United States Environmental Protection Agency, M. E. R. L., Office of Research and Development. *Estimating Water Treatment Costs, Volume 2: Cost Curves Applicable to 1 to 200 Mgd Treatment Plants*; EPA-600/2–79–162b; United States Environmental Protection Agency, **1979**.

41. Vlad, S.; Anderson, W. B.; Peldszus, S.; Huck, P. M. Removal of the Cyanotoxin Anatoxin-a by Drinking Water Treatment Processes: A Review. *J. Water Health.* **2014**, *12* (4), 601–617. https://doi.org/10.2166/wh.2014.018.

42. Keen, O. S.; Dotson, A. D.; Linden, K. G. Evaluation of Hydrogen Peroxide Chemical Quenching Agents Following an Advanced Oxidation Process. *J. Environ. Eng.* **2013**, *139* (1), 137–140.

43. Huang, Y.; Nie, Z.; Wang, C.; Li, Y.; Xu, M.; Hofmann, R. Quenching H_2O_2 Residuals after UV/H_2O_2 Oxidation Using GAC in Drinking Water Treatment. *Environ. Sci. Water Res. Technol.* **2018**, *4* (10), 1662–1670. https://doi.org/10.1039/C8EW00407B.

44. Martijn, A. J. *Impact of the Water Matrix on the Effect and the Side Effect of MP UV/H_2O_2 Treatment for the Removal of Organic Micropollutants in Drinking Water Production*. Ph.D., Wageningen University and Research, The Netherlands, 2015.

45. Zhu, H.; Jia, R.; Sun, S.; Feng, G.; Wang, M.; Zhao, Q.; Xin, X.; Zhou, A. Elimination of Trichloroanisoles by UV/H_2O_2: Kinetics, Degradation Mechanism, Water Matrix Effects and Toxicity Assessment. *Chemosphere.* **2019**, *230*, 258–267. https://doi.org/10.1016/j.chemosphere.2019.05.052.

46. Yang, Y.; Pignatello, J. J.; Ma, J.; Mitch, W. A. Effect of Matrix Components on UV/ H_2O_2 and UV/S2O82− Advanced Oxidation Processes for Trace Organic Degradation in Reverse Osmosis Brines from Municipal Wastewater Reuse Facilities. *Water Res.* **2016**, *89*, 192–200. https://doi.org/10.1016/j.watres.2015.11.049.

47. Metz, D. H.; Meyer, M.; Vala, B.; Beerendonk, E. F.; Dionysiou, D. D. Natural Organic Matter: Effect on Contaminant Destruction by UV/H_2O_2. *J.—Am. Water Works Assoc.* **2012**, *104* (12), E622–E636. https://doi.org/10.5942/jawwa.2012.104.0157.

48. Crittenden, J. C.; Trussell, R. R.; Hand, D. W.; Howe, K. J.; Tchobanoglous, G. *MWH's Water Treatment: Principles and Design*, 3rd ed.; John Wiley & Sons, Inc., **2012**.

49. Wang, C.; Rosenfeldt, E.; Li, Y.; Hofmann, R. External Standard Calibration Method to Measure the Hydroxyl Radical Scavenging Capacity of Water Samples. *Environ. Sci. Technol.* **2020**, *54* (3), 1929–1937. https://doi.org/10.1021/acs.est.9b06273.

10 Current Challenges and Future Prospects in AOPs

Pratibha Biswal, Umit Gunes and Mohamed M. Awad

CONTENTS

10.1 Introduction 215
10.2 Current Status of AOP Research: Worldwide 216
10.3 Established AOP Techniques and Their Applications 216
    10.3.1 Fenton and Photo-Fenton Oxidation Treatment 216
    10.3.2 Ozone-based Oxidation Treatment 219
    10.3.3 Photocatalytic Treatment 221
    10.3.4 UV-based Oxidation Treatment 222
    10.3.5 Sonication/Sonolysis 222
10.4 Current Status of Emerging AOP Techniques 223
10.5 Conclusions and Future Outlook 224
References 224

10.1 INTRODUCTION

Over the past decades, environmental pollution resulting from industrialization and urbanization has attracted significant attention from researchers. One of the major focuses has been on wastewater treatment of water effluents from various sources, including municipal, industrial, and agricultural wastewater. Many works in the literature have discussed wastewater treatment involving physical and chemical mechanisms. However, due to the presence of several newly suspended or dissolved materials attributed to over-industrialization, conventional methods may no longer be efficient for water treatment. Researchers worldwide are thus working on new chemical treatment methods and advanced oxidation processes (AOPs) (Suty et al. 2004).

AOPs refer to the chemical treatment of wastewater in order to remove organic contaminants via the production of highly reactive hydroxyl radicals. AOP processes involve chemical processes that use oxidants (e.g., ozone, hydrogen peroxide, oxygen) and/or light sources (e.g., UV/solar light) and catalysts (e.g., iron ions, electrodes, metal oxides). Earlier works in the literature and on treatment plants have observed AOP's efficiency in reducing contaminant concentrations. Of the earlier discussed methods, processes involving light sources are quite effective for treatment. AOP usage occurs in many industrial applications, such as the pharmaceutical industry (Kanakaraju et al. 2018), pulp and paper industry (Hermosilla et al. 2015), automobile industry

DOI: 10.1201/9781003247913-10

215

(Mudliar et al. 2009), and textile industry (Zazou et al. 2019), to name a few. Implementing AOPs effectively addresses significant limitations observed in conventional wastewater treatment methods. AOPs also provide an effective means for pretreating recalcitrant pollutants, owing to the exceptional reaction rate facilitated by hydroxyl radicals.

This chapter provides a comprehensive examination of AOPs and their significance in the field of wastewater treatment. The initial section explores the utilization of bibliometric analysis to assess the global impact and prominence of AOPs. Subsequently, it presents recent studies focusing on applying different widely recognized AOP methods. Furthermore, this section offers insights into the current research trends and practical implementations of all existing AOP techniques.

10.2   CURRENT STATUS OF AOP RESEARCH: WORLDWIDE

Figure 10.1(a) depicts the number of studies conducted on AOPs, while Figure 10.1(b) presents the prominent organizations engaged in research focused on AOPs. These figures provide compelling evidence for the significance of AOPs and the continually growing interest in their various aspects. Notably, China has emerged as a leading contributor to research and publications on AOPs. The preference for publishing AOP studies in English is evident, as shown in Figure 10.2(a). Furthermore, Figure 10.2(b) indicates a substantial volume of research articles in this field. The remarkable abundance of research endeavors in this area has also led to numerous review articles on the current state of AOPs. Figure 10.3(a) exhibits an exponential growth trend in the number of journal publications, with the count reaching 1,013 in 2021. Similarly, Figure 10.3(b) reflects a similar pattern in the number of citations received by AOP-related works, with an astonishing total of approximately 53,000 citations. Most of the studies on AOPs are concentrated in the environmental sciences, followed by chemistry and chemical engineering, as shown in Figure 10.4(a). Additionally, Chemical Engineering Journal emerges as the primary publication venue for major works on AOPs, closely followed by Chemosphere, as illustrated in Figure 10.4(b).

10.3   ESTABLISHED AOP TECHNIQUES AND THEIR APPLICATIONS

The field of AOP holds immense global significance, as evidenced by the extensive research conducted in this area. The existing literature in this domain showcases noteworthy advancements in AOPs, particularly in relation to their application in diverse industrial sectors. Since their inception, AOPs have evolved into various types, encompassing a range of materials and methodologies, each tailored for specific industrial applications. In the subsequent section, we will delve into the major types of AOPs, emphasizing their industrial relevance and applications.

10.3.1   FENTON AND PHOTO-FENTON OXIDATION TREATMENT

The Fenton reagent (mixture of H_2O_2 and Fe^{2+}) generates a highly reactive hydroxyl radical (OH). This Fenton oxidation process proves to be highly efficient in chemically treating wastewater. Apart from the traditional Fenton process, numerous hybrid methods have been developed, one of which is the light-coupling technique

Number of publications

< 10   10–50   50–100   100–300   300–500   500–1,000   ≥ 1,000

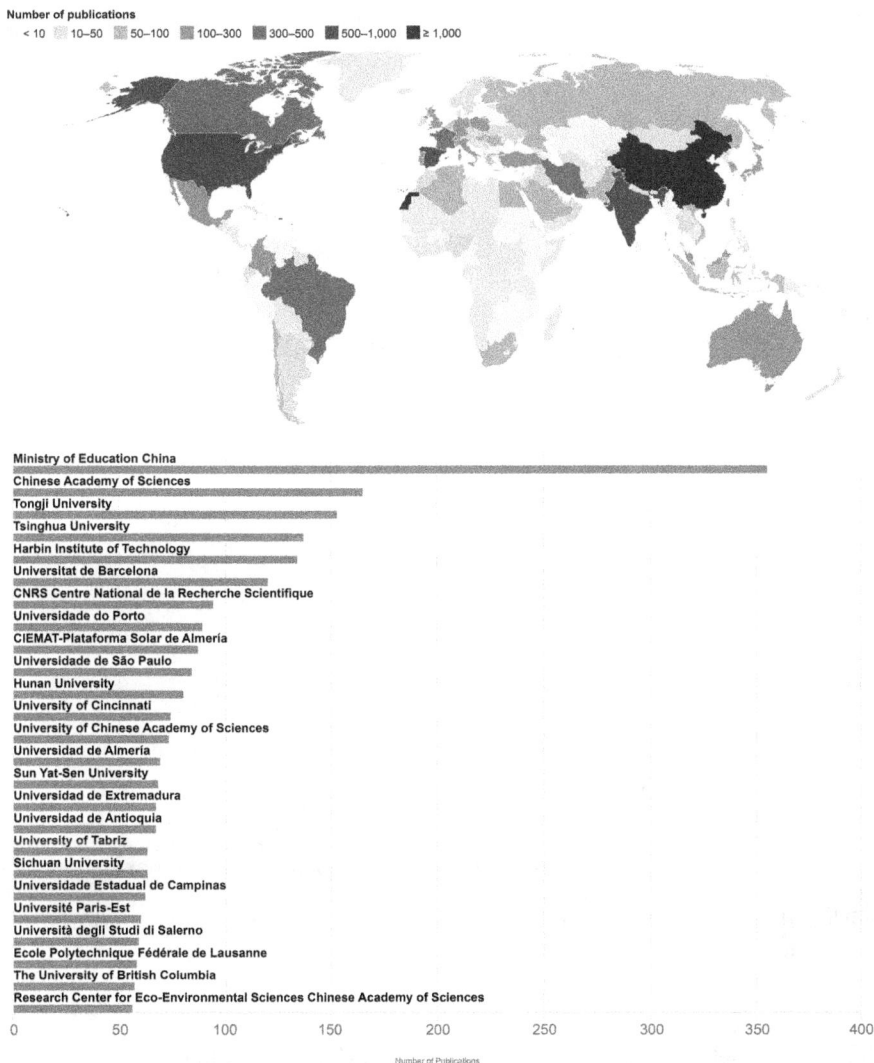

FIGURE 10.1 (a) Current status of the number of publications worldwide on advanced oxidation processes (AOP). (b) Designation of authors worked on top articles published on AOP.

known as photo-Fenton oxidation. The photo-Fenton reaction demonstrates accelerated reaction rates and enhanced levels of mineralization when compared to the conventional Fenton process. Moreover, this reaction can be driven by low-energy photons, making it feasible to utilize solar irradiation as an energy source for the process. Ebrahiem et al. (2017) mentioned that the photo-Fenton process could reduce operational costs and improve kinetics and performance.

Previous studies have acknowledged the efficacy of photo-Fenton processes, specifically in the context of treating wastewater from the textile industry. Pérez et al.

```
English                                                                    7360
Chinese    145
Portuguese  33
Polish     14
Spanish    14
French     12
Japanese   11
Persian     9
German      8
Korean      6
Croatian    4
Turkish     4
Bosnian     3
Italian     2
Moldavian   2
Moldovan    2
Romanian    2
Slovenian   2

0      1000    2000    3000    4000    5000    6000    7000    8000
```

```
Article
                                                                       6200

Review
        699

Conference Paper
       491

Book Chapter
    166

Editorial
 27

Book
 7

Erratum
 6

Letter
 5

Note
 4

Data Paper
 3

Short Survey
 3

0    500   1000  1500  2000  2500  3000  3500  4000  4500  5000  5500  6000  6500

                        Number of Publications
```

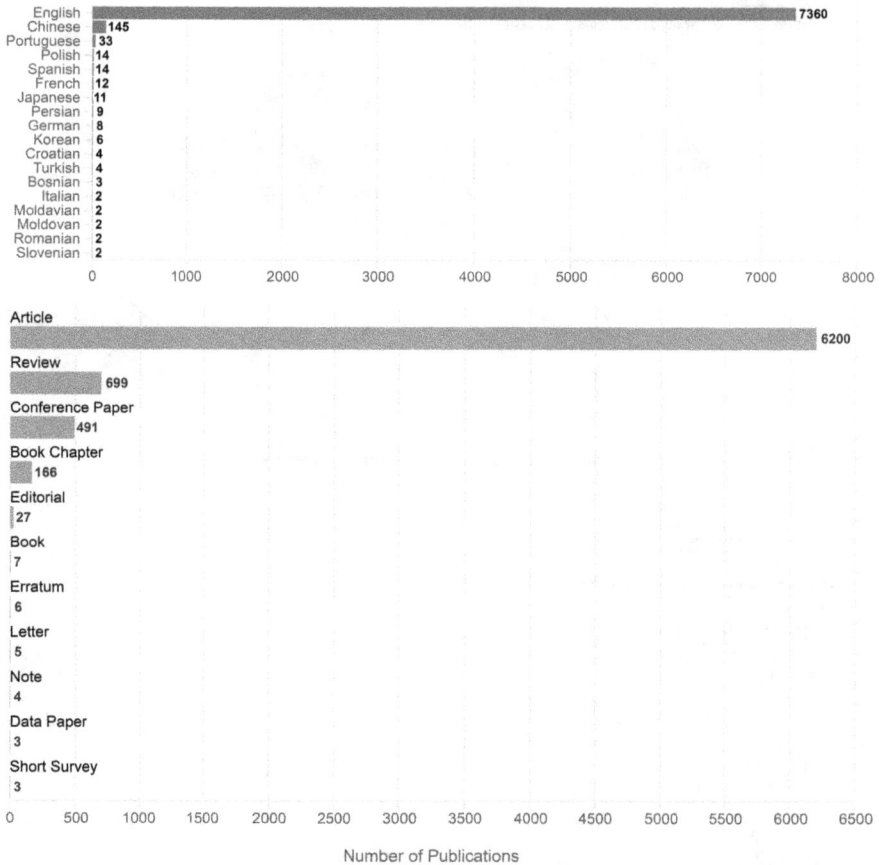

FIGURE 10.2 (a) Publications languages on Advanced Oxidation Processes. (b) Types of articles published on Advance Oxidation Process (AOP).

(2002b) reported on a photo-Fenton-based approach for treating textile wastewaters, while Lucas and Peres (2006) emphasized the significance of these methods in decolorizing azo dyes, such as Reactive Black 5. Furthermore, Fenton and photo-Fenton processes have demonstrated their specialization in textile wastewater treatment and subsequent reuse, as highlighted by (Blanco et al. 2014). GilPavas et al. (2017) presented work on Fenton or photo-Fenton processes as an alternative for industrial textile wastewater treatment. Lebron et al. (2021) recently studied photo-Fenton and membrane-based techniques for textile effluent treatment.

In addition to the textile industry, many other works have also presented the importance of the Fenton process for degrading the organic content of paper industry effluent (Pérez et al. 2002a). This method is also used to treat effluents from the paint industry (Oliveira et al. 2007). Zapata et al. (2010) presented a combined solar photo-Fenton/biological system for the remediation of pesticide-contaminated water. Furthermore, for the treatment of landfill leachates, these AOPs can also be used (Umar et al. 2010). Zhang and Pagilla (2010) effectively employed photo-Fenton

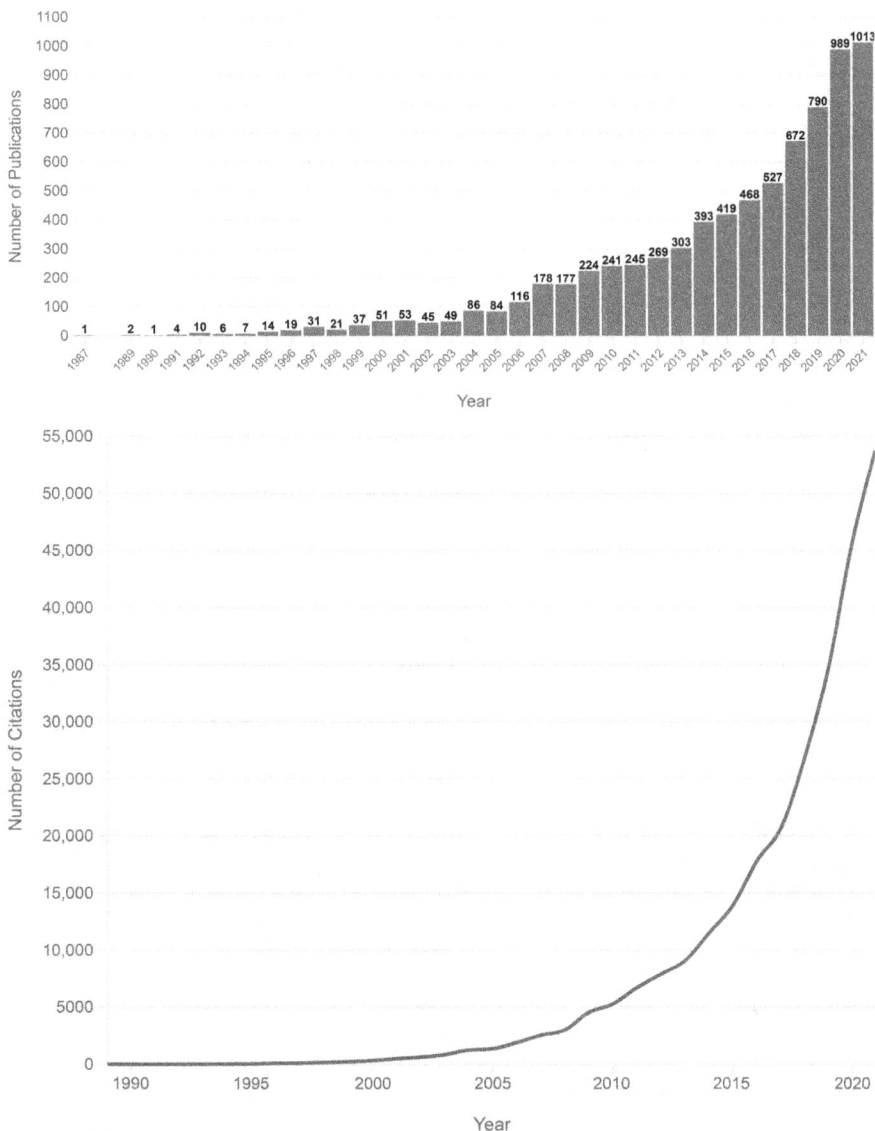

FIGURE 10.3 (a) Number of publications since 1987 on advanced oxidation process (AOP). (b) Number of citations of works related to AOPs.

reagents to treat wastewater generated by the pesticide industry. Fenton and photo-Fenton methods in other industries involving issues with photo-Fenton are the operations at low pH and the presence of high heavy metal ions (Ahmed et al. 2011).

10.3.2 OZONE-BASED OXIDATION TREATMENT

The ozone-based oxidation method generates ozone through electric discharge in a stream of oxygen or air. As mentioned earlier, ozone is a powerful oxidizing agent.

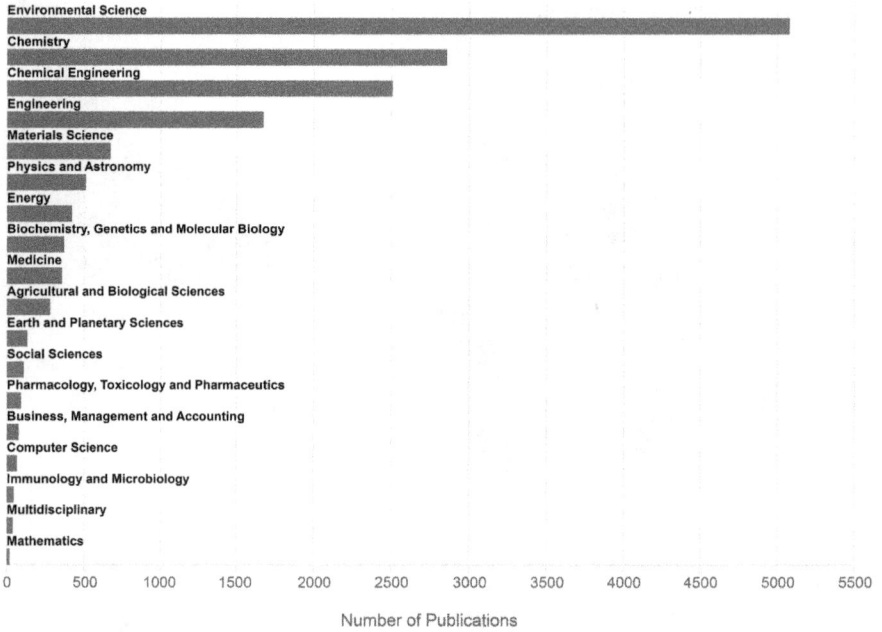

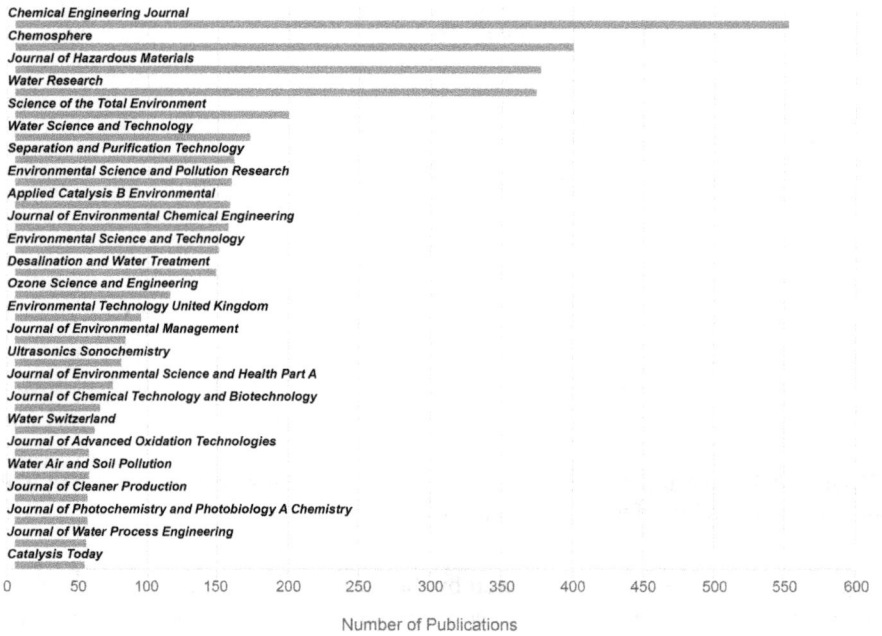

FIGURE 10.4 (a) Importance of advance oxidation process in various areas of STEM. (b) Number of publications in major journals on STEM.

Thus, it can convert organic pollutants into smaller compounds or other materials (carbon dioxide, water, sulphate, and nitrate anions). This process is beneficial in terms of the quality of the final product without odours or tastes. In their study on ozone-based AOPs, Rosenfeldt et al. (2006) examined the efficiency of hydroxyl radical generation during ozonation. They reported that enhancing the exposure to hydroxyl radicals resulted in a reduction of the energy gap between the UV/H2O2-AOP and ozone processes. A recent review presented a history of ozone-based AOPs and advances (Rekhate and Srivastava 2020) and concluded that combined methods are the most cost-effective. Ozone-based AOPs have gained significant popularity and demonstrated successful applications across various fields. The key advantages of these processes stem from the exceptional selectivity of molecular ozone, enabling targeted and efficient reactions. Furthermore, the generation of highly reactive hydroxyl radicals, although less selective, enhances the attractiveness of this method even further.

In a series of studies, Wu et al. (2004) explored the efficacy of various ozone-based AOPs for treating landfill leachates. Similarly, Chin and Bérubé (2005) compared the performance of ozone, ultraviolet irradiation (UV), and their combined AOPs in eliminating disinfectant byproduct precursors from raw surface water. The efficiency of O3/granular-activated carbon and common AOPs was compared by Sánchez-Polo et al. (2006), while Wu et al. (2007) investigated the oxidation of dimethyl sulfoxide using ozone-based methods. Acero and von Gunten (2008) presented a study highlighting the impact of carbonate on the ozone (i.e., hydrogen peroxide) process. Cuiping et al. (2011) employed an ozone-based oxidation process to eliminate rhodamine B from an aqueous solution. Tichonovas et al. (2017) applied ozone-based AOPs to remove 2-naphthol, phenol, oxalic acid, phthalate, methylene blue, and D-glucose. Lastly, Phungsai et al. (2019) investigated the semiquantitative changes that occur during oxidation using ozone and an ozone/hydrogen peroxide mixture.

10.3.3 PHOTOCATALYTIC TREATMENT

Oxidizing agents can be combined with a catalyst and/or light using the photocatalytic treatment. These processes may be of particular interest for treating effluent containing highly toxic compounds and for which the biological processes might not be pertinent. However, this process is ineffective in removing the high organic content of molasses wastewater. Therefore, it has been recommended for application as a post-treatment process for anaerobic digestion. A review of parameters impacting the photocatalytic process involving semiconductors and their physical types was presented for wastewater treatment of effluent from different industries (Gholami and Bahrami 2019).

Kamat and Meisel (2002) have worked on photocatalysis using semiconductor nanoparticles to develop self-cleaning glasses and air purification systems. Černigoj et al. (2007) studied the simultaneous effect of pH and ozone dosage on titanium oxide photocatalytic ozonization. In a comprehensive scientific review, Demeestere et al. (2007) examined the fundamental principles and practicality of heterogeneous photocatalysis as an AOP specifically for treating volatile organic compounds. Likewise, Chan et al. (2011) conducted an in-depth review focusing on using metal oxide semiconductors as photocatalysts within AOPs, specifically for treating dye wastewater. Finally, Brame et al. (2015) investigated the detrimental impacts of natural organic matter on photocatalytic AOPs, shedding light on the potential challenges associated with their implementation.

Further, Fernandes et al. (2020) studied the effect of a titanium-oxide photocatalytic AOP while treating refinery effluents. Zhang et al. (2019) studied the coupling of heterogeneous AOPs and photocatalysis for treating tetracycline hydrochloride, while. Gayathri et al. (2019) studied microwave photocatalysis mediated by ZnO and persulphate to remove rhodamine B dye pollutants. More recently, Sun et al. (2021) attempted to use photocatalysis and persulfate-based AOPs alongside a reaction mechanism and path of photogenerated charges.

10.3.4   UV-BASED OXIDATION TREATMENT

UV direct photolysis is a prominent mechanism employed for the degradation of contaminants through the absorption of incident radiation from UV light. Consequently, this method is primarily applied to target contaminants with high UV absorption capacity.

Mascolo et al. (2008) studied the effectiveness of UV-based AOPs for treating hydrocarbon compounds found in groundwater. Furthermore, UV-based AOPs were found to be effective in controlling biofilm formation in water systems, as demonstrated by Furthermore, UV-based AOPs were found to be effective in controlling biofilm formation in water systems, as demonstrated by Lakretz et al. (2010). Zoscke et al. (2012) presented studies on UV-based oxidation processes for removing geosmin and 2-methyl isoborneol (2-MIB). Deng et al. (2013) investigated the performance of carbamazepine degradation via UV radiation in the presence of different oxidants. Additionally, Tan et al. (2014) studied the treatment of acetaminophen in various UV-based AOPs. Xu et al. (2016) reported the mineralization of sucralose by UV-based AOPs (i.e., peroxydisulfate [PDS], and hydrogen peroxide). The importance and use of UV-based AOPs in algal medium recycling were presented by Wang et al. (2018). Varanasi et al. (2018) studied the degradation of dissolved organic matter via UV-based AOPs. Mangalgiri et al. (2019) employed UV-based AOP to treat trace organic contaminants. Wen et al. (2019) used UV-based AOP to control the photoreactivation of fungal spores in water. Lin et al. (2020) presented a degradation of organosilicon wastewater using a UV-based AOP (UV/H_2O_2, UV/PDS and UV/peroxymonosulfate [PMS]). They showed the COD removal rate to be significant for UV/H_2O_2 compared to other processes. Yea et al. (2021) reviewed iodine and its occurrence in water, followed by the importance of UV-based AOP for their treatment. Additionally, Meng et al. (2021) conducted a study focusing on UV-based AOPs and examined the optimal dosage of oxidants in relation to primary radical concentrations.

10.3.5   SONICATION/SONOLYSIS

Ultrasound has emerged as a promising approach within AOPs for treating waste and sewage water. It employs high-intensity acoustic irradiation to target both chemical and microbiological pollutants effectively. Extensive research has been conducted to investigate the degradation capabilities of ultrasound on various substances, including dyes (Eren and Ince, 2010). Aromatic compounds and pollutants (Goskonda et al. 2002; Jiang et al. 2002), volatile and non-volatile organic compounds (Goel et al. 2004), and pharmaceutical pollutants.

TABLE 10.1

A Summary of AOP Techniques and Their Status in the Current Industrial, Pilot Scale, and Laboratory Applications.

Types	Relevant References	Sub-types	Status
Ozone-based AOPs	• Rekhate and Srivastava (2020) • Suty et al. (2014)	O_3	Established method
		O_3/H_2O_2	Established method
		O_3/Catalyst	Laboratory and pilot-scale tests
Ultraviolet-based AOPs	• Lei et al. (2021) • Matafonova and Batoev (2018)	UV/H_2O_2	Established method
		UV/O_3	Laboratory and pilot-scale tests
		UV/PDS	Laboratory and pilot-scale tests
		UV/Cl_2	Established method
Electrochemical AOPs	• Sires et al. (2014) • Moreira et al. (2017)	BDD-electrode	Laboratory and pilot-scale tests
		SnO_2-doped electrode	Laboratory-scale tests
		PbO_2-doped electrode	Laboratory-scale tests
		TiO_2-electrode	Laboratory scale tests
Catalytic AOPs	• Wang et al. (2021) • Sharma and Feng (2019)	Fenton	Established method
		Photo-Fenton	Established method
		UV/Catalyst	Laboratory and pilot-scale tests
Photo-assisted AOPs	• Serpone et al. (2010) • Subbian et al. (2018)	Electron beam	Laboratory scale tests
		Ultrasound	Laboratory and pilot-scale tests
		Plasma	Laboratory and pilot-scale tests
		Microwave	Laboratory scale tests

10.4 CURRENT STATUS OF EMERGING AOP TECHNIQUES

In addition to the previously mentioned advanced oxidation processes (AOPs), a plethora of new and emerging techniques have been described in recent scientific literature. Table 10.1 presents a comprehensive overview of both traditional and innovative AOP techniques, along with their current status in terms of industrial, pilot scale, and laboratory applications. The table indicates that numerous novel

AOP techniques are still in the early laboratory-scale testing and research stages. Some promising methods, such as catalytic and UV-based processes, have successfully advanced beyond the laboratory and are now being utilized at both bench and pilot scales. However, the primary hurdles faced by less mature processes are their substantial costs and energy requirements. When designing and developing an AOP system, careful consideration must be given to the associated operation and maintenance expenses. As the previous review highlights, the widespread adoption and commercial implementation of many underdeveloped AOPs poses significant challenges in terms of long-term applications and economic viability.

10.5   CONCLUSIONS AND FUTURE OUTLOOK

This study provides a comprehensive overview of the current state of AOPs in both research and industry. While various types of AOPs have been documented in the literature, no single method is universally applicable for treating all types of industrial pollutants. Therefore, integrating different established and emerging AOPs should be considered to address a wider range of pollutants effectively. Incorporating solar and other renewable energy sources into these hybrid treatment approaches can significantly enhance economic and energy efficiency. Moreover, it is important to note that the most innovative AOP systems are still in the early stages of development at the laboratory level. Further endeavors are required to advance these emerging technologies and eventually enable their successful implementation in industrial and real-life applications in the future.

REFERENCES

Acero, J.L., von Gunten, U., (2008) Influence of carbonate on the ozone/hydrogen peroxide based advanced oxidation process for drinking water treatment, The Journal of the International Ozone Association, vol. 22, pp. 305–328.

Ahmed, B., Limem, E., Abdel-Wahab, A., Nasr, B., (2011) Photo-Fenton treatment of actual agro-industrial wastewaters, Industrial Engineering Chemistry Research, vol. 50, pp. 6673–6680.

Blanco, J., Torrades, F., Morón, M., Brouta-Agnésa, M., García-Montaño, J., (2014) Photo-Fenton and sequencing batch reactor coupled to photo-Fenton processes for textile wastewater reclamation: Feasibility of reuse in dyeing processes, Chemical Engineering Journal, vol. 240, pp. 469–475.

Brame, J., Long, M., Li, Q., Alvarez, P., (2015) Inhibitory effect of natural organic matter or other background constituents on photocatalytic advanced oxidation processes: Mechanistic model development and validation, Water Research, vol. 84, pp. 362–371.

Černigoj, U., Štangar, U.L., Trebše, P., (2007) Degradation of neonicotinoid insecticides by different advanced oxidation processes and studying the effect of ozone on TiO_2 photocatalysis, Applied Catalysis B: Environmental, vol. 75, pp. 229–238.

Chan, S.H.S., Wu, T.Y., Juan, J.C., Teh, C. Y., (2011) Recent developments of metal oxide semiconductors as photocatalysts in advanced oxidation processes (AOPs) for treatment of dye wastewater, Journal of Chemical Technology & Biotechnology, vol. 86, pp. 1130–1158.

Chin, A., Bérubé, P. R., (2005) Removal of disinfection byproduct precursors with ozone-UV advanced oxidation process, Water Research, vol. 39, pp. 2136–2144.

Cuiping, B., Xianfeng, X., Wenqi, G., Dexin, F., Mo, X., Zhongxue, G., Nian, X., (2011) Removal of rhodamine B by ozone-based advanced oxidation process, Desalination, vol. 278, pp. 84–90.

Demeestere, K., Dewulf, J., Langenhove, H.V., (2007) Heterogeneous photocatalysis as an advanced oxidation process for the abatement of chlorinated, monocyclic aromatic and sulfurous volatile organic compounds in air: State of the art, Critical Reviews in Environmental Science and Technology, vol. 37, pp. 489–538.

Deng, J., Shaoab, Y., Gao, N., Xia S., Tan, C., Zhou, S., Hu, X., (2013) Degradation of the anti-epileptic drug carbamazepine upon different UV-based advanced oxidation processes in water, Chemical Engineering Journal, vol. 222, pp. 150–158.

Ebrahiem, E.E., Al-Maghrabi, M.N., Mobarki, A.R., (2017) Removal of organic pollutants from industrial wastewater by applying photo-Fenton oxidation technology, Arabian Journal of Chemistry, pp. S1674–S1679.

Eren, Z., Ince, N.H., (2010) Sonolytic and sonocatalytic degradation of azo dyes by low and high frequency ultrasound, Journal of Hazardius Material, vol. 177, p. 10191024.

Fernandes, A., Makoś, P., Wang, Z., Boczkaj, G., (2020) Synergistic effect of TiO_2 photocatalytic advanced oxidation processes in the treatment of refinery effluents, Chemical Engineering Journal, vol. 391, p. 123488.

Gayathri, P.V., Yesodharan, S., Yesodharan, E.P., (2019) Microwave/Persulphate assisted ZnO mediated photocatalysis (MW/PS/UV/ZnO) as an efficient advanced oxidation process for the removal of RhB dye pollutant from water, Journal of Environmental Chemical Engineering, vol. 7, p. 103122.

Gerrity, D., Stanford, B.D., Trenholm, R.A., Snyder, S. A., (2010) An evaluation of a pilot-scale nonthermal plasma advanced oxidation process for trace organic compound degradation, Water Research, vol. 44, pp. 493–504.

Gholami, A., Bahrami, A.S.H., (2019) Application of heterogeneous nano-semiconductors for photocatalytic advanced oxidation of organic compounds: A review, Journal of Environmental Chemical Engineering, vol. 7, p. 103283.

GilPavas, E., Dobrosz-Gómez, I., Gómez-García, M.A., (2017) Coagulation-flocculation sequential with Fenton or Photo-Fenton processes as an alternative for the industrial textile wastewater treatment, Journal of Environmental Management, vol. 191, pp. 189–197.

Goel, M., Hongqiang, H., Mujumdar, A.S., Ray, M.B., (2004) Sonochemical decomposition of volatile and non-volatile organic compounds—a comparative study, Water Research, vol. 38, p. 42474261.

Goskonda, S., Catallo, J.W., Junk, T., (2002) Sonochemical degradation of aromatic organic pollutants. Waste Manage, vol. 22, pp. 351–356.

Jiang, Y., Petrier, C., Waite, T. D., (2002) Effect of pH on the ultrasonic degradation of ionic aromatic compounds in aqueous solution, Ultrasonics Sonochemistry, vol. 9, p. 163168.

Kamat, P.V., Meisel, D., (2002) Nanoparticles in advanced oxidation processes, Current Opinion in Colloid & Interface Science, vol. 7, pp. 282–287.

Kanakaraju, D., Glass, B.D., Oelgemoller, M., (2018) Advanced oxidation process-mediated removal of pharmaceuticals from water: A review, Journal of Environmental Management, vol. 219, pp. 189–207.

Lakretz, A., Ron, E.Z., Mamane, H., (2011) Biofilm control in water by a UV-based advanced oxidation process, Biofouling: The Journal of Bioadhesion and Biofilm Research, Vol. 27, pp. 295–307.

Lebron, Y.A.R., Moreira, V.R., Maia, A., Couto, C.F., Moravia, W.G., Amaral, M.C.S., (2021) Integrated photo-Fenton and membrane-based techniques for textile effluent reclamation, Separation and Purification Technology, vol. 272, p. 118932.

Lei, X., Lei, Y., Zhang, X., Yang, X., (2021) Treating disinfection byproducts with UV or solar irradiation and in UV advanced oxidation processes: A review, Journal of Hazardous Materials, vol. 408, p. 124435.

Lin, H., Ai, J., Li, R., Deng, L., Tan, W., Ye, Z., Wu, X., Zhang, H., (2020) Treatment of organosilicon wastewater by UV-based advanced oxidation processes: Performance comparison and fluorescence parallel factor analysis, Chemical Engineering Journal, vol. 380, p. 122536.

Lucas, M.S., Peres, J.A., (2006) Decolorization of the azo dye reactive Black 5 by Fenton and photo-Fenton oxidation, Dyes and Pigments, vol. 71, pp. 236–244.

Mangalgiri, K.P., Patton, S., Wu, L., Xu, S., Ishida, K.P., Liu, H., (2019) Optimizing potable water reuse systems: Chloramines or hydrogen peroxide for UV-based advanced oxidation process? Environmental Science & Technology, vol. 53, pp. 13323–13331.

Mascolo, G., Ciannarella, R., Balest, L., Lopez, A., (2008) Effectiveness of UV-based advanced oxidation processes for the remediation of hydrocarbon pollution in the groundwater: A laboratory investigation, Journal of Hazardous Materials, vol. 152, pp. 1138–1145.

Matafonova, G., Batoev, V., (2018) Recent advances in application of UV light-emitting diodes for degrading organic pollutants in water through advanced oxidation processes: A review, Water Research, vol. 132, pp. 177–189.

Meng, T., Sun, W., Sua, X., Suna, P., (2021) The optimal dose of oxidants in UV-based advanced oxidation processes with respect to primary radical concentrations, Water Research, vol. 206, p. 117738.

Moreira, F.C., Boaventura, R.A.R., Brillas, E., Vilar, V.J.P., (2017) Electrochemical advanced oxidation processes: A review on their application to synthetic and real wastewaters, Applied Catalysis B: Environmental vol. 202, pp. 217–261.

Mudliar, R., Umare, S.S., Ramteke, D.S., Wate, S. R., (2009) Energy efficient—Advanced oxidation process for treatment of cyanide containing automobile industry wastewater, Journal of Hazardous Materials, vol. 164, pp. 1474–1479.

Oliveira, I.S., Viana, L., Verona, C., Fallavena, V.L.V., Maria, C., Azevedo, N., Pires, M., (2007) Alkydic resin wastewaters treatment by fenton and photo-Fenton processes, Journal of Hazardous Materials, vol. 146, pp. 564–568.

Pérez, M., Torrades, F., Domènech, X., Peral, J., (2002b) Fenton and photo-Fenton oxidation of textile effluents, Water Research, vol. 36, pp. 2703–2710.

Pérez, M., Torrades, F., García-Hortal, J.A., Domènech, X., Peral, J., (2002a) Removal of organic contaminants in paper pulp treatment effluents under Fenton and photo-Fenton conditions, Applied Catalysis B: Environmental, vol. 36, pp. 63–74.

Phungsai, P., Kurisu, F., Kasuga, I., Furumai, H., (2019) Molecular characteristics of dissolved organic matter transformed by O_3 and O_3/H_2O_2 treatments and the effects on formation of unknown disinfection byproducts, Water Research, vol. 159, pp. 214–222.

Rekhate, C.V., Srivastava, J.K., (2020) Recent advances in ozone-based advanced oxidation processes for treatment of wastewater- A review, Chemical Engineering Journal Advances, vol. 3, p. 100031.

Rosenfeldt, E.J., Linden, K.G., Canonica, S., von Gunten, U., (2006) Comparison of the efficiency of radical OH radical formation during ozonation and the advanced oxidation processes O_3/H_2O_2 and UV/H_2O_2, Water Research, vol. 40, pp. 3695–3704.

Sánchez-Polo, M., Salhi, E., Rivera-Utrilla, J., von Gunten, U., (2006) Combination of ozone with activated carbon as an alternative to conventional advanced oxidation processes, The Journal of the International Ozone Association, vol. 28, pp. 237–245.

Serpone, N., Horikoshi, S., Emeline, A.V., (2010) Microwaves in advanced oxidation processes for environmental applications. A brief review, Journal of Photochemistry and Photobiology C: Photochemistry Reviews, vol. 11, pp. 114–131.

Sharma, V. K., Feng, M., (2019) Water depollution using metal-organic frameworks-catalyzed advanced oxidation processes: A review, Journal of Hazardous Materials, vol. 372, pp. 3–16.

Sirés, I., Brillas, E., Oturan, M.A., Rodrigo, M.A., Panizza, M., (2014) Electrochemical advanced oxidation processes: Today and tomorrow. A review, Environmental Science and Pollution Research, vol. 21, pp. 8336–8367.

Sun, H., Guo, F., Pan, J., Huang, W., Wang, K., Shi, W., (2021) One-pot thermal polymerization route to prepare N-deficient modified g-C_3N_4 for the degradation of tetracycline by the synergistic effect of photocatalysis and persulfate-based advanced oxidation process, Chemical Engineering Journal, vol. 406, p. 126844.

Suty, H., De Traversay, C., Cost, M., (2004) Applications of advanced oxidation processes: Present and future, Water Science and Technology, vol. 49, pp. 227–233.

Tan, C., Gao, N., Zhou, S., Xiao, Y., Zhuang, Z., (2014) Kinetic study of acetaminophen degradation by UV-based advanced oxidation processes, Chemical Engineering Journal, vol. 253, pp. 229–236.

Tichonovas, M., Krugly, E., Jankunaite, D., Racys, V., Martuzevicius, D., (2017) Ozone-UV-catalysis based advanced oxidation process for wastewater treatment, Environmental Science and Pollution Research, vol. 24, pp. 17584–17597.

Umar, M., Aziz, H.A., Yusoff, M.S., (2010) Trends in the use of Fenton, electro-Fenton and photo-Fenton for the treatment of landfill leachate, Waste Management, vol. 30, pp. 2113–2121.

Varanasi, L., Coscarelli, E., Khaksari, M., Mazzoleni, L.R., Minakata, D., (2018) Transformations of dissolved organic matter induced by UV photolysis, Hydroxyl radicals, chlorine radicals, and sulfate radicals in aqueous-phase UV-Based advanced oxidation processes, Water Research, vol. 135, pp. 22–30.

Wang, B., Song, Z., Sun, L., (2021) A review: Comparison of multi-air-pollutant removal by advanced oxidation processes—Industrial implementation for catalytic oxidation processes, Chemical Engineering Journal, vol. 409, p. 128136.

Wang, W., Shaacg, J., Lua, Z., Shao, S., Sun, P., Hua, Q., Zhang, X., (2018) Implementation of UV-based advanced oxidation processes in algal medium recycling, Science of the Total Environment, vol. 634, pp. 243–250.

Wen, G., Deng, X., Wan, Q., Xu, X., Huang, T., (2019) Photoreactivation of fungal spores in water following UV disinfection and their control using UV-based advanced oxidation processes, Water Research, vol. 148, pp. 1–9.

Wu, J.J., Muruganandham, M., Chen, S.H., (2007) Degradation of DMSO by ozone-based advanced oxidation processes, Journal of Hazardous Materials, vol. 149, pp. 218–225.

Wu, J.J., Wu, C., Ma, H., Chang, C., (2004) Treatment of landfill leachate by ozone-based advanced oxidation processes, Chemosphere, vol. 54, pp. 997–1003.

Xu, Y., Lin, Z., Zhang, H., (2016) Mineralization of sucralose by UV-based advanced oxidation processes: UV/PDS versus UV/H_2O_2, Chemical Engineering Journal, vol. 285, pp. 392–401.

Ye, T., Zhang, T., Tian, F., Xu, B., (2021) The fate and transformation of iodine species in UV irradiation and UV-based advanced oxidation processes, Water Research, vol. 206, p. 117755.

Zapata, A., Malato, S., Sánchez-Pérez, J.A., Oller, I., Maldonado, M.I., (2010) Scale-up strategy for a combined solar photo-Fenton/biological system for remediation of pesticide-contaminated water, Catalysis Today, vol. 151, pp. 100–106.

Zazou, H., Afanga, H., Akhouairi, S., Ouchtak, H., Rachi, A.A.A., Akbour, A., Assabbane, A., Douch, J., Elmchaouri, A., Duplay, J., Jada, A., Hamdani, M., (2019) Treatment of textile industry wastewater by electrocoagulation coupled with electrochemical advanced oxidation process, Journal of Water Process Engineering, vol. 28, pp. 214–221.

Zhang, Y., Pagilla, K., (2010) Treatment of malathion pesticide wastewater with nanofiltration and photo-Fenton oxidation, Desalination, vol. 263, pp. 36–44.

Zhang, Y., Zhou, J., Chen, X., Wang, L., Cai, W., (2019) Coupling of heterogeneous advanced oxidation processes and photocatalysis in efficient degradation of tetracycline hydrochloride by Fe-based MOFs: Synergistic effect and degradation pathway, Chemical Engineering Journal, vol. 369, pp. 745–757.

Zoschke, K., Dietrich, N., Börnick, H., Worch, E., (2012) UV-based advanced oxidation processes for the treatment of odour compounds: Efficiency and byproduct formation, Water Research, vol. 46, pp. 5365–5373.

Index

A

absorption, 68, 69, 70, 71, 72, 75, 77, 80, 102, 108, 109, 110, 111, 116, 222
activated carbon, 39, 52, 54, 58, 131, 135, 136, 138, 139, 151, 176, 179, 184, 188, 205, 210, 221, 226
additives, 3, 39, 47, 144, 150, 153
adsorption, 14, 39, 41, 48, 52, 54, 60, 62, 63, 90, 118, 120, 139, 140, 143, 147, 148, 158, 163, 204
advanced oxidation processes, 13, 19, 21, 25, 26, 35, 36, 43, 65, 93, 96, 97, 133, 187, 213, 214, 218
agriculture, 47, 133, 140
aluminum, 37, 166, 167, 183
anionic, 32, 37, 40, 68, 86, 110, 128, 142, 148, 152, 155
anode, 18, 19, 20, 21, 30, 31, 36, 41, 44, 78, 79, 173, 174, 175, 176
anodic, 20, 21, 22, 30, 41, 43, 48, 173, 175, 179, 184
antibiotic, 4, 6, 8, 9, 10, 30, 39, 40, 41, 52, 53, 54, 55, 61, 62, 85, 97, 114, 116, 117, 123, 125, 128, 129, 130, 158, 182, 184
AOP, 14, 19, 24, 27, 32, 33, 35, 36, 41, 43, 44, 49, 66, 82, 83, 86, 108, 110, 112, 113, 114, 116, 117, 119, 121, 133, 134, 135, 139, 142, 144, 148, 149, 159, 167, 168, 181, 189, 194, 205, 206, 207, 208, 209, 210, 211, 214, 215, 216, 217, 218, 219, 221, 222, 223, 224
aquatic, 3, 6, 7, 8, 9, 10, 14, 24, 35, 38, 39, 40, 47, 59, 62, 65, 66, 67, 70, 81, 85, 114, 116, 120, 158, 179, 181

B

benzene, 72, 172
bioaccumulation, 3
biodegradability, 7, 29, 50, 99, 136, 207, 210
biological, 14, 28, 39, 40, 41, 42, 43, 47, 48, 66, 68, 86, 110, 123, 125, 149, 159, 218, 221, 227
bisphenol A, 54, 145, 151, 154, 169, 178, 183, 184

C

carbon, 9, 15, 17, 18, 21, 27, 28, 30, 31, 35, 36, 37, 38, 44, 52, 54, 55, 66, 67, 70, 121, 134, 136, 138, 140, 142, 143, 144, 149, 150, 152, 153, 159, 176, 177, 178, 179, 210

carbon dioxide, 9, 28, 70
carbon nanotubes, 52, 54, 143, 178
carcinogenic, 7, 117, 179
catalysts, 14, 18, 19, 24, 25, 27, 31, 36, 37, 38, 39, 48, 49, 50, 51, 52, 53, 54, 55, 56, 57, 58, 59, 61, 62, 63, 64, 75, 76, 77, 78, 80, 81, 117, 118, 120, 127, 130, 134, 135, 136, 138, 139, 140, 141, 142, 143, 144, 145, 147, 150, 151, 152, 159, 160, 163, 164, 165, 166, 169, 171, 172, 178, 181, 183, 184, 215, 221
cathode, 18, 19, 20, 21, 30, 31, 79, 138, 151, 172, 173, 175, 176, 179, 184
cationic, 40, 97
cerium, 53, 139, 166, 167
chemical vapor deposition, 142
chromium, 7, 133, 166
clay, 52, 53, 57, 61, 62, 63
climate change, 35, 70, 191
coagulation, 14, 29, 42, 189, 190, 195, 204, 205, 207
contaminant, 43, 73, 87, 108, 110, 123, 125, 155, 182, 188, 191, 194, 198, 204, 211, 215
copper, 7, 51, 52, 53, 54, 55, 56, 60, 62, 63, 110, 124, 138, 145, 154, 155, 166, 169, 183
cost, 18, 35, 36, 37, 42, 44, 48, 56, 59, 74, 80, 108, 114, 116, 117, 121, 136, 138, 165, 181, 188, 189, 190, 192, 193, 194, 195, 196, 197, 198, 199, 200, 201, 202, 203, 204, 205, 206, 207, 210, 211, 221
cyclohexane, 172

D

decay, 18, 22, 23, 27, 28, 32, 33, 53
degradation, 8, 14, 16, 17, 22, 24, 26, 27, 28, 29, 30, 33, 35, 36, 37, 38, 39, 40, 41, 42, 43, 44, 49, 50, 51, 52, 53, 54, 55, 56, 57, 58, 59, 60, 61, 62, 63, 64, 66, 69, 70, 71, 72, 74, 75, 76, 77, 78, 79, 80, 81, 83, 84, 85, 86, 87, 88, 89, 99, 100, 101, 105, 106, 108, 109, 110, 112, 113, 114, 115, 116, 117, 118, 119, 120, 121, 122, 123, 124, 125, 126, 127, 128, 129, 130, 134, 135, 136, 137, 138, 139, 140, 141, 142, 143, 144, 145, 146, 147, 148, 149, 150, 151, 152, 153, 154, 155, 159, 161, 162, 164, 165, 166, 169, 172, 174, 176, 177, 178, 179, 182, 183, 184, 191, 207, 208, 209, 222, 225, 227
dyes, 18, 23, 27, 31, 41, 43, 79, 86, 108, 110, 113, 122, 126, 127, 128, 138, 141, 144, 146, 147, 148, 150, 155, 179, 218, 222, 225

E

effluents, 9, 10, 14, 20, 22, 27, 28, 29, 30, 31, 36, 39, 40, 41, 44, 45, 47, 50, 52, 53, 54, 55, 57, 59, 66, 72, 74, 85, 114, 115, 116, 117, 118, 120, 121, 123, 125, 126, 129, 130, 136, 138, 139, 146, 158, 182, 184, 210, 215, 218, 221, 225, 226

electrochemical, 18, 19, 20, 21, 22, 31, 38, 43, 44, 48, 50, 54, 60, 79, 136, 158, 172, 173, 176, 184, 227

electrochemical oxidation, 19

electrode, 21, 30, 44, 136, 138, 150, 174, 175, 176, 181, 223

electrolysis, 14, 48, 63

electron, 18, 19, 20, 23, 36, 37, 54, 55, 56, 73, 74, 78, 79, 80, 93, 95, 111, 112, 113, 118, 134, 144, 145, 147, 167, 168, 172, 173, 178, 183, 223

electron transfer, 18, 19, 23, 54, 79, 80, 144, 167, 168, 183

emerging, 9, 28, 29, 36, 44, 47, 59, 61, 62, 63, 66, 99, 108, 114, 120, 122, 123, 124, 128, 134, 145, 149, 177, 179, 181, 182, 190, 191, 210, 223, 224

emerging materials, 25, 36, 133, 135

endocrine, 3, 5, 6, 9, 10, 39, 40, 76, 108, 109, 114, 116, 126, 158, 178

F

Fenton, 13, 14, 15, 17, 18, 19, 21, 22, 24, 25, 26, 28, 29, 30, 31, 32, 33, 36, 38, 39, 40, 41, 42, 43, 44, 45, 47, 48, 49, 50, 51, 52, 53, 54, 55, 56, 57, 58, 59, 60, 61, 62, 63, 64, 66, 69, 79, 90, 95, 96, 106, 134, 135, 136, 137, 138, 139, 140, 141, 147, 150, 151, 152, 155, 157, 158, 159, 160, 161, 162, 163, 164, 165, 166, 167, 168, 169, 170, 171, 172, 173, 174, 175, 176, 177, 178, 179, 180, 181, 182, 183, 184, 185, 215, 216, 217, 218, 219, 223, 224, 225, 226, 227

ferrihydrite, 52, 157, 165, 178, 183, 184

ferrites, 43, 56, 157, 166

fluidized-bed, 170

G

global warming, 67

goethite, 157, 165

graphene, 27, 52, 54, 60, 62, 63, 77, 78, 93, 94, 95, 130, 136, 139, 142, 150, 151, 152, 153

graphene oxide, 52, 62, 78, 130, 139, 151, 153

groundwater, 3, 8, 9, 47, 114, 158, 190, 191, 222, 226

H

H$_2$O$_2$ 13, 14, 15, 16, 17, 18, 19, 20, 21, 22, 23, 24, 25, 26, 27, 29, 30, 31, 32, 33, 37, 39, 40, 41, 43, 57, 58, 62, 65, 69, 70, 71, 72, 76, 79, 80, 83, 84, 90, 100, 102, 103, 106, 108, 109, 110, 111, 113, 114, 115, 116, 117, 120, 123, 124, 125, 126, 128, 129, 131, 134, 135, 136, 139, 140, 141, 143, 144, 146, 147, 148, 149, 150, 153, 174, 182, 183, 188, 192, 193, 194, 195, 199, 200, 201, 202, 203, 204, 207, 208, 209, 210, 211, 212, 213, 214, 216, 222, 223, 226

hazardous, 4, 5, 6, 7, 20, 30, 72, 99, 149, 159, 182

heavy metals, 7, 21, 22, 40, 134, 219

hematite, 56, 157, 165

heterogeneous, 29, 35, 39, 40, 41, 43, 44, 48, 49, 50, 51, 52, 53, 54, 55, 56, 57, 58, 59, 60, 61, 62, 63, 64, 122, 130, 135, 136, 137, 138, 139, 140, 144, 147, 150, 151, 162, 163, 164, 165, 166, 169, 170, 171, 178, 181, 182, 183, 184, 221, 222, 225, 227

homogeneous, 15, 18, 19, 22, 24, 27, 35, 36, 37, 39, 40, 41, 43, 44, 48, 50, 51, 59, 60, 82, 137, 138, 144, 159, 162, 163, 164, 165, 168, 169, 170, 172, 181

hormones, 6, 29, 39, 40

hydrocarbons, 3, 33, 36, 40, 43, 69, 136, 222, 226

hydrogen, 14, 16, 19, 22, 23, 24, 37, 40, 41, 42, 48, 49, 51, 53, 54, 55, 70, 72, 84, 100, 101, 102, 103, 106, 114, 116, 118, 134, 146, 149, 154, 158, 159, 175, 176, 182, 183, 215, 221, 222, 224, 226

hydrogen peroxide, 14, 16, 19, 22, 23, 24, 37, 40, 41, 42, 49, 51, 53, 54, 55, 70, 72, 84, 102, 106, 114, 116, 118, 134, 154, 158, 159, 182, 183, 215, 221, 222, 226

hydrothermal, 53, 128

hydroxides, 51

hydroxyl radicals, 14, 16, 19, 37, 43, 48, 49, 51, 52, 53, 55, 57, 71, 73, 74, 75, 80, 99, 106, 108, 109, 114, 115, 118, 121, 125, 133, 134, 138, 158, 159, 182, 190, 195, 203, 207, 208, 209, 215, 216, 221

I

inflation, 190, 195, 210

iron, 15, 18, 19, 25, 27, 29, 36, 38, 39, 40, 44, 45, 48, 49, 50, 51, 52, 53, 54, 55, 56, 57, 58, 59, 62, 63, 64, 106, 136, 138, 139, 140, 141, 151, 152, 159, 160, 161, 162, 163, 164, 165, 166, 167, 169, 170, 171, 172, 173, 174, 178, 179, 181, 182, 183, 215

irradiation, 14, 27, 28, 29, 32, 33, 37, 38, 42, 56, 57, 58, 59, 68, 69, 74, 75, 76, 78, 81, 82, 83, 84, 85, 87, 100, 105, 108, 110, 111, 114, 115, 117, 127, 129, 130, 134, 135, 150, 151, 153, 171, 172, 178, 181, 184, 217, 221, 222, 225

K

kinetic, 35, 42, 56, 78, 99, 102, 103, 105, 106, 107, 116, 123, 124, 125, 126, 127, 129, 149, 163, 164, 184
kinetic modelling, 35

M

magnetic, 41, 52, 54, 55, 61, 62, 139, 143, 147, 148, 153, 164
magnetite, 63, 135, 157, 164
manganese, 55, 57, 140, 157, 166, 168, 169, 183
membrane filtration, 54, 158, 193, 204
metallic-organic frameworks, 55
metal-organic framework, 145
micropollutants, 4, 5,, 8, 9, 19, 31, 22, 23, 25, 26, 29, 30, 35, 36, 37, 39, 40, 41, 43, 54, 61, 83, 86, 87, 88, 108, 109, 110, 111, 112, 113, 117, 118, 120, 121, 122, 123, 124, 128, 129, 130, 131, 133, 134, 136, 140, 142, 143, 144, 145, 148, 149, 157, 158, 159, 162, 165, 171, 175, 177, 178, 179, 181, 182, 184, 190, 207, 208, 212
mxene, 147, 148, 149, 155

N

nanocomposite, 63, 78, 81, 142, 143, 147, 149, 152, 153
nanoplastics, 81
nitrate, 9, 37, 68, 69, 70, 109, 115, 209, 221

O

organic, 3, 4, 9, 15, 16, 17, 18, 19, 21, 24, 27, 29, 30, 35, 36, 37, 38, 39, 40, 41, 42, 43, 44, 48, 49, 54, 55, 59, 60, 61, 62, 63, 65, 66, 67, 68, 70, 71, 74, 78, 79, 80, 81, 83, 84, 85, 86, 88, 99, 106, 108, 109, 110, 113, 114, 115, 116, 122, 123, 124, 125, 128, 130, 134, 135, 138, 139, 140, 142, 145, 147, 148, 149, 151, 153, 155, 158, 159, 160, 161, 162, 164, 165, 166, 167, 168, 170, 171, 172, 174, 175, 177, 179, 182, 183, 184, 190, 192, 193, 205, 207, 208, 210, 215, 218, 221, 222, 224, 225, 226, 227
oxidants, 3, 22, 27, 38, 48, 49, 70, 71, 72, 76, 83, 159, 179, 195, 206, 207, 208, 209, 215, 222, 226

oxidation, 8, 14, 15, 16, 18, 19, 20, 21, 22, 23, 27, 29, 30, 32, 35, 36, 37, 38, 39, 40, 41, 42, 43, 44, 45, 48, 50, 52, 53, 54, 55, 57, 59, 60, 61, 62, 63, 68, 69, 71, 72, 74, 77, 79, 80, 81, 82, 83, 85, 99, 100, 106, 107, 108, 109, 110, 111, 114, 115, 116, 118, 122, 123, 124, 125, 126, 128, 129, 131, 133, 134, 135, 136, 138, 139, 140, 144, 145, 148, 149, 150, 151, 152, 153, 154, 155, 158, 159, 160, 161, 162, 163, 164, 165, 167, 168, 169, 170, 171, 173, 175, 177, 179, 181, 182, 183, 184, 189, 190, 194, 205, 210, 215, 216, 217, 219, 220, 221, 222, 223, 224, 225, 226, 227
oxides, 18, 39, 51, 52, 53, 54, 56, 57, 60, 62, 74, 77, 136, 151, 165, 166, 170, 172, 183, 215
oxygen, 3, 19, 21, 23, 26, 29, 30, 36, 37, 52, 54, 58, 63, 68, 70, 72, 73, 74, 75, 78, 79, 83, 114, 118, 134, 135, 138, 139, 143, 146, 166, 173, 174, 175, 176, 177, 179, 193, 199, 215, 219
oxyhydroxides, 51, 160, 168
ozonation, 16, 19, 22, 23, 24, 25, 27, 30, 37, 39, 40, 41, 48, 49, 72, 76, 77, 124, 128, 134, 144, 152, 182, 189, 192, 197, 198, 204, 207, 221, 226

P

perovskites, 52, 54
peroxide, 16, 17, 23, 24, 26, 70, 80, 83, 108, 126, 140, 142, 146, 149, 152, 161, 167, 168, 221
peroxydisulfate, 23, 45, 145, 154, 222
pesticides, 3, 7, 9, 11, 18, 29, 32, 34, 35, 36, 39, 40, 47, 49, 62, 71, 72, 74, 75, 76, 78, 79, 86, 90, 91, 92, 94, 108, 118, 122, 130, 151, 158
pharmaceuticals, 5, 6, 10, 19, 30, 39, 50, 52, 53, 57, 58, 60, 61, 62, 63, 78, 79, 82, 84, 85, 106, 108, 110, 113, 114, 115, 116, 117, 118, 119, 120, 122, 124, 125, 126, 129, 130, 133, 139, 143, 158, 177, 178, 179, 184, 191, 215, 222, 225
phenol, 17, 36, 38, 40, 54, 86, 112, 135, 136, 142, 150, 152, 161, 162, 167, 172, 177, 179, 180, 181, 184, 185, 221
photocatalysis, 32, 34, 56, 57, 66, 73, 74, 75, 76, 77, 78, 79, 117, 118, 119, 121, 122, 123, 130, 134, 142, 143, 151, 172, 221, 222, 225, 227
photocatalyst, 59, 61, 65, 72, 73, 74, 75, 76, 77, 78, 79, 80, 81, 108, 111, 112, 114, 117, 118, 121, 129, 130, 131, 139, 143, 148, 153, 155, 224
photocatalytic, 17, 35, 36, 39, 40, 48, 56, 57, 62, 72, 73, 74, 75, 76, 77, 78, 79, 80, 111, 113, 117, 118, 119, 121, 122, 123, 124, 127, 128, 129, 130, 139, 143, 144, 148, 149, 153, 155, 183, 221, 224, 225

photochemical, 32, 43, 48, 66, 68, 70, 71, 72, 77, 88, 107, 128, 142, 149
photochemistry, 66, 70
photodegradation, 35, 76, 78, 81, 100, 105, 106, 107, 114, 118, 119, 126, 127, 131, 143, 151
photoluminescence, 78
photolysis, 16, 27, 40, 48, 51, 55, 66, 67, 68, 70, 71, 72, 89, 100, 102, 108, 109, 113, 115, 116, 123, 124, 125, 129, 146, 171, 181, 208, 209, 222, 227
photosensitizers, 68, 79, 80, 115
plasticizers, 39
pollutants, 3, 8, 9, 14, 17, 20, 21, 22, 23, 27, 28, 29, 35, 38, 40, 42, 44, 47, 48, 49, 50, 54, 59, 60, 61, 62, 63, 65, 66, 67, 69, 70, 72, 73, 74, 80, 83, 85, 86, 89, 99, 106, 108, 113, 116, 117, 118, 121, 122, 124, 131, 135, 136, 142, 144, 146, 147, 149, 152, 153, 161, 162, 166, 178, 182, 183, 216, 221, 222, 224, 225, 226
polymer, 58, 99, 100, 101, 103, 106, 147, 183
pyrite, 157, 166, 178

Q

quenching, 71, 188, 193, 199, 201, 202, 203, 204, 206, 207, 209, 210

R

radicals, 14, 17, 19, 20, 22, 24, 26, 27, 30, 35, 36, 37, 48, 54, 60, 67, 70, 72, 79, 82, 83, 84, 85, 86, 90, 99, 100, 101, 103, 105, 106, 108, 109, 111, 113, 116, 118, 133, 134, 136, 139, 144, 147, 151, 154, 158, 160, 162, 166, 167, 168, 171, 173, 177, 207, 208, 227
refractory, 14, 15, 23, 25, 28, 30, 43, 44, 50, 62, 71, 133, 134, 144, 158
remediation, 7, 9, 19, 21, 27, 60, 66, 106, 121, 125, 149, 191, 218, 226, 227
reverse osmosis, 48, 158, 193
ruthenium, 166, 183

S

sedimentation, 48
sludge, 8, 21, 27, 29, 47, 48, 49, 50, 66, 109, 126, 130, 138, 150, 161, 163, 164, 169, 171, 173, 174, 181, 182
solvents, 3
sonolysis, 48, 51, 66, 82, 83, 84, 85, 86, 88, 89, 134, 146, 222
supramolecular, 69
surfactants, 6, 39, 40, 41, 85, 86, 154, 158

T

techno-economic analysis, 19
techno-economic assessment, 36, 187, 189
three-dimensional, 43, 54, 55, 63, 136

toluene, 52, 172, 184
toxic, 6
toxicity, 3, 5, 6, 7, 9, 16, 36, 42, 51, 52, 56, 62, 112, 118, 121, 123, 125, 126, 127, 129, 131, 135, 158, 164, 168

U

ultrasonic, 19, 22, 44, 82, 84, 85, 87, 88, 144, 225
ultrasonic irradiation, 19, 84
ultrasound, 13, 22, 23, 41, 42, 48, 60, 81, 82, 83, 84, 86, 87, 88, 96, 97, 98, 144, 145, 146, 147, 153, 154, 222, 223, 225
ultraviolet, 14, 15, 16, 17, 19, 40, 59, 79, 81, 124, 126, 127, 129, 142, 143, 155, 221
ultraviolet radiation, 16, 19, 40, 79
UV, 13, 14, 16, 17, 22, 25, 26, 27, 28, 29, 30, 31, 32, 33, 34, 35, 36, 37, 38, 39, 40, 41, 42, 43, 44, 45, 48, 49, 55, 56, 57, 58, 60, 70, 71, 72, 73, 74, 75, 76, 77, 78, 80, 81, 90, 91, 92, 93, 94, 99, 100, 101, 102, 105, 106, 107, 108, 109, 110, 111, 112, 113, 114, 115, 116, 117, 118, 119, 120, 121, 122, 123, 124, 125, 126, 127, 128, 129, 130, 131, 134, 142, 143, 146, 147, 148, 149, 153, 158, 171, 172, 177, 178, 184, 187, 188, 189, 191, 192, 193, 194, 195, 200, 201, 202, 203, 204, 206, 208, 209, 210, 211, 212, 213, 214, 215, 221, 222, 223, 224, 225, 226, 227
UV/H_2O_2 13, 17, 25, 26, 28, 29, 30, 32, 33, 35, 36, 42, 43, 70, 71, 72, 90, 91, 99, 100, 101, 105, 107, 108, 109, 110, 112, 113, 114, 115, 116, 117, 118, 120, 121, 122, 123, 124, 125, 126, 128, 129, 130, 131, 187, 188, 189, 191, 192, 193, 194, 195, 200, 201, 202, 203, 204, 206, 208, 210, 211, 212, 213, 214, 221, 222, 223, 226, 227
UV/TiO_2 25, 28, 36, 44, 45, 108, 111, 112, 113, 114, 117, 118, 120, 121, 122, 123, 124, 125, 126, 127, 128, 130, 131, 146
UV treatment, 14, 17, 31, 107, 111

W

wastewater, 3, 4, 5, 7, 8, 9, 10, 14, 16, 19, 21, 23, 28, 29, 30, 32, 33, 35, 36, 38, 39, 40, 41, 42, 43, 44, 47, 49, 50, 52, 53, 58, 59, 60, 61, 62, 63, 65, 66, 74, 77, 80, 83, 85, 86, 88, 99, 106, 107, 108, 109, 114, 116, 117, 118, 119, 120, 121, 122, 123, 124, 125, 126, 127, 128, 129, 130, 131, 134, 135, 136, 139, 140, 141, 143, 144, 147, 149, 150, 151, 153, 158, 159, 168, 177, 179, 181, 182, 184, 188, 190, 210, 215, 216, 217, 218, 219, 221, 222, 224, 225, 226, 227

Z

zeolites, 51, 52, 53, 62, 63, 112, 127, 151, 178, 184

For Product Safety Concerns and Information please contact our EU
representative GPSR@taylorandfrancis.com
Taylor & Francis Verlag GmbH, Kaufingerstraße 24, 80331 München, Germany

www.ingramcontent.com/pod-product-compliance
Lightning Source LLC
Chambersburg PA
CBHW060358220326
41598CB00023B/2955